Disequilibrium and Self-Organisation

Mathematics and Its Applications

Disequilibrium and Self-Organisation

edited by

C. W. Kilmister

Former Professor of Mathematics, King's College, University of London, England

D. Reidel Publishing Company

A MEMBER OF THE KLUWER ACADEMIC PUBLISHERS GROUP

Dordrecht / Boston / Lancaster / Tokyo

Library of Congress Cataloging in Publication Data

CIP

Disequilibrium and self-organisation.

(Mathematics and its applications)
Edited versions of the contributions to the 2nd and 3rd ISG meetings, held in Klosterneuburg near Vienna in 1983 and in Windsor near London in 1985.
Includes index.
1. Self-organizing systems–Congresses. 2. Information theory–Congresses. 3. Thermodynamics–Congresses. I. Kilmister, C. W. (Clive William) II. International Study Group on Self-Organising Systems and Dissipative Structures. III. Series: Mathematics and its applications (D. Reidel Publishing Company)
Q325.D57 1986 003 86–13933
ISBN-13: 978-94-010-8598-4 e-ISBN-13: 978-94-009-4718-4
DOI: 10.1007/978-94-009-4718-4

Published by D. Reidel Publishing Company
P.O. Box 17, 3300 AA Dordrecht, Holland

Sold and distributed in the U.S.A. and Canada
by Kluwer Academic Publishers,
101 Philip Drive, Assinippi Park, Norwell, MA 02061, U.S.A.

In all other countries, sold and distributed
by Kluwer Academic Publishers Group,
P.O. Box 322, 3300 AH Dordrecht, Holland

Softcover reprint of the hardcover 1st edition 1986

TABLE OF CONTENTS

SERIES EDITOR'S PREFACE

Approach your problems from the right end and begin with the answers. Then one day, perhaps you will find the final question.

'The Hermit Clad in Crane Feathers' in R. van Gulik's *The Chinese Maze Murders*.

It isn't that they can't see the solution. It is that they can't see the problem.

G.K. Chesterton. *The Scandal of Father Brown* 'The point of a Pin'.

Growing specialization and diversification have brought a host of monographs and textbooks on increasingly specialized topics. However, the "tree" of knowledge of mathematics and related fields does not grow only by putting forth new branches. It also happens, quite often in fact, that branches which were thought to be completely disparate are suddenly seen to be related.

Further, the kind and level of sophistication of mathematics applied in various sciences has changed drastically in recent years: measure theory is used (non-trivially) in regional and theoretical economics; algebraic geometry interacts with physics; the Minkowsky lemma, coding theory and the structure of water meet one another in packing and covering theory; quantum fields, crystal defects and mathematical programming profit from homotopy theory; Lie algebras are relevant to filtering; and prediction and electrical engineering can use Stein spaces. And in addition to this there are such new emerging subdisciplines as "experimental mathematics", "CFD", "completely integrable systems", "chaos, synergetics and large-scale order", which are almost impossible to fit into the existing classification schemes. They draw upon widely different sections of mathematics. This programme, Mathematics and Its Applications, is devoted to new emerging (sub)disciplines and to such (new) interrelations as exempla gratia:

- a central concept which plays an important role in several different mathematical and/or scientific specialized areas;
- new applications of the results and ideas from one area of scientific endeavour into another;
- influences which the results, problems and concepts of one field of enquiry have and have had on the development of another.

The Mathematics and Its Applications programme tries to make available a careful selection of books which fit the philosophy outlined above. With such books, which are stimulating rather than definitive, intriguing rather than encyclopaedic, we hope to contribute something towards better communication among the practitioners in diversified fields.

It is clear by now that mathematics as a whole now seems prepared to grapple seriously with the world of nonlinear phenomena. Compared to the linear world this one is much more dicey. Rewards or results can be totally out of proportion and, just as easily, inspired efforts can yield virtually no results. Going from linear to nonlinear mathematics is maybe a little like giving up a nice comfortable salaried position with pension, social security, etc. for a life of speculation and living by one's wits in an uncertain environment full of (nasty) surprises and unexpected happenings.

One of the more interesting nonlinear phenomena is the possibility of spontaneous emergence of patterns: self-organisation, auto-catalysis, self-reinforcement, synergetics, ... This is also clearly a group of phenomena which plays of big role in the natural world around us and it is important to understand and model them.

It would be nice if there were just a few mathematical schemes at the root of all these pattern-generating phenomena. That may be so, but we do not know enough yet to judge. At the present stage we need to understand, describe and compare such phenomena in many different disciplines such as geology, history, geography, chemistry, biology, physics, cybernetics, neurology, ... This makes the field totally interdisciplinary and, recognising this, a number of people organised themselves into the International Study Group on Self-Organizing Systems and Dissipative Structures.

When I invited ISG to contribute a volume to the MIA series I described it as a volume that should contain the best of ISG thinking in all its diversity. That describes this volume, and it should turn out to be a most useful source book for those who consider self-organisation a fascinating and/or important factor in Nature.

The unreasonable effectiveness of mathematics in science ...

Eugene Wigner

Well, if you know of a better 'ole, go to it.

Bruce Bairnsfather

What is now proved was once only imagined.

William Blake

As long as algebra and geometry proceeded along separate paths, their advance was slow and their applications limited.

But when these sciences joined company they drew from each other fresh vitality and thenceforward marched on at a rapid pace towards perfection.

Joseph Louis Lagrange.

Bussum, May 1986 Michiel Hazewinkel

PREFACE

The International Study Group on self-organising systems and dissipative structures (ISG) started life in 1981 at its first Whitsun meeting in Wunsiedel, in the Fichtel-gebirge. Perhaps some appropriateness can be seen in the fact that this was the home of Jean Paul, who was the favourite author of K.F.Gauss. The aim of the ISG is to make precise the notion of self-organisation, a general idea invoked by workers in many different fields from cybernetics and physics through biology to economics and sociology, by a careful study of actual usages in these fields.

It was decided to meet at two-yearly intervals, and the second and third meetings were held in Klosterneuburg near Vienna in 1983 and in Windsor near London in 1985. This volume contains edited versions of most of the contributions to the second and third ISG meetings.

My thanks, as retiring President of the ISG, and as ed
iror of this volume, are due to the members of the organising committee of the first three conferences, Manfred Bayerl, Kamal Islam and Rainer Zimmermann, who have worked hard to make this possible.

C.W.Kilmister.

SELF-ORGANIZATION AND INFORMATION THEORY

Karl-Erik Eriksson
Physical Resource Theory Group, Chalmers University of Technology, University of Göteborg,
S-412 96 Göteborg, Sweden

According to the standard model of cosmology [1], the Universe, including space-time itself and matter/energy, was created in one event nearly point-like in time, the Big Bang.

At the very high temperatures reigning at the beginning, the Universe may have been not only in thermal equilibrium but also very symmetric, obeying nearly exact gauge symmetries. These symmetries are still the basis for interactions through gauge fields, but they must have been broken very early (around 10^{-35} s) with the decrease in temperature accompanying the expansion. In this symmetry breaking the first diversity was created [2]. It seems that ever since, growing complexity, including self-organization, has been a steady occupation of the universe.

The asymmetry which arose between matter and antimatter gave us a small remnant of nuclear matter in an intense sea of leptons and of electromagnetic radiation which has now cooled to the relatively insignificant (energy-wise) 2.7 K cosmic background radiation. Around 10 s after Big Bang the neutrinos had decoupled, of the charged leptons only electrons remained, and hadron matter was already in the form of nucleons. The radiation was a heat bath in which nucleons and electrons were dissolved. All this was in thermal equilibrium.

With the continued expansion and cooling, the equilibrium state of the nucleons changed from free nucleons to nuclei, without the proper reaction channels being open to achieve such an equilibrium. Only some helium was formed during the fourth minute, roughly 26% of all nuclear matter. The disequilibrium that arose [3-6] in the nuclear matter implied that exergy (= available energy) [7-10] was created [11]. The corresponding information-theoretical concept will be called contrast - here contrast between the actual state described by a probability distribution P and an equilibrium state $P^{(o)}$ and we define it formally in terms of Kullback information $K(P^{(o)};P)$ [12].

Later, when the pressure had decreased enough to allow gravitational clumping of matter, also gravitational exergy (and gravitational contrast) came into being. This may be a type of process which can take place on an ever increasing scale [13].

Thus, at first contrast - nuclear and gravitational - was present

C. W. Kilmister (ed.), Disequilibrium and Self-Organisation, 1–10.

without any structural manifestation.

This state could be viewed as a supercooled state at a temperature below nuclear and gravitational "freezing points", but the "freezing" into galaxies, stars and nuclei had not yet taken place. Since the stars are required as nuclear reactors for the nuclear burning, the gravitational clumping had to come first. When this happened, matter also became structured into dense and hot regions and thin and cold regions. Such structuring also took place on various scales.

Later, after nuclear burning and supernova explosions had produced the chemical elements, the gravitational structuring of a solar system also led to the separation of the chemical elements, as we know it in our solar system and on the earth itself.

If it could be described on a galactic scale, there would probably be much randomness in the structure of solar systems. From a human view point however, it is not quite so. To us the structure of our own environment is unique. We have given names to our continents and oceans, to our mountains and valleys, to the planets and to the stars in our neighbourhood within our galaxy. The astronomers have given names also to galaxies and to clusters of galaxies.

Exergy is a physically defined quantity [14,15] . The same is true of the closely related concept of contrast, which has the dimension of information [16,17]. It turns out that structure may be similarly defined, and that the expression for contrast may be divided into one term which is structural contrast, i.e., structure, and another term which is structureless contrast. Even in a primitive self-organizing system structureless contrast may be transformed into structural contrast. At the same time energy is dissipated, i.e. contrast is lost [18].

Thus structureless contrast is the "raw material" for structure. As Schrödinger put it: "We feed on negentropy" [19] . The origin of the structureless contrast is in its turn the cosmic imbalance inherent in the expanding universe itself.

Self-organization requires self-catalytic processes and mass transport within the system. Gravitational clumping obviously has these characteristics. Through star formation gravitation also catalyzes nuclear reactions which lead to a radiation flow from the stellar surface. The contrast between this radiation and interstellar space may then be "utilized" by physico-chemical systems of a planetary surface if it has a suitable localization and composition.

Structure may be of a rather trivial nature such as crystalline structure or it may be quite complex such as the structure of an organism or an ecosystem or the infrastructure of a human society. The amount of complexity in a given structured system can be expressed by its content of algorithmic information [20,21].

Let a system A be described in statistical mechanics terms by a certain macrostate, i.e. a probability distribution P(n) over microstates n. The algorithmic information of A is the minimum information H needed for a program (algorithm) that orders a nondeterministic Turing machine (NDT) to produce as outputs microstates n for A such that they are distributed as described by P(n). The algorithmic information of A is then H(P).

Algorithimic information is a concept which may be applied directly to any system which can be described in statistical mechanics terms. Complexity [22] may then be similarly defined. With $P^{(o)}$ a reference state, the complexity of A in the state P is the minimum information $H(P^{(o)}; P)$ needed in a reprogramming algorithm, which orders an NDT to change its output distribution from $P^{(o)}$ to P.

We shall also consider transformation processes involving matter/energy, contrast, structure, and complexity. When dealing with transformation processes performing certain prescribed tasks we call them functions. Some of the processes considered by Professor Schuster in his lecture [23] are of this kind.

A system will be considered as organized, if it performs a set of coordinated functions. The degree of organization of an organized system may then be defined as the complexity of the set of coordinated functions. (This is a definition which needs to be made more precise.)

A system B which is structured in such a way that its structural information on some level expresses the algorithm for forming another system A may be called a blueprint of A. The algorithmic information content of B then expresses the complexity of A.

For theoretical treatment processes which may be considered as stationary or cyclic are particularly simple. There one may apply input/output descriptions and thermodynamical/economic efficiency measures [24].

Table 1. A ladder formed by various levels of self-organization

↑
Organization
Complexity
Structure
Contrast
Matter/energy in space-time

The ladder in Table 1 shows the hierarchy of concepts discussed here. It should probably be extended. Maybe it will be possible to give sharp definitions of several important concepts of higher complexity, even though they may not often be possible to quantify with any exactness. Table 2 shows a dream of what such an extended ladder may look like. It was inspired by such works as Schrödinger's essay on mind and matter [25], Monod's discussion of evolution, science, and values [26], and H.T. Odum's discussion of energy flows in nature and society [27].

Statistical mechanics is the theoretical basis for all science dealing with matter and energy, unless one is dealing with special systems involving only very few degrees of freedom. Information theory should similarly be the theoretical basis for the sciences dealing with ideas. In the real world matter/energy and ideas influence each other. Matter/energy may exist without ideas, but ideas are not known to exist without matter. Fortunately for us, statistical mechanics is closely related to information theory [17]. Concepts such as probability and entropy are common to them. Therefore information theoretical concepts may be fundamental along the whole extended ladder.

Table 2. A dream of an extended self-organization ladder

↑ Values	Universal
Self-knowledge	Collective
	Individual
Knowledge	
Consciousness	
Evolution	
Know-how	
Organization	
Complexity	
Structure	
Contrast	
Matter/energy in space-time	

The aim of this paper is to discuss only the lowest steps of the ladder stretching from the matter/energy of the expanding universe into the realm of ideas. From the reference list it is clear that the basic ideas are not new; they have been collected from various sources.

Mathematical definitions of structure and complexity

The exergy of a system is the amount of mechanical work that can be extracted from the system ideally in a process through which it approaches equlibrium in a reversible way. It has been realized for some time that if the macrostate of the system is given as a probability distribution $P = \{p_i\}_1^n$, where n is the number of available microstates (usually a huge number) and if the equilibrium state is described by $P^{(o)} = \{p_i^{(o)}\}_1^n$, then the exergy is

$$E = (k \ln 2) T_o \, K(P^{(o)};P) \qquad (1)$$

where T_o is the equilibrium temperature, and k is Boltzmann's constant. K is the so-called Kullback information (or surprisal) of the state P relative to the reference state $P^{(o)}$,

$$K(P^{(o)};P) = \frac{1}{\ln 2} \sum_{i=1}^{n} p_i \ln \frac{p_i}{p_i^{(o)}} \geq 0 \qquad (2)$$

$K(P^{(o)};P)$ vanishes only for $P = P^{(o)}$. Although this quantity has all the mathematical characteristics of an information concept, we do not consider it as information in all cases. For instance, solar radiation (let it be described by P) coming in towards the Earth's surface (which defines a reference state $P^{(o)}$ carries with it a considerable amount of exergy, as given by (1) and (2), but very little that we recognize as information. Clearly a scientist who studies the sun itself can extract valuable information about the solar surface from investigating this radiation. The object of study is then the deviation from homogenous blackbody radiation, i.e. deviations from a maximum entropy photon gas. For estimating the exergy these deviations are totally negligible. Hence the information carried by the solar radiation can only be a minute fraction of (2).

On the other hand, the solar exergy in its passage through all the various chains of conversions on the Earth's surface, gives rise to a lot of structure that may be described in information-theoretical terms. Therefore it seems reasonable to consider (2) as information capacity, which may or may not become coded with information. We shall refer to (2) as contrast.

It has been suggested by Chaitin, that algorithmic information be used as the proper information measure. In computer terminology, if X is a computer output consisting of n(X) binary digits, then its information content in ordinary terms is n(X) bits. Its algorithmic information content H(X) is instead the smallest number of bits in a program for a general-purpose computer to print out X. For most numbers ("random numbers") the two concepts very nearly coincide, but for a small set of numbers (ordered sequences like 11111111111111111 or 10101010101010101010) the program may be substantially shorter than the number of digits, and hence $H(X) < n(X)$.

Two computer outputs are (algorithmically) independent if the best description of them is simply the two separate descriptions added, i.e.

$$H(X,Y) = H(X) + H(Y) \tag{3}$$

The left side here is always smaller than or equal to the right side. Then the deviation from equality

$$H(X:Y) = H(X) + H(Y) - H(X,Y) \tag{4}$$

which is called the mutual algorithmic information, is a measure of the organization of the total output (X,Y). This is easily generalized to an output X consisting of any number of parts,

$$X = (X_1, X_2, \dots, X_k) \tag{5}$$

for which the mutual information is

$$H(X_1:X_2:\dots:X_k) = \sum_{i=1}^{k} H(X_i) - H(X_1,X_2,\dots,X_k) \tag{6}$$

If we consider blackbody radiation in such a context, we find that the mutual information is very low, consisting only of a weak correlation in energy and in spatial position or direction (depending on the distance from the source). This weak correlation is then characteristic of blackbody radiation emitted from a certain surface at a certain temperature.

A statistical mechanics description would rather be based on macrostates, which are probability distributions (ensembles) over microstates. The algorithmic information of a certain macrostate (probability distribution) would then be defined through the physical description needed to specify such a state. The temperature and the location of the surface emitting the photons would constitute this information. A computer analogue would require a generator of random numbers to reproduce the required statistical distribution over microstates. A reduction of the type (6) would then not be needed, since the randomness of the distribution and with it the relatively low level of mutual information would be a direct consequence of the algorithm.

The relevant algorithmic information to be compared to the surprisal (2) is the amount of information in a minimum algorithm for changing a computer program designed to generate the probability distribution $P^{(o)}$ into one that generates P. Let us call this quantity $H(P^{(o)};P)$. Then we have (up to an additive constant, which may be neglected in statistical mechanics)

$$H(P^{(o)};P) \leq K(P^{(o)};P) \tag{7}$$

This is the same as the information content in a minimum algorithm for changing a program for printing $P^{(o)} = \{p_i^{(o)}\}_1^n$ into one for printing $P = \{p_i\}_1^n$. (In statistical mechanics n is so large, that any printing task seems totally intractable; still the algorithm may be quite possible to handle.

Clearly when $P = P^{(o)}$, the required program is of nearly zero length, as it should be. As has been said already, the algorithmic information contained in blackbody radiation as against blackbody radiation of another temperature is very small, because the program is very simple. If instead P describes radiation coded with a message against a background $P^{(o)}$ which still describes blackbody radiation, then $H(P^{(o)};P)$ is very close to the information content of the message. It is thus very reasonable to consider the information content $H(P^{(o)};P)$ of a reprogramming algorithm as the proper measure of information for a macrostate. In those instances where H<K - as in the example of solar radiation on the Earth that we have considered - the distinction between information capacity K and information content H becomes quite essential.

We shall now show how structure may be defined in terms of Kullback information (2). Let $i_1,\ldots,i_r$ be microstates indices for the system of interest, assumed to consist of r subsystems, so that i_k labels the state of the subsystem. From the total probability distribution

$$\begin{cases} P = \left\{p_{i_1 \cdots i_r}\right\}; \; i_k = 1,\ldots,n_k; \; k = 1,\ldots,r. \\ p_{i_1 \cdots i_r} \geq 0 \\ \sum_{i_1=1}^{n_1} \cdots \sum_{i_r=1}^{n_r} p_{i_1 \cdots i_r} = 1 \end{cases} \tag{8}$$

one may derive an individual distribution for each of the subsystems,

$$\begin{cases} P_k = \left\{ p_i^{(k)} \right\}_1^{n_k} \\ p_i^{(k)} = \sum_{i_1=1}^{n_1} \cdots \sum_{i_r=1}^{n_r} p_{i_1 \ldots i_r} \, \delta_{i_k i} \\ \sum_{i=1}^{n_k} p_i^{(k)} = 1 \end{cases} \tag{9}$$

We assume the reference probability distribution to be totally uncorrelated,

$$\begin{cases} P^{(o)} = \left\{ p_{i_1 \ldots i_r}^{(o)} \right\} \\ p_{i_1 \ldots i_r}^{(o)} = \prod_{k=1}^{r} p_{i_k}^{(k,o)} \; ; \; \sum_{i=1}^{n_k} p_i^{(k,o)} = 1 \end{cases} \tag{10}$$

The subsystem distributions are

$$P_k^{(o)} = \left\{ p_i^{(k,o)} \right\}_1^{n_k} \tag{11}$$

The mutual Kullback information corresponding to (6) is now

$$K(P^{(o)};P) - \sum_{k=1}^{r} K(P_k^{(o)};P_k) = K(P';P) \tag{12}$$

where we have introduced the uncorrelated distribution

$$P' = \left\{ \prod_{k=1}^{r} p_{i_k}^{(k)} \right\} \tag{13}$$

We note that $P^{(o)}$ drops out from (12), provided it is uncorrelated as assumed in (10). $K(P';P)$ is the <u>structural contrast</u>.

Let us assume that $P^{(o)}$ is a homogenous state, i.e. that all the r distributions (11) are equal. Further, let $\bar{P}$ be the probability distribution constructed from a similar homogenization of P' (total values of extensive variables remaining unchanged).

$$\bar{P} = \prod_{k=1}^{r} \bar{p}_{i_k} \; ; \; \bar{p}_i = \frac{1}{r} \sum_{k=1}^{r} p_i^{(k)} \tag{14}$$

Then using (12) we may write the total contrast $K(P^{(o)};P)$ as

$$K(P^{(o)};P) = K(P^{(o)};\bar{P}) + K(\bar{P};P') + K(P';P) \tag{15}$$

If we now apply this to a one-dimensional chemical system with k as the position coordinate, then the three terms on the right side of (15) may be interpreted as chemical contrast, contrast from inhomogeneity and structural contrast or structure.

The structural term may be further divided into parts,

$$K(P';P) = K(P';P_2) + K(P_2;P) \geq K(P';P_2) \tag{16}$$

where P_2 is the probability distribution with the same averages and two-point correlations as P but with all higher order correlations smeared out. $K(P';P_2)$ can easily be expressed in terms of the two-point correlation function,

$$\eta_{ii'}(k) = \frac{r^{-1}\sum_{j=1}^{r}\sum_{i \quad i} p_{i \quad i}\,\delta_{ii_j}\,\delta_{i'i_{j+k}}}{p_i p_{i'}} - 1 , \tag{17}$$

$$\left(p_i = \frac{1}{r}\sum_{j=1}^{r} p_{i_j \ldots i_r}\,\delta_{i_j i}\right)$$

as

$$K(P';P_2) = r\sum_{ii'k} p_i p_{i'}\,(1+\eta_{ii'}(k))\,\ln(1+\eta_{ii'}(k)) \tag{18}$$

From the dynamics one may sometimes determine the correlation function that develops in a self-organizing system such as the Brusselator. Then it is also possible (i) to estimate the efficiency in the transition from chemical contrast to structural contrast and (ii) to determine the cost in terms of supplied chemical contrast per unit time to maintain a certain stationary structure.

The algorithmic information contained in the mutual information (12) is smaller than (12) itself, if there exists a simplifying algorithm to change a program that generates P' into one that generates P. The mutual algorithmic information in P is its internal <u>complexity</u>.

$$H(P';P) \tag{19}$$

where P' is given by (13) and (9).

Chaitin [22] has shown how to define structure and complexity on different levels of subsystem size. Adopted to the macrostate description of statistical mechanics that we have been using, the subdivision into subsystems in (8) would have to be such that no subsystem contains more than a certain number, let us say N, of the most elementary subsystems, each having a microstate,

$$n_k \leq a_i^N \; ; \; k = 1, \ldots, r(N) \tag{20}$$

The probability distribution $P = p_{i_1 \ldots i_{r(N)}}$ is the same whatever the subdivision, but

P' as defined in (13) depends on N and on the particular subdivision s applied; hence

$$P' = P'_N (s) \tag{21}$$

The relevant complexity measure for the level N of subsystem size is

$$H_N (P) = \min_s H (P'_N(s);P) \tag{22}$$

It increases with decreasing N and so reveals organization on successive size levels. Chaitin describes the role of $H_N(P)$ as that of a "Fourier transform".

References

1. S. Weinberg: Gravitation and Cosmology: Principles and Applications of the General Principle of Relativity (Wiley, New York 1972).
2. F. Wilczek: Scientific American 243:12 (1980).
3. R.C. Tolman: Relativity, Thermodynamics, and Cosmology (University Press, Oxford 1934).
4. P.C.W. Davies: The Physics of Time Asymmetry (University of California Press, Berkeley 1974).
5. D. Layzer: Scientific American 233:12 56 (1975).
6. P. Landsberg: Thermodynamics and Statistical Mechanics (University Press, Oxford 1978).
7. Z. Rant: Forschung auf dem Gebiete des Ingenieurswesens, 22, 36 (1956).
8. H. Baehr: Energie und Exergie (VDI-Verlag, Düsseldorf 1965).
9. M. Tribus, P.T. Shannon, and R.B. Evans: Am. Inst. Chem. Eng. J. 12, 244 (1966).
10. R.B. Evans: A Proof that essergy is the only consistent measure of potential work (Thesis, Dartmouth College, Hannover, NH 1969).
11. K.-E. Eriksson, S. Islam, and B.-S. Skagerstam. Nature (London) 296, 540 (1982).
12. S. Kullback: Information Theory and Statistics (Wiley, New York 1959).
13. S. Frautschi: Science 217, 593 (1982).
14. M. Tribus and E.C. McIrvine: Scientific American 225:9 (1971).
15. R.B. Evans: A New Approach for Deciding upon Constraints in R.D. Levine and M. Tribus (eds), The Maximum Entropy Formalism (MIT Press Cambridge, Massachusetts 1979).
16. L. Brillouin: Science and Information Theory (Academic Press, New York 1962).
17. E.T. Jaynes: Information Theory and Statistical Mechanics in K.W. Ford (ed.), Statistical Physics, Brandeis Lectures, 1962 (Benjamin, New York 1963).
18. G. Nicolis and I. Prigogine: Self-Organization in Non-Equlibrium Systems (Wiley-Interscience, New York 1977).
19. E. Schrödinger: What is Life? (Univesity Press, Cambridge 1944).
20. G.J. Chaitin: Scientific American 232:5, 47 (1975).

21. A.N. Kolmogorov, work mentioned in ref. 20.
22. G.J. Chaitin: Toward a Mathematical Definition of "Life" in Levine and Tribus (eds), op.cit.
23. P. Schuster. Polynucleotides - A Class of Self-Organizing Biomolecules (these proceedings).
24. P. Salamon, A. Nitzan, B. Andresen and R.S. Berry: Phys. Rev. A 21, 2115 (1980).
25. E. Schrödinger: Mind and Matter (Cambridge University Press 1958).
26. J. Monod: Chance and Necessity (Collins, London 1971).
27. H.T. Odum: Environment, Power and Society (Wiley, New York 1971).

THE MATHEMATICS NEEDED FOR SELF-ORGANISATION

C.W.Kilmister
Department of History and Philosophy of Science
Chelsea College
London SW 7
UK

ABSTRACT. True self-organisation requires a new type of mathematics; the initial stages of its development are described and it is shown to be isomorphic to an earlier algebraic construction.

1. INTRODUCTION

I seek to answer a question raised at the second ISG meeting in 1983, by Professor Muschik as to the sense in which the algebraic system formulated by Bastin and Parker-Rhodes and developed by others (1) is self-organising. This questionis linked with a problem which I discussed in the first ISG meeting (2) in 1981 but which was of longer standing. This problem is the role of randomness in self-organisation. A Turing machine which could be called self-organising would be one which changed its programme according to its state; but such a Turing machine is (by Turing's hypothesis) simply represented by a larger deterministic Turing machine, so that the idea of self-organisation has vanished. The usual answer is that a random element has to be introduced. But of course this is really saying that the Turing machine model has ceased to be an appropriate one for a self-organising system. A different theoretical structure is called for; and since Turing machines are simply a model for classical mathematics, this in turn is called into question.

There are two essential features of self-organisation, which I can illustrate by a simple example. The free motion of cars along a road can be described under normal conditions by giving the motion of each individual car or equally well by looking at the motion of an ensemble. But when congestion develops the individual car description becomes almost useless and one is forced to the collective or higher level one, and one introduces new concepts such as a jam. The two

C. W. Kilmister (ed.), Disequilibrium and Self-Organisation, 11–17.

essential features then are:

(a) The subjective one that a change of descriptive level is forced by the development, and so the system must have such levels;

(b) new features arise and so the mathematical formulation must allow the entry of new symbols.

Classical mathematics does not fit very well with either feature but especially not with (b). Indeed such a creative feature suggests intuitionism and I have flirted with that in the past, but it turns out not to be necessary to go so far. Rather, to adopt an intuitionistic stance has traces of simply taking what is to hand instead of looking for exactly what is needed.

2. GENERATION AND DISCRIMINATION

Let us begin with (b) above. New entities arise. How can this be? There must be some generation process and as each new entity is created it must receive a label. We may use as a set of labels the symbols 0,1,2,3,... but these are not, of course, the cardinal numbers. On the other hand they can be read as ordinal numbers, and if, as we shall see is appropriate, we consider 0 as in a special category, we can read the symbol m as labelling the mth element produced, either by the generation operation or internally inside the mathematical system. This notation emphasises, what is clear from the generation idea, that the resultant system will be either finite or enumerable.

An important question is begged in the preceding paragraph. When an entity is produced, by whatever means, how can we be sure that it is a new one? In one sense it must be because it differs from all the other entities in its stage of production; but to regard the entities in that way is to produce a system with no structure at all. Rather we must have a notion of equivalence between two examples of an entity, so that each potentially new element must be checked to see if it is equivalent to one which has occurred before. At this point it is again necessary to take note of the continuing generation; for it means that we cannot simply ask whether the new element is or is not a member of some given set. This set would be that of the already generated elements and so is not 'given' - whatever that means for the classical mathematician - but is continually changing.

A device which avoids this (it may not be the only possible way to proceed) is to use the nineteenth century idea of a function, that is, a rule which gives a value for each value of the argument. It must not be assumed that any later ideas of domain or range are implied by the use of the word 'function'; there is simply a rule. This rule is then constrained to be such that the values of the function for elements equivalent to already generated elements lie in a fixed set which is disjoint from the set of values for truly new elements. This condition goes some way to determining the rule, as will become clear later. Such an operation will be called discrimination.

3. THE SIMPLE CASE

The simple case in which one compares the 'new' element with a single known one determines much of the structure. Let u be the known element, and x a potentially new one. Denote the function by $f(u,x)$ so as to include reference to the known element. Denote the fixed set of leements which are values of the function when u,x are equivalent by Z. Since we are specifying an equivalence relation, we must have:

(i) $f(x,x) \in Z$ for all x,
(ii) if $f(x,y) \in Z$, then $f(y,x) \in Z$;
(iii) if $f(x,y) \in Z$ and $f(y,z) \in Z$, then $f(x,z) \in Z$.

There is an obvious equivalence relation between such functions f, which holds when they do exactly the same job. That is, one defines f, f^* to be equivalent if, whenever $f(x,y) \in Z$, then also $f^*(x,y) \in Z^*$. It is then straightforward to prove that

THEOREM 1. Each equivalence class of f's contains an element g (say) which is such that:
(a) the Z for g has one element, say 0;
(b) $g(x,x) = 0$ for all x;
(c) $g(x,y) = g(y,x)$ for all x,y;
(d) $g(g(x,y),z) = g(x,g(y,z))$.

It will be noticed that each of (b),(c),(d) are specialisations of the conditions (i),(ii),(iii). But since the conditions on g are exactly those of commutativity and associativity in a field of characteristic 2, it is convenient to use the notation x+y for $g(x,y)$. The proof of the theorem may easily be worked out by the reader, as far as (b) and (c) are concerned. The final step is accomplished by what I call 'Conway's trick' because it is used (in a somewhat different context) in his book (3). This is to take the value of x+y as:

$$x+y = \min z(\ f(z,x'+y) \neq 0,\ f(z,x+y') \neq 0 \text{ for all } x' < x \text{ and all } y' < y).$$

Here f is a member of the equivalence class which has already been chosen to satisfy (b) and (c). It is then easy to see that 0+0 = 0, and indeed 0+x = x. So 1+1 cannot be the same as 1+0 = 1, but can be 0. Then 1+2 cannot be 1 or 0 or 2 so must be a new element 3. Of course, 1+3 = 1+1+2 = 2 and so the set (1,2,3) is 'discriminately closed' (that is, any two different elements x,y are such that x+y is in the set). If now the generation process throws up a new element, it will be labelled 4, and 4+1 = 5, 4+2 = 6, 4+3 = 7 and the set (1,2,3,4,5,6,7) is again evidently discriminately closed. It is clear that every discriminately closed set is of order

$2^r - 1$, for some integer r. Correspondingly, if the set is not yet closed:

THEOREM 2. At every stage the system can be embedded in a discriminately closed one of size $2^r - 1$.

Thus the numbers 3,7,15,31,63,... are going to be of importance in describing self-organising systems.

A useful change of notation is provided by

THEOREM 3. The system of theorem 2 has an injection into V_2/Z_2, where V_r is the vector space of r dimensions, and Z_2 is the field of two elements.

The proof of this is easy as soon as one realises that the injection in question can be taken in the form that the element k, where k = pqr... in the scale of two, is mapped onto the vector of components p,q,r,... . Thus 5 = 101 and so is mapped onto the transpose of (1,0,1) whilst 9 = 1001 and so is mapped onto the transpose of (1,0,0,1). It is of course important to bear in mind that the dimension of the space is not fixed but increases.

4. THE GENERAL CASE

The case considered above gives a guide to the general case. Here our concern is with whether a new element x belongs to a set $S = (u_1, u_2, u_3, ...)$ of already generated ones. We use the same trick of defining a function f(S,x) which vanishes if and only if one of the terms u_i+x vanishes. Because of the symmetry having gone it is more convenient to use the notation of characteristic functions, and write

$$f(S,x) = F_S(x).$$

If we define an addition operation between characteristic functions by the obvious induction:

$$F(x) + G(x) = (F + G)(x)$$

for all x in play up to the point at which the definition is being made, then it is easy to see that this operation has exactly similar properties to the original discrimination operation, so that the set of characteristic functions is itself a discrimination system.

We next make a similar definition of equivalence between characteristic functions to the one made in the last section. That is, we define F,F* to be equivalent if, whenever F(x) = 0, then also F*(x) = 0 and vice versa. Amongst a set of equivalent F's it is obvious that there will be some of the special form

$$F(u_1,u_2,\dots,x\ ) = h(u_1+x,u_2+x,\dots)$$

where h is function which vanishes if and only if one of its arguments vanishes. (Such F's will in fact be those in which the value for those x's which are not one of the u's depends only on the values of the u+x's.) Of course one defines two such functions h, h* as equivalent just when the corresponding F, F* are equivalent. Then, by an argument that closely parallels that for theorem 1, we can prove

THEOREM 4. Every equivalence class of h's contains a particular one with the properties:
(a) $h(x_1,x_2,\dots,x_n)$ depends only on the set $(x_1,x_2,\dots,x_n)$ so is invariant under permutations of the arguments;
(b) $h(x_1,h(x_2,x_3)) = h(h(x_1,x_2),x_3)$; with obvious generalisations to more arguments;
(c) $h(x_1,x_2+x_3) = h(x_1,x_2) + h(x_1,x_3)$.

Here again the proof relies on Conway's trick, but the proof is more intricate than before. Conway uses a least number principle applies to h(x',y) + h(x,y') + h(x',y') to define h(x,y), where as before x' and y' are any smaller elements than x,y. But since (b), (c) are simply the associative rule and distribution over addition, it is intuitively simpler (though more complicated to work in practice) to define h(x,y) = xy as the least element not forbidden, taking account of the commutative and associative rules and distributivity over addition. For example, 0x = 0 and 1x = x are straightforward. Then 2.2 cannot be 0 or 2; could it be 1? No, because if it were, then 2.2 = 1 = 2(3+1) = 2.3 + 2, so that 2.3 = 3 which is forbidden since 1.3 = 3. So 2.2 is in fact 3 (and it is easy to see that this is not forbidden).

The set (1,2,3) is now closed again (2.3 = 1); when 4 is introduced, the distributive law alone allows the completion of the table up to 7.7, so up to the next power of 2, but the table does not now give a closed set. Now the associative law allows the rest of the table up to 15.15 to be filled in, which gives a closed set of size 15. In the same way the next closed set of size 255 arises; and the general conclusion is that amongst the numbers which have already been noticed as being of importance in describing self-organisation, the particular ones 3, 15, 255, ... will be of particular importance.

5. A RETURN TO MUSCHIK'S QUESTION

I can now return to the question raised by Professor Muschik. It is clear that one can describe a set og elements in the system in two ways: either in terms of the individual elements, or as the function at the next level which specifies them. But the characteristic of self-organisation is that one particular description (the higher level one) acquires a great advantage. This happens here in one case. For a general set S it can never be the case that

$$F_S x + F_S y = F_S(x + y),$$

but

THEOREM 5. If and only if S is discriminately closed, the equivalence class of characteristic functions for S contains a member F for S contains a member F for which $F(x) + F(y) = F(x + y)$.

In the corresponding vector space picture such an F is represented by a square matrix, and the addition defined between functions becomes ordinary matrix addition. In the original construction by Parker-Rhodes the matrix specifying a discriminately closed subset is taken as that leaving unchanged all the elements of the subset. These Parker-Rhodes matrices simply differ from those constructed here by the addition of the unit matrix. In the Parker-Rhodes construction one chooses linearly independent functions which then in their turn have as additive closure another discriminately closed subset. Thus, going back to the numbers singled out in the last section, at the first stage 3 linearly independent mambers have to be found in a 3+1 dimensional space; these 3 give rise to 7 d.c.subsets and so 7 operators must be found in 16 dimensions. Then 127 in 256 dimensions and finally $1.7 \; 10^{38}$ in 65536. The last case cannot give rise to linearly independent operators so that this construction stops there and suggests the importance of the numbers 3, 10, 137, 10^{38} which arise as the cumulative totals of elements.

This is evidently a synergetic system of a particularly simple, discrete kind, which yet exhibits the characteristic synergetic property that individual parts cooperate to produce structures (4). As soon as one works out the details of the construction a wealth of numbers arise, which call for identification. For example, the number of automorphisms of a discriminately closed subset generated by n elements is $2^{\frac{1}{2}n(n-1)}F(n)$, where F(n) is defined as n*(n-1)*...3*2*1*, and n* = 2^n - 1. So when n = 3 we have 168, and when n = 4, 20160. This latter number is made up of 2912 conjugate pairs of 'unfixers' that change every element, 11640 operators fixing 1 element, 2590 fixing three elements, 105 fixing 7, together with the identity fixing everything. The 2590 in turn split up into three groups of 1260, 1120 and 210 according to their effect on the remaining elements, the 11640 similarly split into 5760, 3360 and 2520 and the 2912 conjugate pairs split up into 5 groups in a complicated way. The further investigation of the significance of these numbers and those at higher levels is proceeding actively.

REFERENCES

1. T.Bastin,H.P.Noyes,J.Amson and C.W.Kilmister 'On the Physical Interpretation and the Mathematical Structure of the Combinatorial Hierarchy.' Int.Jour.Theor.Phys.18, 445-488, 1979.
2. C.W.Kilmister 'Turing's hypotheses and selforganisation.' Proceedings of the first international Whitsun meeting on self-organisation Ed.R.E.Zimmermann, Berlin 1981.
3. J.H.Conway On Numbers and Games London 1976.
4. H.Haken in W.Weidlich and G.Haag Concepts and models of a Quantitative Sociology, Springer, Berlin 1983.

ENTROPY AND ORDER

Peter T. Landsberg
Faculty of Mathematical Studies
University of Southampton
Southampton
England

ABSTRACT. Definitions which decouple entropy and disorder are given, and are illustrated by a very simple example.

It is often said that

> the statistical entropy is decreased as one gathers more information about a system. (1)

In order to examine (1) take

$$S = -k \sum_{i=1}^{n(t)} p_i \ln p_i \leqslant k \ln n(t) \equiv S_{max} \tag{2}$$

as the (statistical) entropy definition, where p_i is the probability of finding the system in state i at a certain time t and $n(t)$ is the number of available states. Then one can say:

A. Information given at <u>constant</u> $n(t)$.

a. Only the value of $n(t)$ is given. Then $S = S_{max}$.

b. Add information which makes at least one p_i unequal to $1/n(t)$. Then at least one other p_i is also unequal to $1/n(t)$ and the entropy is decreased in accordance with (1).

c. Add information that the system is in a particular definite state. Then $S = 0$. This is also in agreement with (1).

B. Variation in $n(t)$.

a. If the "information" is that $n(t)$ is greater than thought previously, $m(t) > n(t)$ say, but no other information.

C. W. Kilmister (ed.), Disequilibrium and Self-Organisation, 19–21.

Then the entropy is increased from $k \ln n(t)$ to $k \ln m(t)$, in contradiction with (1).

b. If a description of the system is made less coarse (i.e. finer) the entropy is increased, in contradiction with (1).

c. The converse of Bb. holds also.

Following earlier ideas (1,2) which will not be rehearsed here in the written version of the talk, the intensive variable "disorder" will be defined in a technical sense by

$$D = S/k \ln n(t) \ . \tag{3}$$

It is a number which lies between 0 and unity in virtue of equation (2). The order Ω can therefore be defined by the condition

$$\Omega + D = 1 \ . \tag{4}$$

A simple application of this idea will now be made in order to elucidate the definitions (3) and (4), and also in order to explain that the contradiction between (1) and the propositions Ba and Bb arises from the variability of $n(t)$. This is not normally allowed for when the statement (1) is made.

Let us therefore consider the important proposition Bb. Suppose a system has two levels: one at energy E_1 and another at $E_2 = 2E_1$ and these are equally populated. Additional information leads us to regard the lower level to be a doublet of energies $E_1 \pm \varepsilon$. Let us examine the changes in a number of interesting parameters which are brought about by this additional information (Table 1). We can see here very simply that, indeed, entropy and order can increase together.

These ideas are applicable in cosmology (3) and biology (4) and indeed to any expanding or growing system.

Table 1

Effect of the additional information on the "two-level" system

	Initial values	Final values
Probabilities	$(\frac{1}{2},\frac{1}{2})$	$(\frac{1}{4},\frac{1}{4},\frac{1}{2})$
Energy	$\frac{1}{2} E_1 + \frac{1}{2}(2E_1) = \frac{3}{2} E_1$	$\frac{1}{4}(E_1-\varepsilon) + \frac{1}{4}(E_1+\varepsilon) + \frac{1}{2}(2E_1)$ $= \frac{3}{2} E_1$
Entropy	$-2k(\frac{1}{2} \ln \frac{1}{2}) = k \ln 2$	$-k(2\frac{1}{4} \ln \frac{1}{4} + \frac{1}{2} \ln \frac{1}{2})$ $= \frac{3}{2} k \ln 2$
Disorder D	$k \ln 2/k \ln 2 = 1$	$S/k \ln 3 = \frac{3}{2}\frac{\ln 2}{\ln 3}$ $= 0.05$
Order Ω	0	0.05

References

(1) P.T. Landsberg, Thermodynamics and Statistical Mechanics (Oxford: University Press, 1978, p.366).

(2) P.T. Landsberg, "Can entropy and 'order' increase together?" Physics Lett. 102A 171 (1984).

(3) S. Frautschi, "Entropy in an expanding universe", Science 217, 593 (1982).

(4) D.R. Brooks, P.H. Leblond and D.D. Cumming, "Information and entropy in a simple evolution model", J. Theoret. Biol. 109, 77 (1984).

PATTERN FORMATION FOR A ONE DIMENSIONAL EVOLUTION EQUATION BASED ON THOM'S RIVER BASIN MODEL.

Michiel Hazewinkel,
C.W.I. Amsterdam, The Netherlands
Johan F. Kaashoek,
Econometric Inst., Erasmus Univ.,
P.O. Box 1738, Rotterdam, The Netherlands
Bart Leynse,
Dept. Chem. Path., Erasmus Univ.,
Rotterdam, The Netherlands

Abstract. A one component, one dimensional diffusion model is presented in which spatial structure is generated by means of a density dependent diffusive mechanism such that for some density values mass flow is proportional to the mass density gradient. Although stability and attractivity properties of a set of analytic periodic stationary solutions are not strong enough, the analytical and numerical work reported here, show that this evolution equation in one space variable with zero-flux boundary conditions will have stationary attracting periodic limit distributions.

C. W. Kilmister (ed.), Disequilibrium and Self-Organisation, 23–46.

1. Introduction

Pattern formation of one kind or another occurs in many systems, c.q.: star clustering (astronomy), amoebae concentrations (chemotaxis; biology), periodic precipitation (chemistry), population concentrations in cities, (spatial economy), self-fulfilling prophycies (economic behaviour). This tendency to order is even seen in some physical systems as the Bénard convection shows clearly.
Since a macrosystem is described by the average density of the constituents, such cases of pattern formation can be seen as the evolution of an initial uniform distribution function to a non-uniform, well profiled function defined on the space of possible outcomes of the process (e.g.: a price distribution function in the case of self-fulfilling prophycies; spatial distribution of rising water in the case of Bénard convection).
Such an evolution process can be modeled mathematically by nonlinear partial differential systems, in which "almost anything can happen", reflecting "the beauty and great variation of manifestations of the nonlinear in the bio-, geo- and other spheres around us" [5].
We mention here the so called activator-inhibitor models of Meinhardt [4]. These models are based on the possibility to distinguish between slow diffusing, growth (self-) enhancing components and fast diffusing, growth inhibiting components in the system. Since the inhibitor substance is almost uniformly present, only areas with a high activator concentration can grow further.
However, not in all cases "growth" is inhibited by some substance. For instance, an aggregation centre grows by attracting substance, at the same time causing low concentration in the neighbourhood of the centre. So the

rise of new centres becomes impossible in the surroundings of already existing centres.

Apart from this depletion effect, these evolution processes are characterised by self-enhancement, also called autokatalysis. Local higher density areas themselfes are the source of amplyfiing fluctuations of an initial uniform ordening, for instance by gravitational instability, or economies of scale.

In this paper we shall introduce a diffusion equation with density dependent diffusion coefficient, such that for some values of the density, this coefficient will be negative. This means that mass flow will be proportional to the mass concentration gradient and as such opposed at the flow direction of a Fickian diffusion. In other words: mass flow is directed towards higher concentration areas. Instability and self-amplyfication of fluctuations are due to this reverse diffusion. (Section I). In section II we shall give some stationary solutions of such an equation.

As far as we know it is mathematically still an open question to show rigorously that an evolution in one space variable with zero-flux boundary conditions can have stationary attracting periodic limit distributions. There are analytic periodic stationary solutions in a number of cases whose stability and attractively properties are not strong enough. However, analytical and numerical work reported in this note, show existence of stable stationary solutions. (Section III).

I.1. Thom's river basin model

In [6] the following situation is described by Thom. Steadily rain is falling on a sandy hill; at the top brooklets are formed and destroyed almost continuously. Down the hill, the slope is less and erosion is less strong. The

pattern of watersheds and brooklets becomes more stable. Remaining brooks compete with each other over the available space. The result will be an almost regular pattern at the bottom of the hill. Such pattern can be observed in nature, e.g. in Death Valley in California.
Let $s_n(t)$ denote the position of the n-the watershed (at time t). Suppose the eroding power of a stream is proportional to its basin width, then the position s_n will be governed by the following differential equation:

$$\dot{s}_n = -c(s_{n+1}-s_n) + c(s_n-s_{n-1}); \; c > 0 \tag{I.1}$$

(where . denotes derivative with respect to t).
Any equidistant distribution with basin width a for all streams is a stationary solution of (I.1).
The character of equation (I.1) becomes clear by doing some linear stability analysis. Consider two streams with watersheds at +a and -a, and at u near o on ℝ. Assuming c depends also on the basin width, we get:

$$\dot{u} = 2c(a)u + 2ac'(a)u + u^2(\ldots)+\ldots \tag{I.2}$$

We will have stability if $c(a) + ac'(a) < 0$.
Since erosion power will diminish at greater values of the basin width, a reasonable graph of c would be as depicted in figure (1.1):

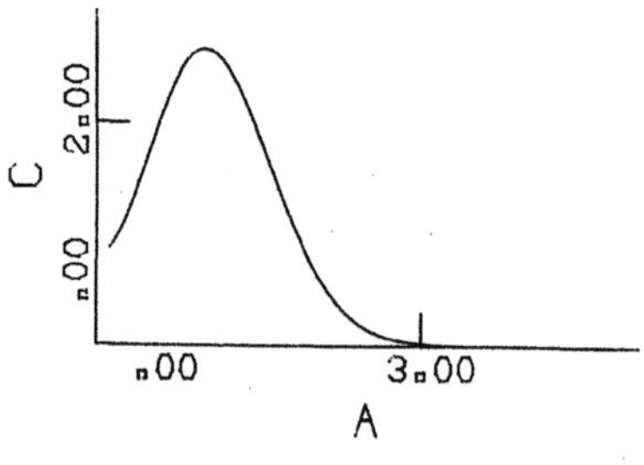

figure 1.1

Then, say for $a > a_o$ with $c(a_o) + a_o c'(a_o) = 0$, stability will be obtained.
On the contrary, if stream width is in the range where $c(a) + ac'(a) > 0$, broader streams will grow at the cost of smaller ones.
As smaller streams coincide with a higher density of the watersheds, at least in the instability range of the model, local maxima of the watersheds spatial distribution function will grow. Now at least two questions are wide open: (i) how does a more or less regular spacing of watersheds at characteristic distance a_o arise from the initially homogeneous situation, and (ii) how does the model select between different possible a_o. Indeed, what is observed e.g. in the Death Valley pictures alluded to above is a characteristic wave length for the spacing of the watersheds. The "equilibrium restoring force" for a spacing a_o is $2c(a_o) + 2a_o c'(a_o)$ (<0) and one could agree that there would be a natural tendency towards that spacing width a_o for which this quantity $|c(a_o)+a_o c'(a_o)|$ is maximal. Other arguments favour the "largest" a_o for which $c(a_o) + a_o c'(a_o)$ is still negative. We shall return to this question in section III.

I.2. The small amplitude scaling continuous limit of the Thom's river basin equation

Let $a_n(t) = s_n(t) - s_{n-1}(t)$, $v_n(t) = \dot{s}_n(t)$ and $\phi(a) = ac(a)$, then (I.1) becomes:

$$v_n(t) = -\phi(a_{n+1}) + \phi(a_n) \qquad \text{(I.3)}$$

Assume ϕ is such that
for $0 < \varepsilon << 1$, $\phi(\varepsilon) > \phi(a)$ for all $a > \varepsilon$

then there will be functions v and a, defined on $I \times \mathbb{R}^+$, $I = [0,L] \subset \mathbb{R}$ such that: [7]

$$v(s_n(t),t) = v_n(t) \tag{I.4.1}$$

$$a\left(\frac{s_n(t)+s_{n-1}(t)}{2},t\right) = a_n(t) \tag{I.4.2}$$

Denoting the partial derivative with respect to s_n by $\frac{\partial}{\partial_n}$, we get:

$$a_{n+1}(t) = \left(\exp \tfrac{1}{2}a_{n+1} \frac{\partial}{\partial n}\right)a(s_n,t)$$
$$a_n(t) = \left(\exp - \tfrac{1}{2}a_n \frac{\partial}{\partial n}\right)a(s_n,t)$$

This gives the following approximations (suppressing the arguments of a):

$$a_{n+1}-a_n = a\frac{\partial a}{\partial n} + \tfrac{1}{4}a\left(\frac{\partial a}{\partial n}\right)^3 + \tfrac{1}{4}a^2\left(\frac{\partial a}{\partial n}\right)\left(\frac{\partial^2 a}{\partial n^2}\right) + \ldots \tag{I.5.1}$$

$$a_{n+1}+a_n = 2a + \tfrac{1}{2}a\left(\frac{\partial a}{\partial n}\right)^2 + \ldots \tag{I.5.2}$$

Take a_o such that $\phi'(a_o) > 0$ (unstable equidistant distribution) and expand ϕ as a MacLaurin series in $a = a_o$, to find:

$$v(s_n,t) = -\phi'(a_o)(a_{n+1}-a_n) - \tfrac{1}{2}\phi''(a_o)(a_{n+1}-a_n)(a_{n+1}+a_n-2a_o) - \\ - \frac{1}{3!}\phi'''(a_o)(a_{n+1}-a_n)(a_{n+1}^2+a_{n+1}a_n+a_n^2-3a_o(a_{n+1}+a_n)+3a_o^2) - \ldots \tag{I.6}$$

Now using (I.5) to get:

$$v(s_n,t) =$$

$$-\phi'(a_o).[a\frac{\partial a}{\partial n}+\frac{1}{4}a(\frac{\partial a}{\partial n})^3 + \frac{1}{4}a^2(\frac{\partial a}{\partial n})(\frac{\partial^2 a}{\partial n^2}) + \frac{1}{24}a^3\frac{\partial^3 a}{\partial n^3} +\ldots] -$$

$$-\frac{1}{2}\phi''(a_o)[2a(a-a_o)\frac{\partial a}{\partial n} + \frac{1}{2}a(a-a_o)(\frac{\partial a}{\partial n})^3 +\ldots] -$$

$$-\frac{1}{6}\phi'''(a_o)[3a(a-a_o)^2\frac{\partial a}{\partial n} + \frac{3}{2}a^2(a-a_o)(\frac{\partial a}{\partial n})^3 +\ldots] - \ldots \quad (I.7)$$

Again using (I.4) and (I.5) we find:

$$\frac{\partial a}{\partial t} + \frac{\partial a}{\partial n}.v = \frac{da}{dt} = \frac{\dot{a}_{n+1}+\dot{a}_n}{2} = \frac{v(s_{n+1},t)-v(s_{n-1},t)}{2} \simeq \frac{\partial v}{\partial n}.a$$

So the continuity equation for a becomes:

$$\frac{\partial a}{\partial t} = a^2\frac{\partial(\frac{v}{a})}{\partial n} \quad (I.8)$$

Inserting (I.7) in (I.8) gives:

$$\frac{\partial a}{\partial t} = a^2\{-\phi'(a_o).[\frac{\partial^2 a}{\partial n^2} + \frac{1}{24}\frac{\partial^4 a^3}{\partial n^4} + \ldots] -\tfrac{1}{2}\phi''(a_o).[\frac{\partial^2(a-a_o)^2}{\partial n^2}$$

$$+ \tfrac{1}{2}\frac{\partial}{\partial n}((a-a_o)(\frac{\partial a}{\partial n})^3) + \ldots] -$$

$$-\frac{1}{6}\phi'''(a_o)[\frac{\partial^2(a-a_o)^3}{\partial n^2} + \frac{3}{2}\frac{\partial}{\partial n}(a(a-a_o)(\frac{\partial a}{\partial n})^3)+ \ldots] - \ldots\}$$

$$(I.9)$$

Now introduce a scaling factor ℓ and write $x = \ell s_n$, so that $\frac{\partial}{\partial n} = \ell\frac{\partial}{\partial x}$.

Given the definition of the function a in (I.4.2), a mass (watershed) distribution function ρ can be defined:

$$\rho(x,t) = \frac{\ell}{a(x,t)}.$$

Let $\rho_o = \frac{\ell}{a_o}$ and $U(x,t) = \rho(x,t) - \rho_o$, then $a-a_0 \sim -\frac{\ell}{\rho_o^2}U$

and $a^3 \sim a_o^3 - 3\frac{\ell^3 U}{\rho_o^4}$

By neglecting higher than seventh order terms in ℓ, and noting that $\phi'(\frac{\ell}{a_o}) > 0$, our final equation will be of the form (after an additional time scaling):

$$\frac{\partial U}{\partial t} = [\frac{\partial^2}{\partial x^2}(-U + r_1 U^2 - r_2 U^3) - \gamma\frac{\partial^4 U}{\partial x^4}] \qquad (I.10)$$

with $r_1 \sim \ell$, $r_2 \sim \ell^2$, $\gamma \sim \ell^4$ and $\gamma > 0$.

We take no-flux boundary conditions.

In the linear approximation (I.10) is a diffusion equation with negative diffusion coefficient; mass flux is proportional to the gradient of U. As such, equation (I.10) can be called a anti-diffusion equation stressing the fact that the flux is in the opposite direction compared to the usual normal, Fickian diffusion.

It is interesting to note that based on the Landau-Ginzburg free energy model, the same form of diffusion equation as (I.10) can be derived. In this case, the linear diffusion coefficient depends on the diffusing substance environment (e.g. Temperature) and becomes negative near the aggregative state. [2]

II. Stationary solutions and stability

We seek stationary solutions of (I.10)

in $H^* = \{u \in L^2[0,L] \,|\, \langle u,1\rangle_2 = 0\}$

where $u(t) = U(\cdot,t)$ and $\langle u,1\rangle_2 = \int_0^L U(x)dx$.

II.1. Uniform distributions:

In the Hilbert space H^*, equation (I.10) has the following form:

$$\frac{du}{dt} = A(\gamma)u + N(u), \qquad \text{(II.2)}$$

where $A(\gamma)$ is a linear operator and N is non-linear operator defined for $u = U(.)$. In the sequel we suppress the variable t.

The linear operator $A(\gamma)$ is specified by:

$$[A(\gamma)u](x) = \{-\frac{\partial^2 U(x)}{\partial x^2} - \gamma\frac{\partial^4 U(x)}{\partial x^4}\} \qquad \text{(II.3)}$$

Stability of the null solution $u = 0$ depends on the eigenvalues $w(\gamma)$ of $A(\gamma)$, which are entirely given by:

$$w(\gamma)U = \{-\frac{\partial^2 U(x)}{\partial x^2} - \gamma\frac{\partial^4 U(x)}{\partial x^4}\} \qquad \text{(II.4)}$$

Then the eigenvalues $w(\gamma)$ are:

$$w(\gamma) = \{\frac{k^2\pi^2}{L^2} - \gamma\frac{k^4\pi^4}{L^4}\}, \quad k \in \mathbb{N}^+ \qquad \text{(II.5)}$$

with eigenvectors proportional to $\cos \frac{k\pi}{L}x$.

So the linear system would be stable if $\gamma > \frac{L^2}{\pi^2}$, and in this case the nonlinear system is conditionally stable ([2]).

For $\gamma < \frac{L^2}{\pi^2}$, the system is unstable in one of more modes; the fastest growing mode would be given by k such

that $\frac{k^2\pi^2}{L^2} = \frac{1}{2\gamma}$.

However, solutions (non-constant) of the linear system are

proportional to $\cos\frac{k\pi}{L}x$ with k such that $\frac{k^2\pi^2}{L^2} = \frac{1}{\gamma}$.

Although every constant function U is a solution of (I.10), we restrict ourselves to the null solution. Since $\rho(x,t) = \rho_o + U(x,t)$, with $\rho_o > 0$, constant, equation (I.10) is the evolution equation of a deviation U from a uniform distribution $\rho(x,.) = \rho_o, \forall x \in I$. So stability of a uniform distribution ρ_o is given by the stability of the null solution of (I.10).

Note that the minus sign of the term $\frac{\partial^2 U}{\partial x^2}$ was given by assuming $\phi'(a_o) > 0$ which coincide with unstability of the original discrete Thom equations. However, with the term $\frac{\partial^4 U}{\partial x^4}$, the unstability range becomes more restricted, reflecting viscosity-effects.

II.2.1. Non-constant stationary solutions

There are stationary solutions of

$$\frac{\partial U}{\partial t} = \{\frac{\partial^2}{\partial x^2}(-U+r_1U^2-r_2U^3) - \gamma\frac{\partial^4 U}{\partial x^4}\} \qquad \text{(II.6)}$$

with $r_1 \sim \ell$; $r_2 \sim \ell^2$; $\gamma \sim \ell^4$, which can be obtained as follows. Consider:

$$U(x) - r_1U^2(x) + r_2U^3(x) + \gamma\frac{d^2U}{dx^2} = \text{Constant} \qquad \text{(II.7)}$$

If U(x) also satifies the boundary conditions, then it will be a stationary solution of (II.6).
There are solutions of the form:

$$U_o(x) = \frac{1}{\alpha+\beta\cos\frac{k\pi}{L}x} \qquad (II.8)$$

with $r_1 = 3\alpha$; $r_2 = 2(\alpha^2-\beta^2)$ and $\gamma.(\frac{k\pi}{L})^2 = 1$ (II.9)

These yields bounded solutions only if $r_2 > 0$. Note that one can solve for α,β,k in terms of r_1,r_2,γ. (We shall mention other solutions at the end of this section). Since we conceive of equation (I.10) as the evolution equation of a disturbance U of a uniform distribution, (II.8) can not be taken as a solution of (I.10). Given the boundary conditions, one must have:

$$\int_0^L U_o(x)dx = 0.$$

Let $V_o(x) = U_o(x) - d$, with $U_o(x)$ as in (II.8) and

$$d = \frac{1}{L}\int_o^L U_o(x)dx.$$

Then $V_o(x)$ will be a solution of:

$$V(x) - r_1^* V^2(x) + r_2^* V^3(x) - \gamma^* \frac{d^2V}{dx^2} = \text{Constant} \qquad (II.10)$$

with
$$r_1^* = \frac{r_1-3r_2d}{1-2r_1d+3r_2d^2}$$

$$r_2^* = \frac{r_2}{1-2r_1d+3r_2d^2}$$

$$\gamma^* = \frac{\gamma}{1-2r_1d+3r_2d^2}$$

Since $d = \frac{1}{\sqrt{\alpha^2-\beta^2}} \sim \frac{1}{\ell}$, all the coefficients of (II.10) are of the same order in ℓ, as in equation (I.10). So $V_o(x)$ will be a proper solution of equation (I.10) with coefficients r_1^*, r_2^* and γ^*. This solution has a different wave length then a solution in the linear case (see section II.1 with $\gamma^* \frac{k^2\pi^2}{L^2} = 1$).

And definitely, the term $\frac{\partial^4 U}{\partial x^4}$ models "viscosity"-effects.

If γ^* tends to zero, the wave length becomes infinitely small and there is no coherence at all between the mass particles (watersheds!). In the other limit case, $\gamma^* \to \infty$, V_o tends to the null solution (Any uniform distribution is stable !).

Before reporting on the stability of the solution $V_o(x)$, we mention other stationary solutions of (II.6):

i) $U(x) = \frac{1}{\alpha+\beta\cos^2(\frac{k\pi}{L}x)}$ with: $r_1 = \frac{3}{2}\alpha$; $r_2 = 2\alpha(\alpha+\beta)$;

$$\gamma \cdot \frac{k^2\pi^2}{L^2} = \frac{1}{4}$$

(This solution belongs to the above mentioned family of solutions.)

ii) $U(x) = \frac{1}{\alpha+\beta_1\cos\frac{k\pi}{L}x+\beta_2\cos^3(\frac{k\pi}{L}x)}$ (II.11)

$$r_1 = 3\alpha;\; r_2 = 2(\alpha^2 - \tfrac{1}{9}\beta_1^2);\; \gamma\,\frac{k^2\pi^2}{L^2} = \frac{1}{9};\; \beta_2 = -\tfrac{4}{3}\beta_1. \quad (II.12)$$

iii) $$U(x) = \frac{1}{\alpha + \beta_1\cos^2(\frac{k\pi}{L}x) + \beta_2\cos^4(\frac{k\pi}{L}x)} \quad (II.13)$$

with $r_1 = 3\alpha + \frac{3}{8}\beta_1$; $r_2 = 2(\alpha^2 + \frac{1}{4}\alpha\beta_1)$; $\gamma.\frac{k^2\pi^2}{L^2} = \frac{1}{16}$; $\beta_2 = -\beta_1$.

These solutions are bounded only if $r_2 > 0$.
Of related interest are solutions of:

$$U(x) + r_2U^3(x) + r_4U^5(x) + \gamma\frac{d^2U}{dx^2} = 0.$$

which are of the form

$$U(x) = \frac{1}{\sqrt{\alpha + \beta\cos\frac{k\pi x}{L}}}, \quad (II.14)$$

with $r_2 = -4\alpha(< 0!)$, $\gamma.\frac{k^2\pi^2}{L^2} = 4$ and $\alpha^2 > \beta^2$ if $r_4 > 0$.

II.2.2. Instability of a non-constant solution

In this section we investigate the stability of a solution $V_o(x)$ of equation (II.10) as derived in II.2.1.
Since $V_o(x) = U_o(x) - d$, with $U_o(x)$ is a solution of (II.7), resp.a stationary solution of (II.6), and the connection between the coefficients of (II.10) and (II.6), stability of $V_o(x)$ follows from stability of $U_o(x)$ and vice versa.
Let $W(x,t)$ be a disturbance of $U_o(x)$, then:

$$\frac{\partial W}{\partial t} = \{\frac{\partial^2}{\partial x^2}(-1+2r_1U_o(x)-3r_2U_o^2(x))W - \gamma\,\frac{\partial^4 W}{\partial x^4}\} + \text{non-linear terms in } W \quad (II.15)$$

with no-flux boundary conditions.
And $U_o(x)$ will be stable if the null solution $W(.,.) \equiv 0$ is

a stable solution of (II.15.1.)

As in section II.1, we can write down (II.15.1) in the Hilbert space H^*, and now the linear operator $A(\gamma)$ is specified as:

$$[A(\gamma)w](x) = \{\frac{\partial^2}{\partial x^2}(-1+2r_1U_o(x) - 3r_2U_o^2(x))W(x) - \gamma\frac{\partial^4 W(x)}{\partial x^4}\}$$

with $w(.) \in H^*$; $w(.) = W(x,.)$.

However, eigenvalues of $A(\gamma)$ are not found, so linear stability cannot be established along this way.

Denoting $2r_1U_o(x) - 3r_2U_o^2(x)$ by $f(x)$, then

$$f(x) = \frac{6\delta}{\delta+\cos\frac{k\pi x}{L}} - \frac{6(\delta^2-1)}{(\delta+\cos\frac{k\pi x}{L})^2} \qquad \text{(II.16)}$$

where $\delta = \frac{\alpha}{\beta}$ and $\frac{k\pi}{L} = \sqrt{\frac{1}{\gamma}}$ (see (II.9)).

Define in H^* the functional $F(t)$ by:

$$[F(t)]w = \int_o^L W^2(x,t)dx = \|w\|_2 \qquad \text{(II.17)}$$

Then $[F(t)]w = 0 \rightarrow w = 0$ and $\frac{dF}{dt}(w) = \int_o^L 2W(x,t)\frac{\partial W}{\partial t}dx.$

Now using only the linear part of (II.15.1), then by partial integration,

$$\frac{dF}{dt}(w) = 2[\int_o^L (\frac{\partial W}{\partial x})^2 dx - \gamma \int_o^L (\frac{\partial^2 W}{\partial x^2})^2 dx + \int_o^L f(x)W(x,t)\frac{\partial^2 W}{\partial x^2}dx] \qquad \text{(II.18)}$$

If we take $W(x,.)$ proportional to $\cos\frac{k\pi x}{L}$ (with $\frac{k\pi}{L} = \sqrt{\frac{1}{\gamma}}$) then (II.18) reduces to:

$$\frac{dF}{dt}(w) = 2\int_o^L f(x)W(x,t)\frac{\partial^2 W}{\partial x^2}dx \qquad \text{(II.19)}$$

And in this case (W proportional to $\cos\frac{k\pi x}{L}$), $\frac{dF}{dt} > 0$ if

$$\int_0^L f(x)\cos^2\left(\frac{k\pi x}{L}\right)dx < 0 \qquad \text{(II.20)}$$

(II.20) is equivalent with:

$$6 - 12\delta^2 + 12\delta\sqrt{\delta^2-1} < 0 \qquad \text{(II.21)}$$

which holds for any δ, with $\delta^2 \geq 1$ ($\delta^2 = \frac{\alpha^2}{\beta^2} > 1$ if $r_2 > 0$).

Since F gives the norm of w, and w with $w(.) = W(x,.)$ proportional to $\cos\frac{k\pi x}{L}$ can be taken as close in norm to $w = 0$ as one wishes, $\frac{dF}{dt} > 0$ means linear unstability of the solution $U_o(x)$ (in L_2-norm).
So the solution $V_o(x)$ is unstable.

III <u>Existence of stable non-constant solutions and numerical simulations</u>

We have made numerical simulations of

$$\frac{\partial U}{\partial t} = \frac{\partial^2}{\partial x^2}\phi(U) - \gamma\frac{\partial^4 U}{\partial x^4} \qquad \text{(III.1)}$$

where $\phi(U) = -U + r_1U^2 - r_2U^3$ (III.2)

defined on the interval $[0,L] \subset \mathbb{R}$ with no-flux boundary conditions.
For $r_2 < 0$, we shall show the existence of non-constant stable solutions.

III.1 <u>Case $r_2 > 0$</u>

The stationary solutions of section II.2 are never found numerically. Even if the initial value is a discretization of such a non-constant solution, in time the solution becomes unbounded. Since these solution are unstable, another outcome could not be expected

III.2 Case $r_2 < 0$

III.2.1. Existence of stable non-constant solutions.

Stationary solutions of (III.1) must satisfy:

$$\phi(U) - \gamma\frac{\partial^2 U}{\partial x^2} = \beta \text{ and } \int_0^L U(x)dx = 0 \qquad \text{(III.3)}$$

with $\beta = \int_0^L \phi(U)dx$

We seek solutions of (III.3) in the Hilbert space $H^{1,*} = \{u\in H^1 | \langle u,1\rangle_1 = 0\}$ with H^1 Sobolev:

$$\langle f,g\rangle_1 = \int_0^L (fg+f_x g_x)dx \qquad (f_x = \frac{\partial f}{\partial x})$$

Then the variational formulation of (III.3) is:

$$\int_0^L (\phi(U) - \gamma\frac{\partial^2 U}{\partial x^2})h\,dx = 0 \qquad \forall h \in H^{1,*} \qquad \text{(III.4)}$$

Let $V(u) = \int_0^u \phi(s)ds$

Define on $H^{1,*}$ a functional F:

$$F(u) = \int_0^L V(u) + \tfrac{1}{2}\gamma(\frac{\partial U}{\partial x})^2 dx \qquad \text{(III.5)}$$

$$(u(t) = U(\cdot,t))$$

Then solutions of (III.4) are aequivalent with the critical points of F.

Since $\frac{dF}{dt}(u) = \int_0^L - (\frac{\partial}{\partial x}(\phi(U) - \gamma\frac{\partial^2 U}{\partial x^2}))^2 dx \leq 0$, F is a Lyapunov functional for the system (III.1).

Returning to the specification of $\phi(U) = -U + r_1U^2 - r_2U^3$, gives that F is bounded from below if $r_2 < 0$. Since F is C^1, following the argumentation in [1], F attains its minimum. Now, if $U \equiv 0$ is unstable, then $U \equiv 0$ is not a minimum of F. So, there exists a non-zero solution U of (III.3) where F attains its minimum and U is at least stable.
Since $\int_0^L U dx = 0$, , U is non-constant.

III.2.2. Numerical simulations

III.2.2.1 $\gamma = 0$

Using notations of (III.2.1), stationary solutions of (III.1) are given by:

$$\phi(U) = \beta; \quad \int_0^L U(x)dx = 0 \qquad \text{(III.6)}$$

$\phi(U) = \beta$ is aequivalent with $V(U) = \beta U + \alpha$. So, in this case the functional F (III.5) is minimal if α is minimal. For $r_2 < 0$, this is the case, if $\beta U + \alpha$ is tangent at $V(U)$ in $U = U_o$ and $U = U_1$ $(U_o \neq U_1)$
Using ϕ is cubic in U, the values U_o and U_1, must satisfy:

$$\phi(U_o) = \phi(U_1)$$

$$\phi'(U_o) = \phi'(U_1) > 0 \qquad \text{(III.7.)}$$

The final pattern is characterised by these two values U_o and U_1 (under restriction: sum U is zero).
In figure 2.1, $r_1 = 0$, $r_2 = -1/3$ and $U_o = \sqrt{3}$, $U_1 = -\sqrt{3}$.
In figure 2.2, $r_1 = -4/5$, $r_2 = -1/5$ and
$U_o = \frac{4 + \sqrt{93}}{3}$, $U_1 = \frac{4 - \sqrt{93}}{3}$.
See also figure 2.3b.

III.2.2.2 $\gamma \neq 0$

In this case we can choose γ such that the null solution is unstable in modes with wavelength bounded away from zero. The pattern evolution is dominated initially by the

mode $\cos\frac{k\pi x}{L}$ with $\frac{k^2\pi^2}{L^2} = \frac{1}{2\gamma}$ (see section II.1).

In figure 2.3a, the pattern evolution is shown for $r_1 = 0$, $r_2 = -\frac{1}{3}$ and $\gamma = 0{,}00482$; the largest positive eigenvalue is for k = 16 which corresponds to a wavelength ~ 0,6 (e.h.) in figure 2.3. The final pattern shows a wavelength which is at least twice as long. The same holds for other values of γ; in figure 2.4 we have taken $\gamma = 0{,}00241$ (largest eigenvalue for k = 24, wavelength ~ 0,4 (e.h.); figure 2.5: $\gamma = 0{,}01563$ (largest eigenvalue for k = 9, wavelength ~ 1,1 (e.h.).
In figure 2.6, $r_1 = -\frac{4}{5}$, $r_2 = -\frac{1}{5}$, $\gamma = 0{,}01563$. Compared with figure 2.5, the difference must be a consequence of the fact that the number of gridpoints with $U_0 = \frac{4+\sqrt{93}}{3}$ (figure 2.2) is far less than the number of gridpoints with $U_0 = \sqrt{3}$ (figure 2.1) which are the corresponding figures if $\gamma = 0$.
In figure 2.3b, we have set $\gamma = 0$ after reaching the final pattern as depicted in III.3a. The final block form is totally given by the values U_0 and U_1 as derived in section III.2.2.1.

III.3. Figures

Figure 2.1: $\phi(U) = -U + \frac{1}{3}U^3$; $\gamma = 0$

Figure 2.2: $\phi(U) = -U - \frac{4}{5}U^2 + \frac{1}{5}U^3$; $\gamma = 0$

Figure 2.3: $\phi(U) = -U + \frac{1}{3}U^3$; $\gamma = 0{,}00482$

2.3a: dotted curve initial evolution pattern; continuous curve final pattern.

2.3b: after setting $\gamma = 0$, the final pattern of figure 2.3a becomes block-like.

Figure 2.4: $\phi(U) = -U + \frac{1}{3}U^3$; $\gamma = 0{,}00241$.

Figure 2.5: $\phi(U) = -U + \frac{1}{3}U^3$; $\gamma = 0{,}01563$

dotted curve: initial evolution pattern;

continuous curve: final pattern.

Figure 2.6: $\phi(U) = -U - \frac{4}{5}U^2 + \frac{1}{5}U^3$; $\gamma = 0{,}01563$

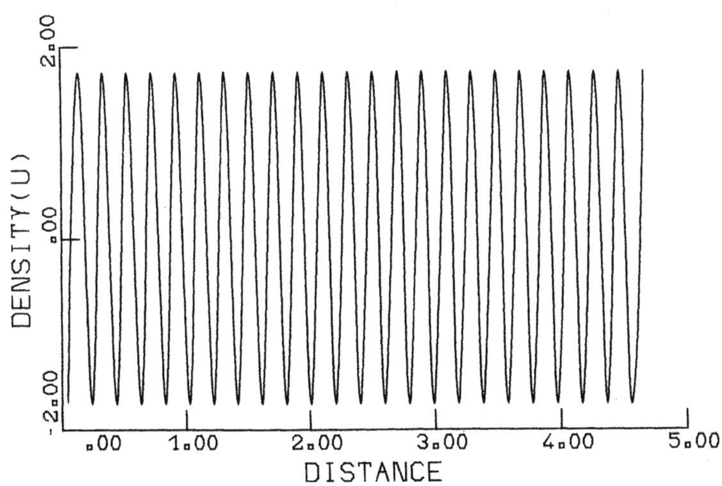

Figure 2.1: $\phi(U) = - U + \frac{1}{3}U^3$; $\gamma = 0$

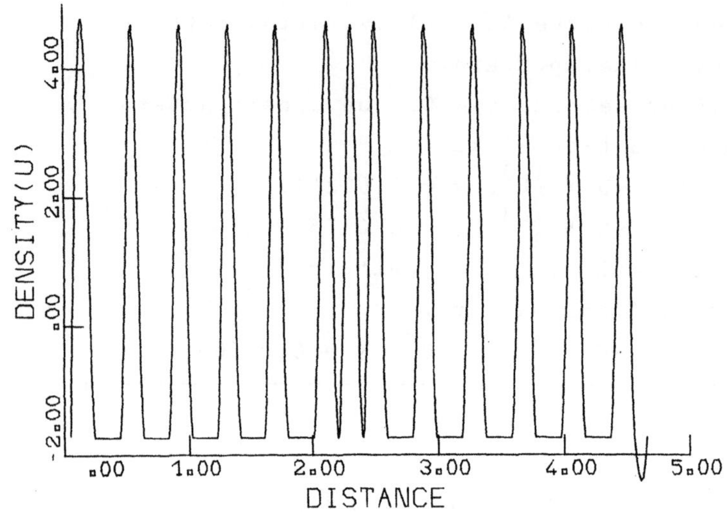

Figure 2.2: $\phi(U) = -U - \frac{4}{5}U^2 + \frac{1}{5}U^3$; $\gamma = 0$

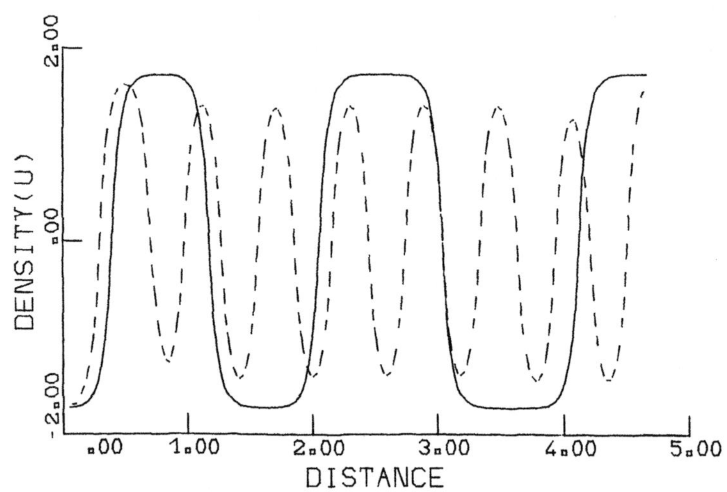

Figure 2.3a:pattern evolution for $\phi(U) = -U + \frac{1}{3}U^3$;
$\gamma = 0{,}00482$

Final pattern: continuous curve

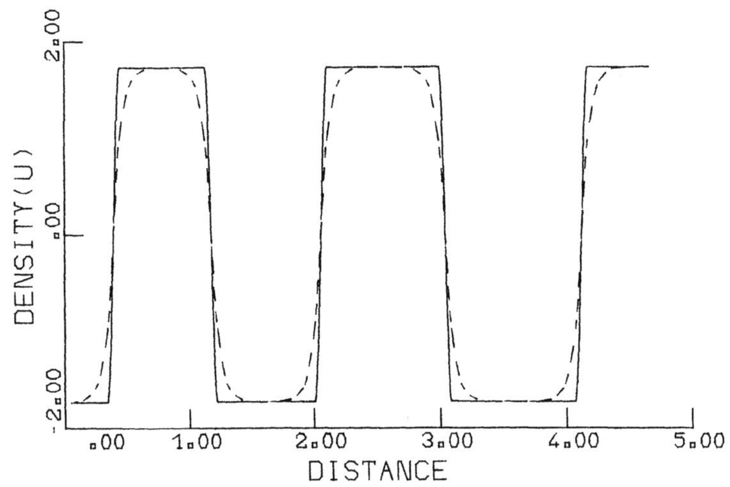

Figure 2.3b: Continuation of final pattern of (2.3a) after setting $\gamma = 0$.

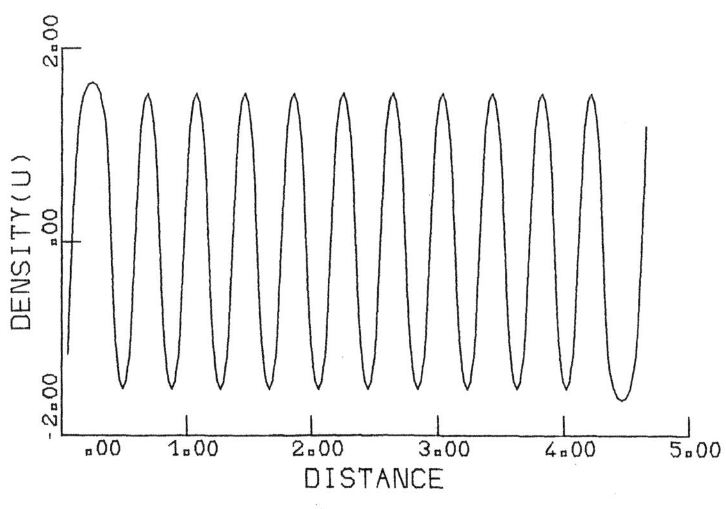

Figure 2.4: $\phi(U) = -U + \frac{1}{3}U^3$; $\gamma = 0{,}00241$.

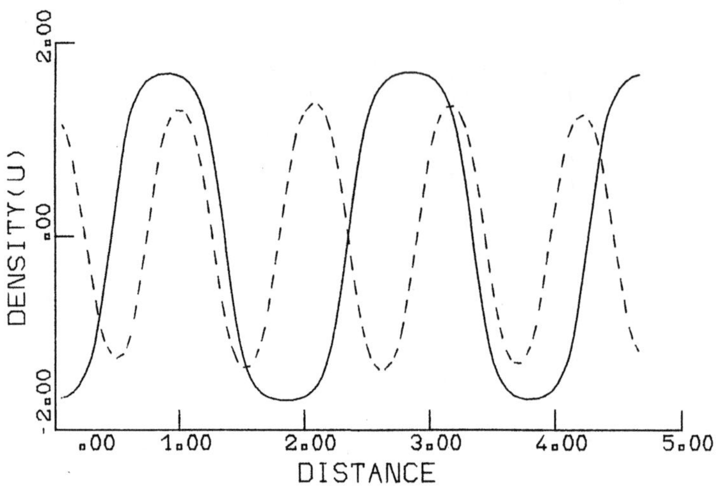

Figure 2.5:pattern evolution for
$\phi(U) = -U + \frac{1}{3}U^3$; $\gamma = 0,01563$
Final pattern: continuous curve

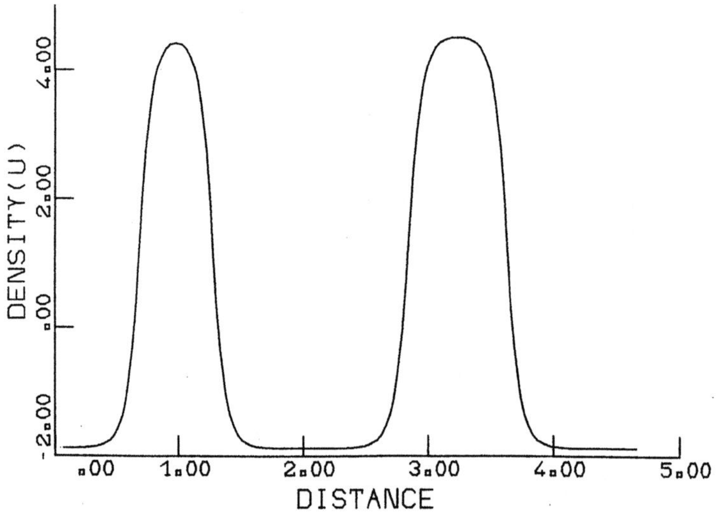

Figure 2.6: $\phi(U) = -U - \frac{4}{5}U^2 + \frac{1}{5}U^3$; $\gamma = 0,01563$.

Conclusion

Thom's river basin model describes the situation in which for a certain basin width, large basins grow at the cost of smaller ones. In defining a continuous distribution function for the watersheds of the basins, we have derived an evolution equation of pure diffusion type. The diffusion coefficient is density dependent and negative for some range of density values. In deriving this continuous form of Thom's discrete model, it is quite natural to incorporate a fourth order derivative term which can be interpreted as modelling viscosity effects.

We have found a class of stationary solutions of this so-called anti-diffusion equation, which are, however, not stable. If the diffusion term is of cubic form with two stable branches, numerically we have found bounded non-constant solutions. Existence of stable non constant solutions is analytically established. By deleting the fourth order derivative term, the solution becomes block form like and is totally characterised by two unique density values.

As such, this anti-diffusion equation seems a respectable candidate for modelling evolution processes which form patterns in the case that there is only one substance involved.

Acknowledgements

The research of the second author is supported by the Netherlands Organization for Advancement of Pure Research (Z.W.O.).

References

1. Chow, S.N. and Hale, J.K.: Methods of Bifurcation Theory, Springer-Verlag, New York, 1982.

2. Cohen, D.S. and Murray, J.D.: 'A generalized diffusion model for growth an dispersal in a population', Journal of Mathematical Biology 12 (1981), 237-249.
3. Iooss, G. and Joseph, D.D.: Elementary stability and bifurcation theory, Springer-Verlag, New York, 1980.
4. Meinhardt, H.: 'The Spatial Control of Cell Differentiation by Autocatalysis and Lateral Inhibition', in H. Haken (ed), Synergetics, a workshop, Springer-Verlag Berlin, 1977.
5. Ortoleva, P.: 'The Multifaceted Family of the Nonlinear: Waves and Fields, Center Dynamics, Catastrophes, Rock Bands and Precipitation Patterns', in A. Pacault and C. Vidal (eds.), Synergetics, Far from Equilibrium, Springer-Verlag Berlin, 1979.
6. Thom, R.: 'Symmetries gained and lost', in K. Maurin and R. Raczka (eds.), Mathematical physics and physical mathematics, Reidel, Dordrecht, 1976.
7. Whitman, G.B.: Linear and Nonlinear Waves, Wiley-Interscience New York, 1974.

ON THE MODELLING OF VEHICULAR TRAFFIC FLOW

Alfred E. Beylich
Technische Hochschule
5100 Aachen
Federal Republic of Germany

ABSTRACT. Vehicular traffic flow belongs to the category of transport processes, and it can be studied at different levels: by phenomenological or continuum description, by a microscopic approach or kinetic theory, or by direct numerical experiment. In the present study we try to find interrelations between the different levels using, as an example, a numerical simulation of steady/homogeneous flow and of unsteady wave propagation. Similarities are found between the present results and solutions of the Burgers-Korteweg-deVries equation. Numerical simulation is also useful as a guide for the development of a kinetic theory. Such a kinetic model is presented which considers clustering and includes the desired speed of the ensemble of vehicles.

1. INTRODUCTION

Vehicular traffic flow is a special type of transport process: the objects to be transported are persons and goods. The problem is characterized by one-dimensional motion on lanes with provisions for passing by means of additional lanes.

In thermodynamics we speak of driving "forces" that cause the "fluxes". In vehicular traffic our main question is: What are the forces or what are the attractors ? If we want to model the process, we are somewhat in the situation of a member of a space ship from another star looking from far distance at the complicated network of roads and wondering why vehicles move.

Although we could know why an individual vehicle moves and we are able to describe its kinematics in every detail, this does not help us much, since we are basically interested in averaged quantities and general behaviour. Yet, we always will be aware of the fact that the number of "particles" we are dealing with is small (compared to the luxuriously high numbers of molecules, for instance in a gas), and therefore all averaged quantities will have some fuzziness.

In order to proceed, one may have a look at a simplified version of the problem, see Fig. 1. We note that we should distinguish two groups. First there exist complicated networks of lanes with many nodes. We may very roughly divide them into (modern) chessboard-type

C. W. Kilmister (ed.), Disequilibrium and Self-Organisation, 47–63.

cities and (medieval) spiderweb-type cities. The characteristic length s_A between two nodes in units of vehicle length σ_0 is of the order of 10 to 100. There exist speed limits, hindrance due to high density, and at the nodes regulation by traffic lights. We may call this "type A" traffic.

Second we notice that between the cities we find speedways (Autobahnen) with multilane unidirectional flow. Here the characteristic length s_B is of the order of 10^3 to 10^4 σ_0. We call this "type B" traffic.

There is no doubt that type B seems to belong to the less complicated problems. It is not only one-dimensional, but also the motivation for driving is more easily modelled. We can say that a vehicle attempts to get for instance from city C to city D in an "optimum" way, or more precisely: a vehicle attempts to drive at a desired speed w(x,t) and, trying to do so, it is only hindered by the other vehicles. As mentioned before, up to this point the problem is still a simple one, complications arise when we try to average and introduce a "desired speed" distribution function in which the individuality is lost.

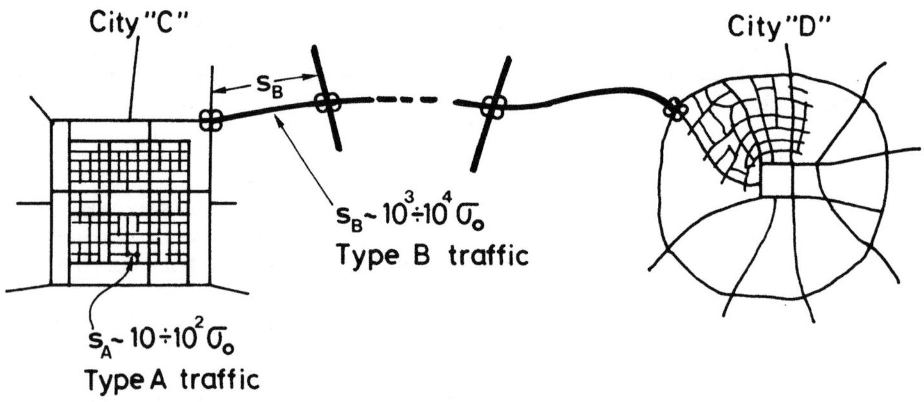

Figure 1: Types of traffic flow

In this paper, I want to restrict myself to attempts of modelling two-lane uni-directional flow without sources and sinks. Since it has always improved understanding of a problem to approach it from different levels and with different techniques, I want to report on some work for which the scheme of Fig. 2 may be a guideline.

In the following, first a Monte Carlo method will be described, then some useful relations from continuum theory will be presented. Some results from computer experiments will be discussed and, finally, a kinetic model will be developed.

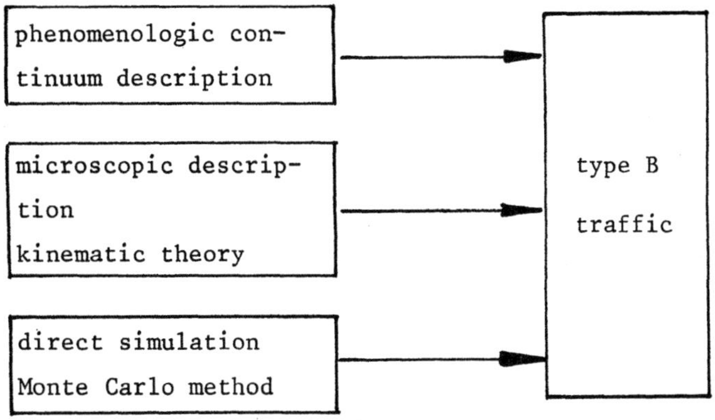

Figure 2: Hierarchy of methods of approach

2. MONTE CARLO MODEL

Intuitively, direct or crude simulation [1,2] seems to be the most straightforward method to obtain results from a microscopic description by averaging. Initially, the desired speed of vehicle i is picked by a random number out of a distributio Φ (w; x,t); then the vehicle is moved according to an acceleration law

$$\frac{dv_i}{dt} = \begin{cases} \dfrac{w_i - v_i}{\tau} & \text{, for free vehicles ,} \qquad (1a) \\ b_- & \text{, for } (x_j - x_i) < \sigma_i \text{ ,} \qquad (1b) \end{cases}$$

where τ is a relaxation time, b_- a deceleration, σ_i the safety distance of vehicle i, and j the index of a leading vehicle. Here we have

$$\sigma_i = \sigma_0 + \tilde{\alpha}\, v_i + \tilde{\beta}\, (v_i^2 - v_j^2). \qquad (2)$$

with σ_0 as the vehicle length, $\tilde{\alpha}$ a reaction time, and $\tilde{\beta} = 1/2\, b_-$.

The numerical procedure splits into two steps. First, movement according to Eq. (1) with condition (2); second, check for change of lanes and book-keeping. The local environment of a vehicle is completely described by a small set of parameters:

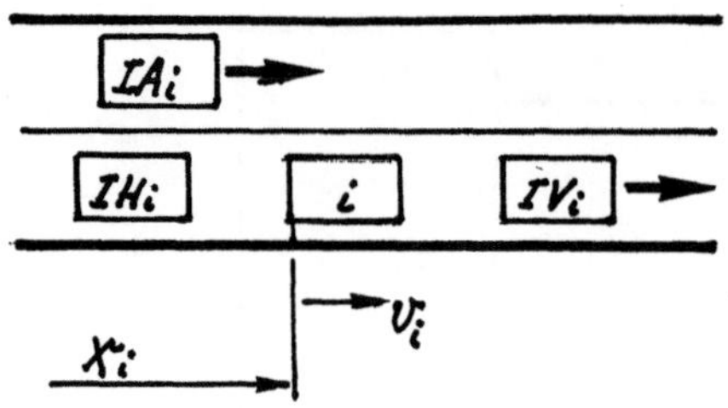

Figure 3: Local environment

i = number of vehicle
w_i = desired speed
v_i = actual speed
x_i = location of vehicle i
IV_i = number of leading vehicle
IH_i = number of trailing vehicle
IA_i = trailing vehicle on other lane
$IZ_i \begin{cases} \geq 1 & \text{right lane} \\ \leq -1 & \text{left lane} \end{cases}$

with $|IZ_i| = 1$ for a free vehicle and $|IZ_i| > 1$ for condensed vehicle. The following rules for the change of lanes hold:

Change to left for free vehicles, if

(1) $\dfrac{x(IV(I)) - x(I)}{V(I) - V(IV(I))} < t_v$ and $V(IV(I)) < V(I)$

(2) a free gap on the left lane exists (this means that IA_i is far enough behind so that it does not have to start an emergency stop and $IV(IA_i)$ is sufficiently far ahead).

Change to left for condensed vehicle) is attemped continuously and condition (2) holds.

Change to right is only performed by free vehicles (IZ = -1) and if the time gap (the time it would spend on the right lane at the moment of decision) is larger than a minimum time t_{vw}.

There exists a lower limit for the deceleration b_- if accidents have to be excluded.

$$b_- \geq \frac{(v_i - v_j)^2}{2\,[\tilde{\alpha}\, v_i + \tilde{\beta}(v_i^2 - v_j^2]}. \qquad (3a)$$

Since it is simpler to work with a fixed value, one may use

$$b_- \simeq \frac{w_{max}}{2\tilde{\alpha}}. \qquad (3b)$$

All quantities are non-dimensionalized using the length σ_0 and the maximum speed w_{max}.

The desired speed distribution $\Phi(w)$ is, for simplicity, independent of space and time; the individual speeds are produced with help of a random generator. For a given random number R_i, w_i, with $0 \leq R_i \leq 1$, the desired speed will, in our case, become

$$w_i = w_u + (1 - w_u) R_i, \tag{4}$$

since we have chosen a constant (boxcar) distribution between the lower speed w_u and $w_{max} = 1$. Any other distribution may be produced by integrating $\Phi(w)$ and inversion.

The ensemble of vehicles (a typical number is 1000) is randomly distributed on the two-lane system of length L which is closed to a ring. This avoids creation and annihilation of further vehicles.

The finite size of time steps, Δt, introduces an additional dispersion into the safety distance σ. After a number of time steps, say 100, average values (moments) may be taken by proper summation. For this purpose, one has to define sample widths in space Δx and in speed Δv.

3. SOME USEFUL TOOLS FROM CONTINUUM THEORY

Similar to approaches used for instance in the physics of fluids, one may under certain conditions ignore the fact that a system consists of discrete units, and change to (smooth) field variables. Although the averaging process and its theoretical foundations are in many cases extremely complicated, once the smoothing has taken place (more often than not by some crude procedures), the new system is relatively simple or rather its problems lie on another (a higher) level. The ad hoc starting from a continuum level is such common practice that someone may wonder why others go through the trouble to start on a microscopic level. But even if a microscopic description may not lead to many practically useful results, it always leads to much insight into the limitations of a continuum theory.

In the present problem of type B traffic one may for instance profit from the achievements of wave theory [3,4] when describing the flux of vehicles on roads. It seems realistic to use only one balance equation: the conservation of number density n(x,t) or of concentration $\varrho(x,t)$. If we restrict ourselves to problems with continuous changes, the continuity equation is written in differential form

$$\frac{\partial}{\partial t} \varrho(x,t) + \frac{\partial}{\partial x} Q(x,t) = 0 \tag{5}$$

where Q is the flux of concentration. In the spirit of wave theory one may assume that Q is, in first order, only dependent upon ϱ. Further, it may depend also on the gradients of ϱ with respect to space and time. Then, one may write down the following expansion

$$Q = Q_0 + \left(\frac{\partial Q}{\partial \varrho}\right)\varrho + \left(\frac{1}{2}\frac{\partial^2 Q}{\partial \varrho^2}\right)\varrho^2 + \dots$$
$$+ \left(\frac{\partial Q}{\partial \varrho_x}\right)\frac{\partial \varrho}{\partial x} + \left(\frac{\partial^2 Q}{\partial \varrho_{xx}}\right)\frac{\partial^2 \varrho}{\partial x^2} + \dots$$
$$+ \left(\frac{\partial Q}{\partial \varrho_t}\right)\frac{\partial \varrho}{\partial t} + \dots \quad . \tag{6}$$

Putting this expansion into Eq.(5) (neglecting the last term), we obtain

$$\varrho_t + \left(\frac{\partial Q}{\partial \varrho}\right)\varrho_x + \left(\frac{\partial^2 Q}{\partial \varrho^2}\right)\varrho\varrho_x + \left(\frac{\partial Q}{\partial \varrho_x}\right)\varrho_{xx} + \left(\frac{\partial Q}{\partial \varrho_{xx}}\right)\varrho_{xxx} = 0 \tag{7}$$

If we restrict ourselves to the first two terms, we obtain the differential equation for a *simple wave*; for a simple dependence $Q(\varrho)$ then $\partial Q/\partial \varrho = c_0$ is the wave speed, i.e. the speed by which perturbations are transported:

$$\varrho_t + c_0\, \varrho_x = 0. \tag{8}$$

This equation allows to construct, for general inhomogenous and unsteady phenomena, the concentration at x,t from the initial distribution $\varrho(x,t = 0)$; c_0 is the tangent to $Q(\varrho)$ (this curve is the so called fundamental diagram).

One may proceed to *quasi-simple waves*, when taking into account the other terms. For this reason it is useful to transform to a frame moving with c_0,

$$\left.\begin{aligned} x' &= x - c_0\, t\,, \\ t' &= t \quad . \end{aligned}\right\} \tag{9}$$

We further introduce

$$\eta = -\left(\frac{\partial^2 Q}{\partial \varrho^2}\right), \qquad \nu = -\left(\frac{\partial Q}{\partial \varrho_x}\right), \qquad \beta = \left(\frac{\partial^2 Q}{\partial \varrho_{xx}}\right).$$

Then we obtain for Eq.(7)

$$\eta_t - \eta\eta_x - \nu\eta_{xx} + \beta\eta_{xxx} = 0 \,, \tag{10}$$

which is, without the third term, the Korteweg-deVries (KdV) equation. From the dispersion relation for the linearized equation

$$\omega = c_0\, k - \beta\, k^3 - i\, \nu\, k^2 \,, \tag{11}$$

where k is a wave number, one may notice that β introduces dispersion into the system and the (friction) term ν damping (dissipation). The total Eq. (10) is called Burgers-Korteweg-deVries (BKdV) equation, which has been found in various very different fields. Its solutions are characterized by the parameters ν and β:for β being positive and $\nu > \nu_{crit}$, one obtains smooth profiles. For $\nu < \nu_{crit}$ oscillatoric structures appear in the tail of an initially smooth step. Below a critical value of β, with ν very small an initially smooth wave decays into a finite number of solitons. For $\beta < 0$, oscillatoric structures develop as forerunners. The nonlinear (second) term in Eq.(10) causes steepening and, interacting with the third term, allows the existence of steady wave profiles.

These properties should be kept in mind when interpreting the results from numerical experiments in the next chapter.

4. RESULTS FROM NUMERICAL EXPERIMENTS

The usefulness of numerical simulation stems from the fact that it allows one to perform inexpensive experiments under clean conditions. the present investigation consists of two parts: first, so-called steady flow is investigated by starting at very low densities and then, step by step, "shrinking" the ring lane system with allowance for relaxation (1000 to 2000 time units) after each step. In the second part, after a relaxation time of 1000 units, provided to produce a homogeneous distribution, suddenly a free strip is inserted into the ring lane in order to study unsteady (wave) behaviour.

Parameters, used in the numerical experiments, are

$$w_u = .5 \qquad t_v = 72$$
$$\alpha = 8.0 \qquad t_{vw} = 216$$
$$\tau = 72 \qquad \beta = -12.5 .$$

The total number of vehicles was 1000. (For a maximum speed $w_{max} = 50$ m/s and a vehicle length of $\sigma_0 = 5$ m, the characteristic time $t_0 = .1$ sec. Thus, the reaction time α is .8 sec and the relaxation time τ corresponds to 7.2 sec).

First, the "fundamental diagram" $Q(\rho)$ was constructed, Fig. 4. One notices the deviation from linear behaviour for small ρ(due to mutual hindrance), a maximum a $\rho \approx .4$, and a further decay of flux at higher ρ. the "close packing" upper limit $(2-\rho)/2\alpha$ is not reached.

Further information about the behaviour of the ensemble is gained by looking at the rate of lane changes (R_w = number of changes/(1000 veh. and 1000 ΔT)). The rapid decay of R_w with increasing ρ, Fig. 5, makes it impossible for vehicles to diffuse, and free movement becomes impossible. We may use another quantity to characterize the quality of the flow, the "entropy" E, which can be defined as follows

$$E = - \sum_{j=1}^{j_{max}} f(v_j) \ln f(v_j). \qquad (12)$$

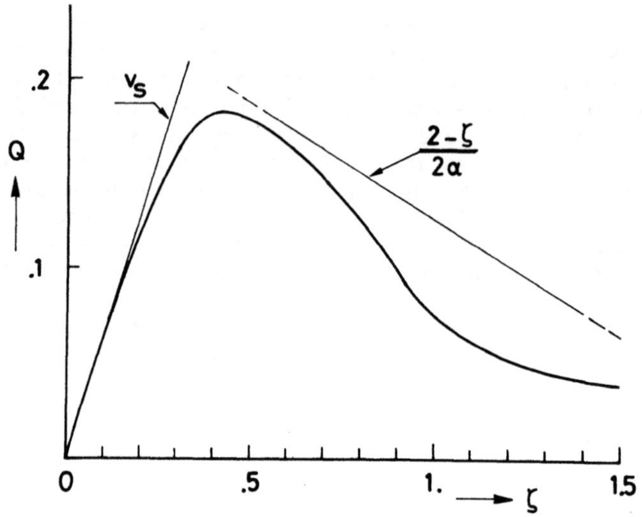

Figure 4: Flux-concentration diagram Q(ϱ) for a so-called steady condition

Velocity space is divided into j_{max} cells, and $f(v_i)$ is the number of vehicles with speed at v_i divided by the total number N. For low ϱ, E decreases. Then, at ϱ ~ .4, E has a peak and decreases for increasing ϱ, until at ϱ ~ .8 a second peak appears. Beyond ϱ = .9, E decreases monotonically. Before this behaviour is interpreted, it may be useful to take a look at the speed distributions for the right and left lane in Fig. 6. The parameter is the concentration ϱ. One notices up to ϱ = 0.3, a narrowing of the distribution which is also evident from the growing peak height; this behaviour (also verified in kinetic theory work [2]) corresponds to a decrease of E. At ϱ = .35, most vehicles have speeds $v > w_u$.

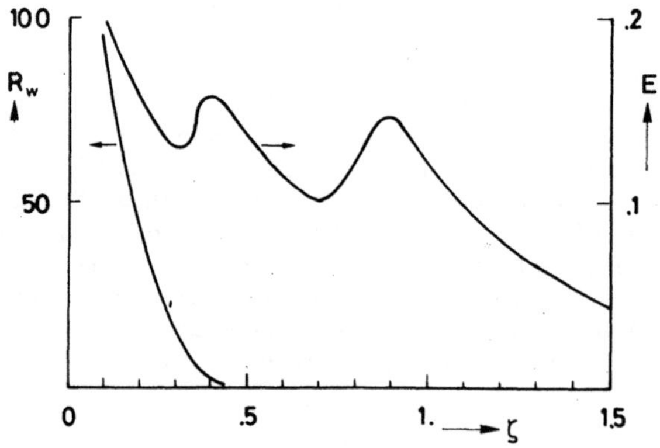

Figure 5: Rate of lane changes R_w and entropy as function of concentration ϱ for "steady" condition

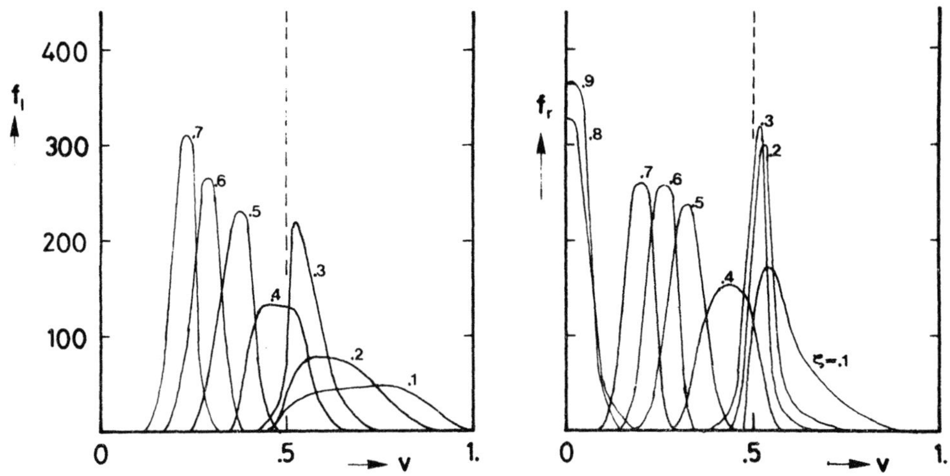

Figure 6: Speed distribution functions $f_{r,l}(v)$ for right and left lanes for concentrations from ϱ = .1 up to .9

Then, at $\varrho \sim .4$ a "break through" occurs, since vehicles are forced to take speeds below w_u due to the prescriptions of the safety distance $\sigma(v)$, Eq.(2). Yet this does not explain the strange effect of broadening of the distribution to which also corresponds the increase in E. An explanation may come from the equation for simple waves, Eq. (8), which tells us that at $\varrho \sim .4$ the wave speed c_0 is practically zero (see also Fig. 4). Thus, perturbations necessary to relax the system to an equilibrium state, are not transported. In fact, attempts to increase the relaxation time did not yield better results, but rather instabilities. In Fig. 6 it can be seen that beyond the "critical" point at $\varrho \sim .4$ the distribution again shows further narrowing which is rewarded by decrease in entropy E. But then, at about $\varrho = .8$, a new effect appears: A certain fraction of vehicles is standing still. This "stop-and-go" traffic causes the second peak of E.

Much further insight can be gained when looking at the behaviour of the vehicles in phase space. In Fig. 7 projections on the v-w plane are shown for ϱ from .1 to 1.2 in steps of .1. At very low ϱ, when there are no interactions at all, the vehicles would all be represented by points on the line attractor (in our example between (v,w) = (.5,.5) and (1., 1.)). When ϱ is increased, the points move away from this diagonal (the fast-moving vehicles suffer most). Further increase in ϱ moves the ensemble of points to lower values of v. Beginning at $\varrho = .8$, where stop-and-go traffic starts, we may observe a structure in the distribution. It is caused by the finite time step size: vehicles which have come to a stop are restarted only at discrete time levels. The structure could easily be randomized, but it is, on the other hand, a clear indicator of stop-and-go traffic, and it does no harm to the system.

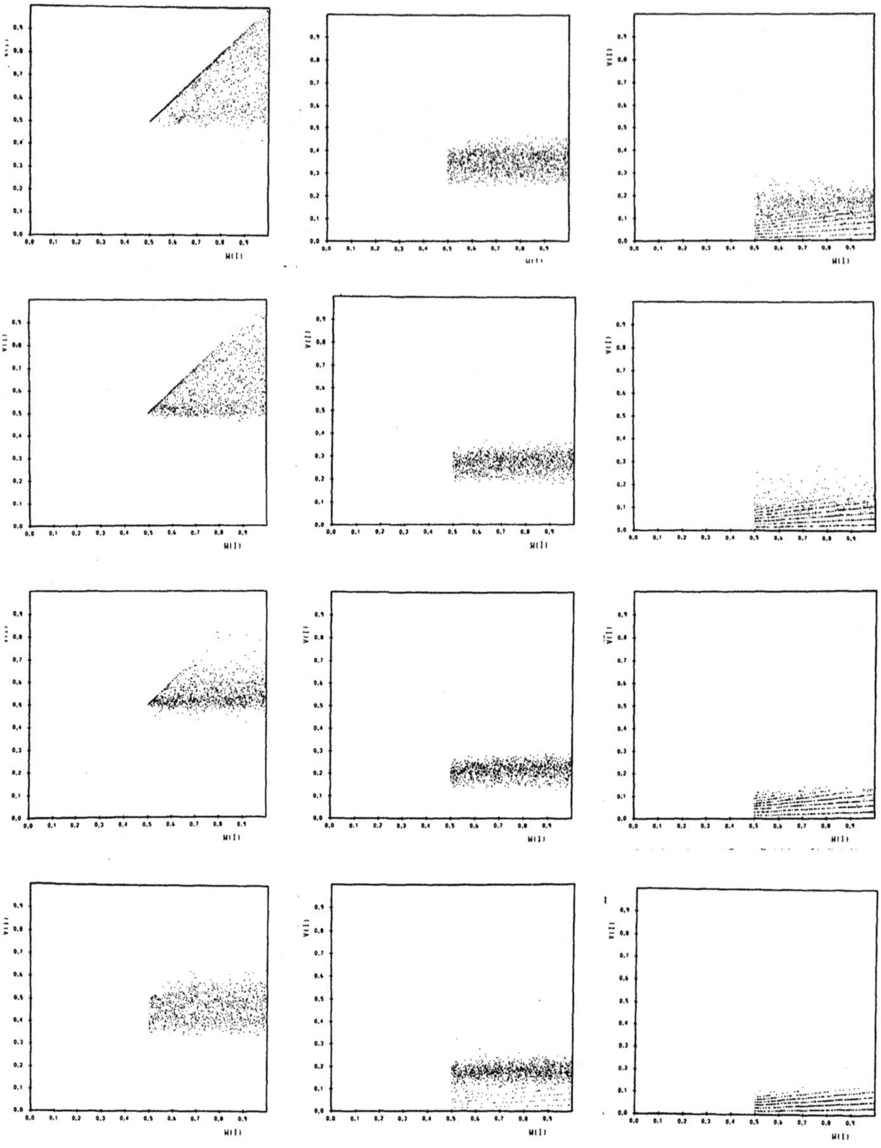

Figure 7: Phase space projection on the v-w plane, Upper left: ζ = .1 increasing in steps of $\Delta\zeta$ = .1 from top to bottom, upper middle : ζ = .5, upper right: ζ = .9

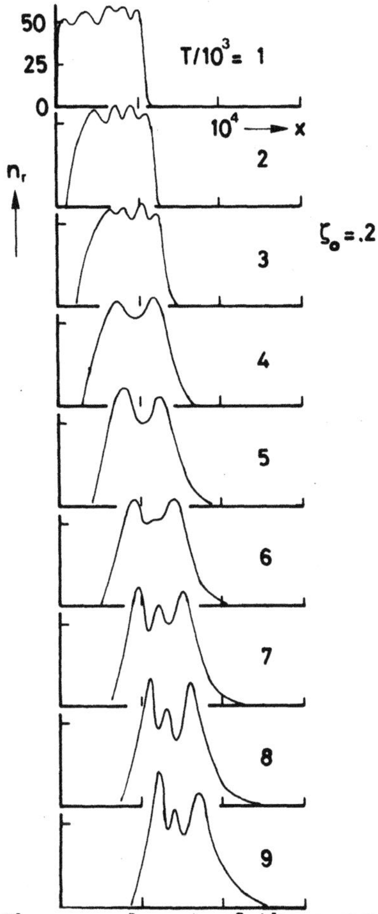

Figure 8:
Development of a boxcar density distribution. Initial concentration ζ_0 = .2. Time steps ΔT = 1.000. Only right lane density is shown. Sample length = 500

The second part of these numerical experiments is concerned with unsteady (wave) phenomena. Figure 8 shows the development of an originally constant concentration between x = 0 and 5000. For an initial concentration of ζ_0 = .2 (which is "subcritical"), we see a wave moving

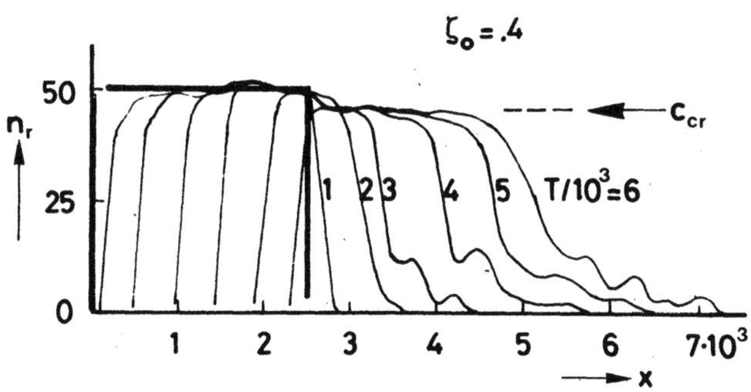

Figure 9: Same as in Fig. 7, but for ζ_0 = .4. Sample length = 250

to the right and developing a structure which strongly reminds us of the soliton solutions of Eq.(10): apparently the damping coefficient ν is small;there exists a dispersion, as one may notice from the broadening of the leading "shock" wave. Simple-wave theory tells us that the trailing rarefaction should not broaden, which is true in this case.

Figure 9 shows a similar situation, but for ζ_0 = .4, which is just about the critical value. Here we notice a small decrease of the plateau (denoted by c_{cr}) downstream of the shock wave moving to the right. Again the shock wave broadens but now, oscillations appear only in the shock front; the rarefaction wave remains steep.

The characteristic wave phenomena are even more clearly seen in Fig. 10, where ζ_0 = .6. One is reminded of the Riemann problem (shock tube problem [9]): at the leading edge, a shock wave travels to the right and a rarefaction wave to the left. The height of the constant plateau is, for all ζ_0 > .4, the same (note the different sample lengths), and independent of ζ_0. Also, the shock speed is always the

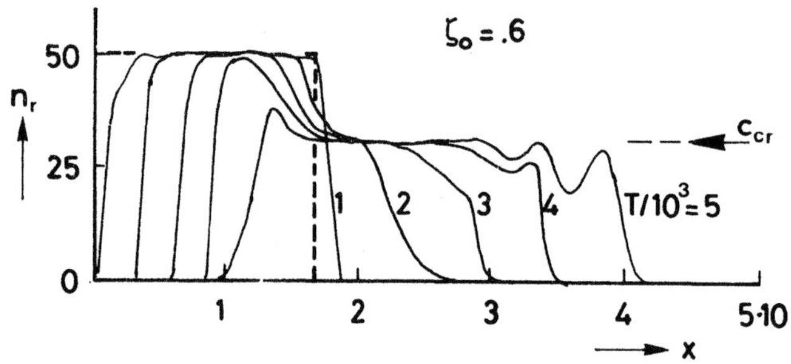

Figure 10: Same as Fig. 7, but for ζ_0 = .6. Sample length = 167

same in Fig. 7-9 and equal to v_s in Fig. 3. Again, the trailing rarefaction remains steep. For large times, oscillations develop at the tail of the shock wave. From these results one can deduce that ν and β must be small quantities; it should be possible to obtain upper and lower limits for ν and β from further systematic studies.

An equivalent description is shown in Fig. 11. The actual speed distribution f(v), here for the right lane, is originally centred at v $\sim$ .1. Then, gradually, a peak at v $\sim$ .5 is filled up, whereas the low speed peak dissappears. We may imagine how the total entropy E will change. In Fig. 12, first E increases due to the creation of a two-peak distribution. For T > 6000 practically a constant profile has developed, and only a slight increase in E is seen over a few 10^4 time units. The constant plateau wave profile, which always develops when starting with ζ_0 > .4, is rather stable over a very long time.

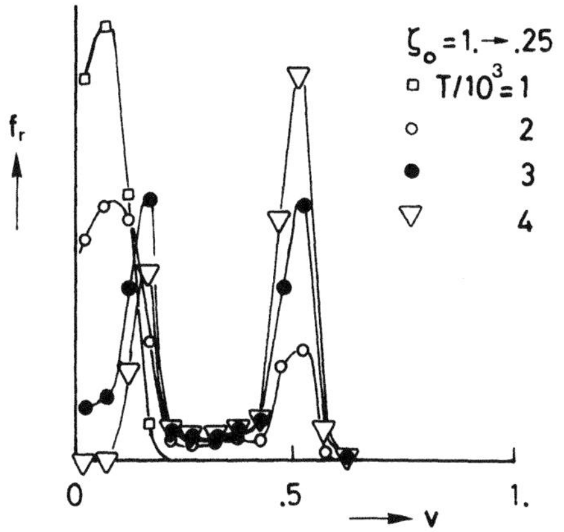

Figure 11: Speed distribution f(v) for right lane. Unsteady (wave) problem with $\zeta_0 = 1$

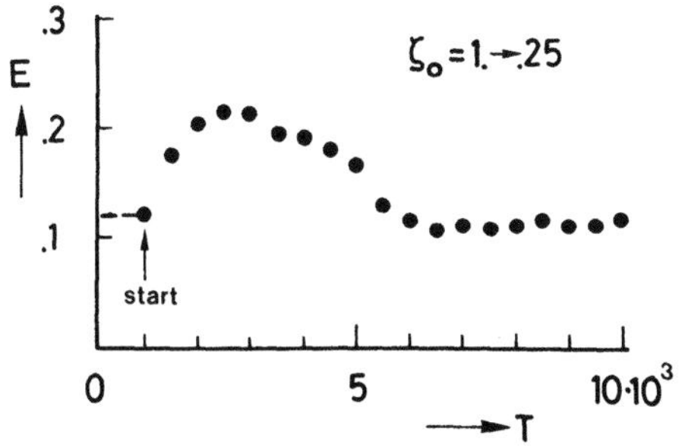

Figure 12: Entropy E as function of time T for problem as described in Fig. 11

5. KINETIC MODEL

Kinetic theory is well established for (not too dense) gases of simple structure (noble gases) [5]. If one wants to transfer the methods and achievements of this field to traffic flow, it is necessary to consider the differences between a gas moving in three-dimensional space and an ensemble of vehicles moving on a lane. In the case of a gas, we are interested in problems where the density changes over many orders of magnitude; at sufficient distance from the critical point, the gas may still be called rarefied (in the sense that the time between two collisions is very much larger than the interaction time) even when condensation starts. In the case of traffic flow, we are concerned only with one decade of density change (say, ζ = .05 to 1); however, this small change in density spans the flow regimes from (what we would call in a gas) "free molecular flow" to "condensed flow" or "liquid state flow". Besides this, we have to deal with particles which have a memory with respect to their desired speed; this memory is transported along particle paths.

If we want to model a kinetic equation for traffic flow which describes the evolution of a group of particles, we have to ask first how many dimensions we have to open up in order to be able to adequately formulate all moments, transition rates, etc. The effect of clustering can be taken into account by introducing distributions for clusters. One may start with formulating the following probability density functions (for more details, see [6]).

$$\psi^k_{ij}(w_j, v,w; x,t)\, dw_j\, dv\, dw\, dx \quad , \tag{13}$$

which can be interpreted as the probability that at time t a vehicle in lane k in a cluster of length i at position j has a desired speed around w_j, if the leader of the cluster has a position at v,w,x in phase space.

We can immediately reduce moments and projections from this distribution. For the leading vehicles the w_j dependence does not exist; we have

$$\psi^k_{i1}\ (v,w; x,t)\, dv\, dw\, dx. \tag{14}$$

Important projections are the actual speed distribution f and the desired speed distribution Φ:

$$f^k_i\ (v; x,t)\, dv = dv \int_0^1 \psi^k_{i1}\ (v,w; x,t)\, dw \quad , \tag{15a}$$

$$\Phi^k_i\ (w; x,t)\, dw = dw \int_0^1 \psi^k_{i1}\ (v,w; x,t)\, dv \quad , \tag{15b}$$

The cluster density c is obtained from integration over v and w:

$$c^k_i\ (x,t) = \int_0^1 \psi^k_{i1}\ (v,w; x,t)\, dv\, dw \ , \tag{16}$$

the total density is

$$\rho(x,t) = \sum_k \sum_i i\, c^k_i(x,t) \quad . \tag{17}$$

Another important moment (as an example) is the occupation of lane (which plays a role in the transition probabilities)

$$s^k(x,t) = \sum_i i \int_0^1 dv\ \sigma(v)\ f^k_i\ (v,x,t) \ . \tag{18}$$

In order to describe the evolution of, say, ψ^k_{i1} it is necessary to write down the balance equation for a volume element in phase space. We have (transforming indices ilk into one index i)

$$\frac{\partial \psi_i}{\partial t} = \left(\frac{\partial \psi_i}{\partial t}\right)_{div} + \left(\frac{\partial \psi_i}{\partial t}\right)_{W} + \left(\frac{\partial \psi_i}{\partial t}\right)_{T} . \tag{19}$$

The change in time of element ψ_i is caused by three groups of terms: the divergence of the fluxes through the element, interaction on the lane which causes gains and losses, and, finally, transitions due to lane changes. With $\underline{\theta} = (x,v,w)$ we can formally write the divergence term as

$$\left(\frac{\partial \psi_i}{\partial t}\right)_{div} = - \underline{\nabla}_\theta \cdot (\underline{\dot{\theta}}\, \psi_i)$$

$$= - v \frac{\partial \psi_i}{\partial x} - \frac{\partial}{\partial v} (\dot{v}\, \psi_i) - \frac{\partial}{\partial w} (\dot{w}\, \psi_i). \tag{20}$$

Since the third term in Eq.(20) does not make much sense, we should replace it by a "spontaneous" interaction type term. We introduce a function $\omega(w,w'; x,t)\, dw\, dw'$, which is the rate of vehicles changing from w, w + dw to w', w' + dw'. Then the third term is replaced by

$$\left(\frac{\partial \psi}{\partial t}\right)_D = \int_0^1 dw' \left[\psi_i(v,w')\, \omega(w',w) - \psi_i(v,w)\, \omega(w,w') \right] . \tag{21}$$

The fundamental equation finally will read as follows

$$\frac{\partial \psi_i}{\partial t} + v \frac{\partial \psi_i}{\partial x} + \frac{\partial}{\partial v} \left[\dot{v}\, \psi_i \right] = \left(\frac{\partial \psi_i}{\partial t}\right)_D + \left(\frac{\partial \psi_i}{\partial t}\right)_W + \left(\frac{\partial \psi_i}{\partial t}\right)_T . \tag{22}$$

The acceleration $\dot{v}$ can be modelled according to Eq.(1) and is, at least in the case of leading vehicles, a function of v. In the absence of the interactions, caused by the right-hand side terms, the ensemble is driven to the attractor at v = w. For increasing densities, the interaction terms keep the ensemble away from the diagonal v = w.

Limited space does not allow to develop the many details of the model in this paper further. Only some final remarks may be apt. It is important to notice that the fundamental equation only describes the evolution of the real and desired speed distribution for given boundary and initial conditions. Besides conservation of particles, there is no further summational invariant.

If one looks at the most general problem of inhomogeneous-unsteady flow, then one needs ψ_i and ω_i at the boundaries. It should be noted that the total desired speed distribution

$$\Phi(w;\ x,t) = \frac{1}{\varrho}\left[\sum_{\substack{ijk\\ j\neq 1}} \int_0^1 dv\, dw\, \overset{*k}{\psi_{ij}}\ (w_j|v,w) + \sum_{i,k} \int_0^1 dv\, \psi^k_{ij}\ (v,w)\right] \quad (23)$$

is not an invariant quantity of the system, even not in the case $\omega_i = 0$ (see for discussion also [7] with respect to [8]).

A somewhat more simple class of problems can be found in the homogeneous - unsteady or the inhomogeneous - steady case. the first group is characterized by some sort of "chemical reaction" connected with a relaxation process due to changes in the desired speed distribution, i.e. $\omega_i(t) \neq 0$.

Practical examples would be: change of Φ due to darkness, rain, fog, etc., or processes from non-equilibrium to steady (Equilibrium) states. For the second group a typical example would be a speed limit at some point or an obstacle on one lane.

Of some practical importance is the simplest case of homogeneous - steady flow, which we might call equilibrium. The fundamental diagram is supposed to be a sequence of equilibrium states for different concentrations ϱ. It may be noted that for this case the boundary/initial conditions are replaced by a global condition:

$$\varrho\ \Phi\ (w) = \sum_{\substack{ijk\\ j\neq 1}} \int_0^1 dv\, dw\, \overset{*}{\psi}{}^k_{ij} + \sum_{i,k} \int_0^1 dv\, \psi^k_{i1}\ . \quad (24)$$

The integral equation connects the ψ functions with the given Φ-total-desired-speed distribution function.

Many simplified versions of the very general fundamental equations have been studied (integral iteration [2], discrete ordinate method [10]). However, much further work needs to be done in order to obrtain an efficient interaction with the continuum approach and with the computer experiments.

REFERENCES

[1] R. Wiedemann: Schriftenreihe des Instituts für Verkehrswesen der Universität Karlsruhe, Heft 8 (1974).

[2] A.E. Beylich: *Forschungsberichte des Landes Nordrhein-Westfalen*, Nr. 2662 (Westdeutscher Verlag, 1977).

[3] G.B. Whitham: *Linear and nonlinear waves* (John Wiley & Sons, New York, 1974).

[4] V.I. Karpmann: *Nichtlineare Wellen in dispersiven Medien* (Vieweg, Braunschweig, 1977).

[5] J.H. Ferziger and H.G. Kaper: *Mathematical theory of transport processes in gases* (North-Holland, Amsterdam, 1972).

[6] A.E. Beylich: In *Neue Wege in der Mechanik*. Festschrift zum 75. Geburtstag von f. Schultz-Grunow (VDI-Verlag, Düsseldorf, 1981).

[7] S.L. Paveri-Fontana: Transportation Research 9, 225 (1975).

[8] I. Prigogine and R. Herman: *Kinetic theory of vehicular traffic flow* (Elsevier, New York, 1971).

[9] A.H. Shapiro: *The dynamics and thermodynamics of compressible fluid flow* (The Ronald Press Co., New York, 1954).

[10] H.J. Poethke: Ein Vierphasenmodell des Verkehrsflusses auf Autobahnen (Dissertation, TH Aachen, 1982).

ALTERNATIVE EXPLOITATION OF DISSIPATION INEQUALITY DEMONSTRATED FOR HYPERBOLIC HEAT CONDUCTION

W. Muschik
Technische Universität Berlin
Institut für Theoretische Physik
Hardenbergstr. 36
D-1000 Berlin 12
Germany

ABSTRACT. Two different methods exist for exploiting dissipation inequalities: The COLEMAN-NOLL technique /1/ which does not take into account the balance equations because in this procedure the body force field and the heat supply are taken for being arbitrary which contradicts experience. The second method /2/ takes balance equations into account by using LAGRANGE multipliers and representation theorems for the constitutive equations. In contrast to the COLEMAN-NOLL technique the second one will be physically correct, if the presuppositions of LIU's theorem /3/ which is used in the second method can be interpreted physically. Such an interpretation is possible. The idea is demonstrated to exploit the dissipation inequality without using special constitutive equations at first taking them into account at second. As a simple example hyperbolic heat conduction in a rigid body is treated.

For demonstrating the method we start out with the choice of a great state space /4/

$$\underline{z} := (\rho,\ 1/\Theta,\ \nabla\rho, (1/\Theta)^{\bullet}\nabla(1/\Theta),\ \nabla\underline{v}\ ) \tag{1}$$

Belonging to an accelleration independent fluid (ρ = mass density, Θ = contact temperature /5/, $\underline{v}$ = material velocity, $\bullet$ = material time derivative). Other choices of great state spaces are possible /6/. For the first derivatives of the equilibrium variables we use the abbreviation

$$\underline{g} := (\nabla\rho, (1/\Theta)^{\bullet}, \nabla\ (1/\Theta)). \tag{2}$$

The first order higher derivatives not included in $\underline{z}$ are

C. W. Kilmister (ed.), Disequilibrium and Self-Organisation, 65–74.

$$\underline{y}:= (\nabla \underline{g},\ \dot{\underline{g}}\ ,\ \underline{v},\ (\nabla \underline{v})^{\bullet}\ \nabla\nabla \underline{v}). \qquad (3)$$

The balance equations of mass, internal energy and momentum

$$\dot{\rho} + \rho \underline{\underline{E}}:\nabla \underline{v} = 0 \qquad (4)$$

$$\rho\dot{\varepsilon} + \underline{\underline{E}} :\nabla \underline{q} - \underline{\underline{P}}:\nabla \underline{v} - r = 0, \qquad (5)$$

$$\rho\dot{\underline{v}} + \nabla\cdot \underline{\underline{P}} - \rho \underline{f} = 0, \qquad (6)$$

are understood as system of equations which determines the directional derivative in state space. (ε = specific internal energy, $\underline{g}$ = heat flux density, $\underline{\underline{P}}$ = CAUCHY's stress tensor, r = heat supply, $\underline{f}$= imposed acceleration field). Because of the choice (1) and (2) of the state space the balance equations yield

$$0 = -\rho \frac{\partial \varepsilon}{\partial \rho} \underline{\underline{E}}:\nabla \underline{v} + \frac{\partial \varepsilon}{\partial (1/\Theta)} \left(\frac{1}{\Theta}\right)^{\bullet} + \rho \frac{\partial \varepsilon}{\partial \underline{g}} \cdot \dot{\underline{g}} + \rho \frac{\partial \varepsilon}{\partial \nabla \underline{v}} :(\nabla \underline{v})^{\bullet} \qquad (7)$$

$$+ \frac{\partial \underline{q}}{\partial \rho} \cdot \nabla\rho + \frac{\partial \underline{q}}{\partial (1/\Theta)} \cdot \nabla \frac{1}{\Theta} + \frac{\partial \underline{q}}{\partial \underline{g}} : \nabla \underline{g} + \frac{\partial \underline{q}}{\partial \nabla \underline{v}} :\nabla\nabla \underline{v} + \underline{\underline{P}}:\nabla \underline{v} - r$$

and

$$0 = \rho\dot{\underline{v}} + \frac{\partial \underline{\underline{P}}}{\partial \rho} \cdot \nabla\rho + \frac{\partial \underline{\underline{P}}}{\partial (1/\Theta)} \cdot \nabla \frac{1}{\Theta} + \frac{\partial \underline{\underline{P}}}{\partial \underline{g}} : \nabla \underline{g} + \frac{\partial \underline{\underline{P}}}{\partial \nabla \underline{v}} : \nabla\nabla \underline{v} - \rho \underline{f} \qquad (8)$$

As we do so with the dissipation inequality

$$\rho\dot{s} + \nabla\cdot (\underline{k} + \underline{q}/\Theta) - \frac{r}{\Theta} =: \sigma \geq 0 \qquad (9)$$

(s = specific entropy, $\underline{k}$ = entropy flux density minus reduced heat flux density) we get

$$0 \leq \rho \frac{\partial s}{\partial \rho} \dot{\rho} + \rho \frac{\partial s}{\partial (1/\Theta)} \left(\frac{1}{\Theta}\right)^{\bullet} + \rho \frac{\partial s}{\partial \underline{g}} \cdot \dot{\underline{g}} + \rho \frac{\partial s}{\partial \nabla \underline{v}} : (\nabla \underline{v})^{\bullet}$$

$$+ \left(\frac{\partial \underline{k}}{\partial \rho} + \frac{1}{\Theta} \frac{\partial \underline{q}}{\partial \rho}\right) \cdot \nabla\rho + \left(\frac{\partial \underline{k}}{\partial (1/\Theta)} + \underline{q} + \frac{1}{\Theta} \frac{\partial \underline{q}}{\partial (1/\Theta)}\right) \cdot \nabla \frac{1}{\Theta}$$

$$+ \left(\frac{\partial \underline{k}}{\partial \nabla \underline{v}} + \frac{1}{\Theta}\ \frac{\partial \underline{q}}{\partial \nabla \underline{v}}\right) \vdots \nabla\nabla \underline{v} + \left(\frac{\partial \underline{k}}{\partial \underline{g}} + \frac{1}{\Theta}\ \frac{\partial \underline{q}}{\partial \underline{g}}\right) : \nabla \underline{g} - r/\Theta \tag{10}$$

If we now use the following abbreviations

$$\underline{\underline{A}} := \left(\frac{\partial \underline{\underline{P}}}{\partial \underline{g}} , \underline{0}, \quad \rho, \underline{0}, \quad \frac{\partial \underline{\underline{P}}}{\partial \nabla \underline{v}} \right) \tag{11}$$

$$\underline{C} := - \frac{\partial \underline{\underline{P}}}{\partial \rho} \cdot \nabla \rho - \frac{\partial \underline{\underline{P}}}{\partial (1/\Theta)} \cdot \nabla \frac{1}{\Theta} + \rho \underline{f} \tag{12}$$

$$\underline{A} := \left(\frac{\partial \underline{q}}{\partial \underline{g}} , \rho \frac{\partial \varepsilon}{\partial \underline{g}} , 0, \rho \frac{\partial \varepsilon}{\partial \nabla \underline{v}} , \frac{\partial \underline{q}}{\partial \nabla \underline{v}} \right) \tag{13}$$

$$C := + \rho \frac{\partial \varepsilon}{\partial \rho} \underline{\underline{E}} : \nabla \underline{v} - \frac{\partial \varepsilon}{\partial (1/\Theta)} \left(\frac{1}{\Theta}\right)^{\bullet} - \frac{\partial \underline{q}}{\partial \rho} \cdot \nabla \rho - \frac{\partial \underline{q}}{\partial (1/\Theta)} \cdot \nabla \frac{1}{\Theta}$$

$$- \underline{\underline{P}} : \nabla \underline{v} + r \tag{14}$$

$$\underline{B} := \left(\frac{\partial \underline{k}}{\partial \underline{g}} + \frac{1}{\Theta}\ \frac{\partial \underline{q}}{\partial \underline{g}} , \rho \frac{\partial s}{\partial \underline{g}} , 0, \rho \frac{\partial s}{\partial \nabla \underline{v}} , \frac{\partial \underline{k}}{\partial \nabla \underline{v}} + \frac{1}{\Theta}\ \frac{\partial \underline{q}}{\partial \nabla \underline{v}} \right) \tag{15}$$

$$D := + \rho \frac{\partial s}{\partial \rho} \underline{\underline{E}} : \nabla \underline{v} - \rho \frac{\partial s}{\partial (1/\Theta)} \left(\frac{1}{\Theta}\right)^{\bullet} - \left(\frac{\partial \underline{k}}{\partial \rho} + \frac{1}{\Theta}\ \frac{\partial \underline{q}}{\partial \rho}\right) \cdot \nabla \rho$$

$$- \left(\frac{\partial \underline{k}}{\partial (1/\Theta)} + \underline{q} + \frac{1}{\Theta}\ \frac{\partial \underline{q}}{\partial (1/\Theta)} \right) \cdot \nabla \frac{1}{\Theta} + r/\Theta \quad . \tag{16}$$

We get short forms of the balance equations of momentum and internal energy and of the dissipation inequality

$$\begin{pmatrix} \underline{\underline{A}} \\ \underline{A} \end{pmatrix} \cdot \underline{y} = \begin{pmatrix} \underline{C} \\ C \end{pmatrix} , \quad \underline{B} \cdot \underline{y} \geq D. \tag{17}$$

For each state this system of relations determines the higher directional derivatives, if the coefficients $\underline{\underline{A}}$ to D in (17) were known. These coefficients are fixed by the constitutive equations of the material which are to be left open until the dissipation inequality is taken into account.

Now the question arises how to exploit the dissipation inequality? First the dissipation inequality limits the higher derivations so that processes into certain directions of the state space are excluded. Therefore states exist having about themselves allowed and forbidden directions due to the dissipation inequality. Then it is easy to be seen /6/ that among these directions there are reversible ones with vanishing entropy production. Second we postulate the

Axiom: Except in equilibrium-subspace reversible directions in state space do not exist

Accepting this axiom consequently all non-equilibrium states have either allowed or forbidden directions in state space about themselves. Because we are dealing with allowed processes the axiom yields the implication: All considered solutions of balance equations satisfy the dissipation inequality. This statement restricts the constitutive equations $\underline{\underline{A}}$ to D. We can prove /7/ the following

Proposition I: In each great state space the constitutive equations satisfy the relation

$$\underline{B} = \underline{\lambda} \cdot \underline{\underline{A}} + \lambda \underline{A} \tag{18}$$

with suitable state functions $\underline{\lambda}$ and λ, the residual dissipation inequality

$$\underline{\lambda} \cdot \underline{C} + \lambda C \geq D \tag{19}$$

is valid, and the entropy production density σ is independent of the higher derivatives $\underline{y}$.

Corollar: The theorem of I.S. LIU /3/ is valid

$$\underline{B} \cdot \underline{y} - \underline{\lambda} \cdot (\underline{\underline{A}} \cdot \underline{y} - \underline{C}) - \lambda(\underline{A} \cdot \underline{y} - C) \geq D. \tag{20}$$

Another statement /7/ being independent of the axiom cited above is

Proposition II: If the transformation

$$\underline{Z}:\ \underline{X} \rightarrow \underline{Z}(\underline{X}) \tag{21}$$

is continuous in the vicinity of $\underline{X} = \underline{0}$, and if

$$\underline{Z}(\underline{X}) \cdot \underline{X} \geqq 0\ , \quad \Lambda X \tag{22}$$

is valid, $\underline{Z}$ is homogeneous in its arguments

$$\underline{Z}(\underline{X}) = \underline{\underline{a}}(\underline{X})\cdot\underline{X}, \quad \underline{\underline{a}} \text{ positive semidefinit.} \tag{23}$$

We now use proposition I and the abbreviations (11) to (16) equation (18) yields

$$\frac{\partial \underline{k}}{\partial \underline{g}} + \frac{1}{\Theta}\ \frac{\partial \underline{q}}{\partial \underline{g}} = \underline{\lambda}\cdot\frac{\partial \underline{\underline{P}}}{\partial \underline{g}} + \lambda\ \frac{\partial \underline{q}}{\partial \underline{g}} \tag{24}$$

$$\rho\ \frac{\partial s}{\partial \underline{g}} = \lambda\ \frac{\partial \varepsilon}{\partial \underline{g}}\ \rho \tag{25}$$

$$\underline{0} = \underline{\lambda}\rho \qquad \rightarrow \underline{\lambda} = \underline{0} \tag{26}$$

$$\rho\ \frac{\partial s}{\partial \nabla \underline{v}} = \lambda\ \frac{\partial \varepsilon}{\partial \nabla \underline{v}}\ \rho \tag{27}$$

$$\frac{\partial \underline{k}}{\partial \nabla \underline{v}} + \frac{1}{\Theta}\ \frac{\partial \underline{q}}{\partial \nabla \underline{v}} = \underline{\lambda}\cdot\frac{\partial \underline{\underline{P}}}{\partial \nabla \underline{v}} + \lambda\ \frac{\partial \underline{q}}{\partial \nabla \underline{v}}\ . \tag{28}$$

The dissipation inequality (19) becomes

$$\left(-\ \lambda \underline{\underline{P}} - \rho\left(\frac{\partial s}{\partial \rho} - \lambda\frac{\partial \varepsilon}{\partial \rho}\right)\underline{\underline{E}}\right):\nabla \underline{v} + \rho\left(\frac{\partial s}{\partial (1/\Theta)} - \lambda\frac{\partial \varepsilon}{\partial (1/\Theta)}\right)\left(\frac{1}{\Theta}\right)^{\bullet} \tag{29}$$

$$+ \left[\frac{\partial \underline{k}}{\partial \rho} + \left(\frac{1}{\Theta} - \lambda\right)\frac{\partial \underline{q}}{\partial \rho}\right]\cdot\nabla\rho + \left[\underline{q} + \frac{\partial \underline{k}}{\partial (1/\Theta)} + \left(\frac{1}{\Theta} - \lambda\right)\frac{\partial \underline{q}}{\partial (1/\Theta)}\right]\cdot\nabla\,\frac{1}{\Theta}$$

$$+ \left(\lambda - \frac{1}{\Theta}\right) r \geqq 0.$$

Using now proposition II the equations (24) to (28) and the dissipation inequality (29) yield in equilibrium (E)

$$\lambda\Big|_E = \frac{1}{\Theta} \quad , \tag{30}$$

$$\underline{\underline{P}}\Big|_E = p\,\underline{\underline{E}} \,, \qquad p := -\,\Theta\rho\left(\frac{\partial s}{\partial \rho}\Big|_E - \frac{1}{\Theta}\,\frac{\partial \varepsilon}{\partial \rho}\Big|_E\right) , \tag{31}$$

$$\frac{\partial \underline{k}}{\partial \rho}\Big|_E = \underline{0}, \qquad \frac{\partial \underline{k}}{\partial (1/\Theta)}\Big|_E = \underline{0} \tag{32}$$

$$\frac{\partial \underline{k}}{\partial \nabla \underline{v}}\Big|_E = \underline{\underline{0}} \,, \qquad \frac{\partial \underline{k}}{\partial \underline{g}}\Big|_E = \underline{\underline{0}} \,, \tag{33}$$

$$\frac{\partial s}{\partial (1/\Theta)}\Big|_E = \frac{1}{\Theta}\,\frac{\partial \varepsilon}{\partial (1/\Theta)}\Big|_E \tag{34}$$

$$\frac{\partial s}{\partial \nabla \underline{v}}\Big|_E = \frac{1}{\Theta}\,\frac{\partial \varepsilon}{\partial \nabla \underline{v}}\Big|_E \quad , \tag{35}$$

$$\frac{\partial s}{\partial \underline{g}}\Big|_E = \frac{1}{\Theta}\,\frac{\partial \varepsilon}{\partial \underline{g}}\Big|_E \quad . \tag{36}$$

The equation (30) yields immediately

$$\lambda - \frac{1}{\Theta} = \underline{\underline{\ell}}^1 : \nabla \underline{v} + \ell^2 \left(\frac{1}{\Theta}\right)^{\cdot} + \underline{\ell}^3 \cdot \nabla \rho + \underline{\ell}^4 \cdot \nabla \frac{1}{\Theta} \tag{37}$$

where the ℓ^j are state space functions. Consequently we get

$$\left.\frac{\partial\lambda}{\partial(1/\Theta)}\right|_E = 1, \qquad \left.\frac{\partial^n\lambda}{\partial(1/\Theta)^n}\right|_E = 0, \quad n\geq 2. \tag{38}$$

Taking proposition II into account, the dissipation inequality (29) yields

$$-\lambda\underline{\underline{P}} - \rho\left(\frac{\partial s}{\partial\rho} - \lambda\frac{\partial\varepsilon}{\partial\rho}\right)\underline{\underline{E}} = \underline{\underline{f}}^1\left(\nabla\underline{v},\ \left(\frac{1}{\Theta}\right)^{\bullet},\ \nabla\rho,\ \nabla\frac{1}{\Theta}\right)$$

$$\rho\left(\frac{\partial s}{\partial(1/\Theta)} - \lambda\frac{\partial\varepsilon}{\partial(1/\Theta)}\right) = f^2(\text{———})$$

$$\frac{\partial\underline{k}}{\partial\rho} + \left(\frac{1}{\Theta} - \lambda\right)\frac{\partial\underline{q}}{\partial\rho} = \underline{f}^3\ (\text{———})$$

$$\underline{q} + \frac{\partial\underline{k}}{\partial(1/\Theta)} + \left(\frac{1}{\Theta} - \lambda\right)\frac{\partial\underline{q}}{\partial(1/\Theta)} = \underline{f}^4\ (\text{———})$$

$$r = f^5\ (\text{———}) \tag{39}$$

$$f^j(\underline{\underline{0}},\ 0,\ \underline{0},\ \underline{0}) = 0, \quad j = 1,\ldots,5. \tag{40}$$

Especially we now consider the case of mechanical equilibrium, i.e.

$$\underline{\underline{f}}^1(\text{———}) \equiv \underline{\underline{0}}\ , \qquad \underline{f}^3(\text{———}) \equiv \underline{0} \tag{41}$$

and we get for the pure heat conduction problem

$$\rho\left(\frac{\partial s}{\partial(1/\Theta)} - \lambda\frac{\partial\varepsilon}{\partial(1/\Theta)}\right) = \hat{\alpha}\left(\frac{1}{\Theta}\right)^{\bullet} + \hat{\underline{\beta}}\cdot\nabla\frac{1}{\Theta} \tag{42}$$

$$\underline{q} + \frac{\partial \underline{k}}{\partial(1/\Theta)} + \left(\frac{1}{\Theta} - \lambda\right) \frac{\partial \underline{q}}{\partial(1/\Theta)} = \hat{\underline{a}} \left(\frac{1}{\Theta}\right)^{\bullet} + \hat{\underline{\underline{b}}} \cdot \nabla \frac{1}{\Theta} \tag{43}$$

$$\frac{1}{\Theta} - \lambda = h \left(\frac{1}{\Theta}\right)^{\bullet} + \underline{m} \cdot \nabla \frac{1}{\Theta} \tag{44}$$

The linearized energy balance equation is by use ~~of~~ (43)

$$\rho \left.\frac{\partial \varepsilon}{\partial(1/\Theta)}\right|_E \left(\frac{1}{\Theta}\right)^{\bullet} + \rho \left.\frac{\partial \varepsilon}{\partial(1/\Theta)^{\bullet}}\right|_E \left(\frac{1}{\Theta}\right)^{\bullet\bullet} + \rho \left.\frac{\partial \varepsilon}{\partial \nabla(1/\Theta)}\right|_E \cdot \left(\nabla \frac{1}{\Theta}\right)^{\bullet}$$

$$= - \nabla \cdot \underline{q} = \tag{45}$$

$$= - \nabla \cdot \left[\frac{\partial \underline{k}}{\partial(1/\Theta)} + \left(\hat{\underline{a}} - h \frac{\partial \underline{q}}{\partial(1/\Theta)}\right) \left(\frac{1}{\Theta}\right)^{\bullet} + \left(\hat{\underline{\underline{b}}} - \underline{m} \frac{\partial \underline{q}}{\partial(1/\Theta)}\right) \cdot \left(\nabla \frac{1}{\Theta}\right)\right]\Bigg|_E$$

The equations (42) to (44) allow to prove the following

<u>Proposition III:</u>

$$\left.\frac{\partial \lambda}{\partial(1/\Theta)^{\bullet}}\right|_E = - h|_E \,, \quad \left.\frac{\partial \lambda}{\partial \nabla(1/\Theta)}\right|_E = \underline{m}|_E \tag{46}$$

$$\rho \left.\frac{\partial \varepsilon}{\partial(1/\Theta)^{\bullet}}\right|_E = \hat{\alpha}\Big|_E - \rho h\Big|_E \left.\frac{\partial \varepsilon}{\partial(1/\Theta)}\right|_E \tag{47}$$

$$\rho \left.\frac{\partial \varepsilon}{\partial \nabla(1/\Theta)}\right|_E = \hat{\underline{\beta}}\Big|_E - \rho \underline{m}\Big|_E \left.\frac{\partial \varepsilon}{\partial(1/\Theta)}\right|_E \tag{48}$$

With aid of proposition III we are able to write down the linearized heat conduction equation whose type matrix becomes

$$\begin{pmatrix} \hat{\alpha}|_E - \rho h|_E \dfrac{\partial \varepsilon}{\partial(1/\Theta)}\Big|_E & \hat{\underline{\beta}}|_E - \underline{m}|_E \dfrac{\partial \varepsilon}{\partial(1/\Theta)}\Big| \\ \hat{\underline{a}}|_E - h|_E \dfrac{\partial \underline{q}}{\partial(1/\Theta)}\Big|_E & \hat{\underline{\underline{b}}}|_E - \underline{m}| \dfrac{\partial \underline{q}}{\partial(1/\Theta)}\Big|_E \end{pmatrix} \tag{49}$$

A comparison with the positive semidefinite matrix

$$\begin{pmatrix} \hat{\alpha}|_E & \hat{\underline{\beta}}|_E \\ \hat{\underline{a}}|_E & \hat{\underline{\underline{b}}}|_E \end{pmatrix} \geqq 0 \tag{50}$$

of the bilinear form of the entropy production density

$$\hat{\alpha} \left[\left(\frac{1}{\Theta}\right)^{\bullet}\right]^2 + \hat{\underline{\beta}}\cdot\nabla \frac{1}{\Theta} \left(\frac{1}{\Theta}\right)^{\bullet} + \hat{\underline{a}}\left(\frac{1}{\Theta}\right)^{\bullet}\cdot\nabla \frac{1}{\Theta} + \hat{\underline{\underline{b}}}:\nabla \frac{1}{\Theta} \nabla \frac{1}{\Theta} \geqq 0 \tag{51}$$

shows

Proposition IV: If the linearized heat conduction equation should be hyperbolic, not both of the quantities

$$h|_E \neq 0 \quad \vee \quad \underline{m}|_E \neq 0$$

can vanish, i.e. for arbitrary $\underline{k}$

$$\lambda \not\equiv \frac{1}{\Theta} \tag{53}$$

must be valid.

As (38) shows all derivatives of λ are zero in equilibrium. Therefore the TAYLOR series of λ has the value $1/\Theta$ for arbitrary deviations out of equilibrium. This shows that near equilibrium the dependence of λ from $1/\Theta$ is extremely weak, such as for instance

$$\lambda\left(\frac{1}{\Theta}\,,\;\left(\frac{1}{\Theta}\right)^{\bullet},\;\nabla\frac{1}{\Theta}\right) \;=\; \frac{1}{\Theta} + \exp\left[-1/f\left(\frac{1}{\Theta}\,,\;\left(\frac{1}{\Theta}\right)^{\bullet},\nabla\frac{1}{\Theta}\right)\right]$$

$$f \geqq 0, \quad f\left(\frac{1}{\Theta}\,,\,0\,,\,\underline{0}\right) \;=\; 0 \tag{54}$$

Therefore the hyperbolicity of heat conduction is a "small" effect which is of course theoretically very important because of causality.

References

1. B.D. Coleman, W. Noll: Arch. Rat. Mech. Anal. 13, 167 (1963)
   B.C. Coleman: Arch. Rat. Mech. Anal. 17, 1 (1964)
2. I. Müller: Thermodynamics, Boston 1985, 5.4.3
3. I.S. Liu: Arch. Rat. Mech. Anal. 46, 131 (1972)
4. W. Muschik: ZAMM 61, T213 (1981)
5. W. Muschik: Arch. Rat. Mech. Anal. 66, 379 (1977)
   W. Muschik, G. Brunk: Int. J. Eng. Sci. 15, 377 (1977)
   W. Muschik,: J. Non-Equilib. Thermodyn. 4, 277, 377 (1979)
6. W. Muschik: 'Derivation of GIBBS' Fundamental Equations by Dissipation Inequalities (Algebraic Approach)', Proc. Intern. Conf. Nonlinear Mech. (ICNM) Shanghai 1985, Ed. CHIEN Wei-zang, p. 155-162.
7. W. Muschik: 'Fundamental Remarks on Evaluating Dissipation Inequalities', Proc. Recent Developments in Nonequilibrium Thermodynamics, Barcelona 1983, Lecture Notes in Physics 199, p. 387-397.

A NEW APPROACH TO THERMODYNAMICS OF IRREVERSIBLE PROCESSES BY MEANS OF LAGRANGE-FORMALISM

K.-H. Anthony
Universität-Gesamthochschule-Paderborn
Fachbereich Physik, Abteilung Theoretische Physik
(4790) Paderborn, FR Germany

ABSTRACT. This paper is concerned with Lagrange-Formalism as applied to phenomenological thermodynamics of irreversible processes. The general method is exemplified by the special process of heat conduction. The total information concerning the irreversible processes is included into a Lagrange-density function. From this all fundamental field and balance equations and all constitutive equations are derived by means of straightforward methods of field theory. In this respect universal invariance requirements and Noethers theorem play a most important rule. The entropy concept is included into the formalism in quite a natural way. Especially this theory allows for a methodical unification of dissipative and non-dissipative physics within Lagrange-Formalism. Onsagers linear theory can be reproduced completely, which is demonstrated for the special case of heat conduction.

1. INTRODUCTION

Hamilton's Principle and the associated Lagrange-Formalism are leading principles in many branches of theoretical physics /1,2,3/. There are several reasons for this fact:

The whole information concerning the processes of a particular system is included into its Lagrangian which is a function, or more generally, a functional of the process variables. Thus the Lagrangian can be considered as the simplest starting point for describing a system. As a particularly prominent example I refer to modern gauge theories which in fact start with appropriate Lagrangians in order to create new physical forces.

The definitions of observables such as energy, linear momentum, mass a.s.o. are related with particular invariance properties of the Lagrangian. The associated invariance principles are universally defined for all systems and thus, from a field-theoretical point of view, these principles are responsible for the

C. W. Kilmister (ed.), Disequilibrium and Self-Organisation, 75–92.

fact, that energy, linear momentum, mass a.s.o. are universal observables which obey conservation laws and which, as a consequence, are capable of exchange processes between systems of different nature.

Hamiltons's Principle, which on the basis of a Lagrangian is a variational principle with free and independent variations of all process variables, has a fundmental meaning with respect to the unification of physical cognition as well as to practically solving of problems. Each real process of a system which can be associated with a Hamilton's Principle, is distinguished by a minimum (or more generally by an extremum) of a quantity called "action". On the other hand the solving of the fundamental equations of a system can be replaced by the solving of Hamilton's variation principle by means of direct Ritz-methods.

Thus in theoretical physics a theory is often considered to be complete only, if its variational principle in the sense of Hamilton is known.

Unfortunately in the past nobody succeeded in describing irreversible and dissipative processes on the phenomenological level by means of Hamilton's Principle and Lagrange-Formalism. - In this context I have to mention, that indeed a lot of variational principles have been established in the past /4/. However, to my knowledge, they altogether are related to restricted variations and are thus no Hamilton's Principles in the original sense. - So research has been done along the lines of Onsager's theory /5/ or, more generally, along the lines of rational thermodynamics /6/. Onsager's theory being an extrapolation of thermostatics has to be restricted to a neighborhood of equilibrium. Both Onsager's theory and rational thermodynamics are based on the standard balance equations (energy, linear momentum, mass a.s.o.) and on a number of ad hoc constitutive equations for those densities and fluxes which occur in the balance equations. Here the question arises, if these constitutive equations are compatible with the existence of a Lagrangian. In general they are not, at least at a first glance.

To summarize my present view of the affair I would like to give the following statements:

For reasons of methodical unification as well as for practical needs thermodynamics of irreversible processes must be included into the scheme of Hamilton's Principle and Lagrange-Formalism.

In the past the non-existence of a Lagrangian for irreversible processes is due to an insufficient choice of the fundamental process variables. A priori it is not clear, that traditionally used and "directly measurable" quantities such as temperature or mass densities are the real fundamental variables. A revision of the definition of fundamental process variables must be done.

In the following I shall give a realization of my statements. Within the scheme of Hamilton's Principle and Lagrange-Formalism a new approach to thermodynamics of irreversible processes is presented. The theory is a phenomenological one. It belongs to the same level as traditional thermostatics. Thermostatics is included as a special class of processes. From the very beginning the theory allows for a quite general dynamics without any restrictions to the neighborhood of equilibrium. However, in order to clarify the general structure of the theory, Onsager's approach to thermodynamics of irreversible pocesses is embedded into Lagrange-Formalism at the first attempt. Afterwards I disregard Onsager's approach and a pure field-theoretical structure is left over, which defines a general thermodynamics of irreversible processes without restrictions to the neighborhood of equilibrium.

In this paper I restrict myself to heat conduction as the most fundmental problem. It will be shown, that all fundamental laws of thermodynamics, all balance and constitutive equations are related with fundamental structures of Lagrange-Formalism. Especially the observable entropy, its density, flux and production rate are a quite natural consequence of simple structures of Lagrange-Formalism, i.e. the second law of thermodynamics is embedded in quite a natural way.

2. FUNDAMENTAL PRINCIPLES

Confining myself to the simplest but nevertheless the most important case the total information of a system is contained in a Lagrange-density-function*)

$$\ell(\psi_i(x,t),\ \partial_t\psi_i(x,t),\ \partial_\alpha\psi_i(x,t)) \tag{1}$$

which generally depends on all field process variables $\psi_i(x,t)$, $i = 1,\dots,N$, of the system and on their first order derivatives with respect to time $(\partial_t = \frac{\partial}{\partial t})$ and space variables $(\partial_\alpha = \frac{\partial}{\partial x^\alpha}$, $\alpha = 1,2,3$). This function is the kernel of the action integral

$$J = \int_{t_1}^{t_2} \int_{V(t)} \ell(\psi_i,\ \partial_t\psi_i,\ \partial_\alpha\psi_i)\ d^3x\ dt \tag{2}$$

which is associated with a process of the system in the fixed time interval $[t_1,t_2]$. I indicate the material volume of the system by $V(t)$ which in general takes part in the process and thus depends on time.

*) In a larger context this theory is called a local theory: The response of the system at point x is due to the local state at this point only. Generalizations to "non-locality" are possible. Furthermore the first order Lagrangian (1) can be generalized to higher orders, taking account of higher derivatives of the process variables.

As a methodical principle the real processes of a system are distinguished as solutions of Hamilton's variational Principle:

$$J = \text{extremum} \tag{3}$$

by free and independent variation of all field variables $\psi_i(x,t)$ with fixed values at the starting and end points t_1 and t_2 of the process:

$$\delta\psi_i(x,t_1) = \delta\psi_i(x,t_2) = 0 \tag{4}$$

Variations of the fields at the boundary and of the shape of the boundary are allowed.

By definition the action integral (2) is associated with an instantaneous response of the system*). Going along the lines of rational thermodynamics it could easily be extended in such a way as to take account of memory effects. However, memory effects are always due to hidden internal variables which in the Lagrange-Formalism should explicitly be taken account of. According to this methodical statement I emphasize, that the Lagrangian is always assumed to be instantaneous, i.e. all relevant internal variables are collected into the set of field variables ψ_i. Furthermore in the Lagrangian (1) there is no explicit dependence on the space and time variables, which means, that we are dealing with a closed system.

According to the independent variation of the N independent fields ψ_i, Hamilton's Principle (3,4) results in N independent Euler-Lagrange-equations

$$\partial_t \frac{\partial \ell}{\partial(\partial_t\psi_i)} + \nabla \cdot \frac{\partial \ell}{\partial(\nabla\psi_i)} - \frac{\partial \ell}{\partial\psi_i} = 0 \quad , \quad i = 1,\dots,N \tag{5}$$

which are the fundamental field equations of the system. The variations at the boundary give rise to a

set of boundary conditions. (6)

It should be emphasized, that the field equations (5) are more fundamental than the usual set of balance equations which in Onsager's theory or in rational thermodynamics are taken as fundmental equations. In general both sets of equations are even different. However, the balance equations can always be derived from the Euler-Lagrange-equations, whereas in general the opposite is not true. At this point the difference between "measurable" and "fundamental" process variables gets important (see below).

Within the framework of Lagrange-Formalism balance equations of a system are due to invariance properties of the Lagrangian with respect to a continuous group of process transformations. A process transformation $T_{\varepsilon_1 \dots \varepsilon_n}$ which is assumed to belong to an n-parameter continuous group with group parameters $\varepsilon_1,\dots,\varepsilon_n$, transforms one real

*) The Lagrange function ℓ at time t depends on the process variables evaluated at the same time.

process into another real process:

$$\{\bar{x},\bar{t},\bar{\psi}\} = \underline{T}_{\varepsilon_1 \dots \varepsilon_n} \{x,t,\psi\} \tag{7}$$

- The arguments x,t in {...} indicate, that generally the transformation $\underline{T}$ applies to the functional variables ψ_i as well as to the coordinate variables x,t. - Both processes have to fulfill Hamilton's Principle simultaneously, which for the Lagrangian leads to an invariance criterium

$$\ell(\bar{\psi},\bar{\partial}\bar{\psi}) = \ell(\psi,\partial\psi) + \partial_t F(\psi,\varepsilon) \tag{8}$$

This has to be satisfied identically in the process variables ψ_i and the group parameters ε. - The last term is a total time derivative of some function F depending on the process variables ψ_i and the group parameters ε only. On the left the symbol $\bar{\partial}$ indicates, that the derivatives are affected by the transformation $\underline{T}$. Of course the left side of eq.(8) depends on the group parameters via eq.(7). -

Using the field equations (5) and the invariance principle (8) the following statement can be proved (Noether's Theorem /1/):

Each group parameter ε_κ of the group of process transformations $\underline{T}_{\varepsilon_1 \dots \varepsilon_n}$ is associated with an observable ω_κ of the first kind which is defined by a homogeneous balance equation:

$$\varepsilon_\kappa \Rightarrow \omega_\kappa \Rightarrow \partial_t a_\kappa + \operatorname{div} \vec{J}_\kappa = 0 \tag{9}$$

The density fuction $\qquad a_\kappa = a_\kappa(\psi,\partial\psi) \qquad (10)$

and the flux density vector $\qquad \vec{J}_\kappa = \vec{J}_\kappa(\psi,\partial\psi) \qquad (11)$

are uniquely determined from the Lagrangian and thus depend on the same variables as ℓ does. Due to eq. (9) the observables of the first kind are conservative quantities.

Now there arise two different problems with respect to process transformations and the invariance criterium:

1. Given a Lagrangian which defines a particular system. Find the complete group of process transformations (7) by means of eq. (8). After that, find all observables ω_κ associated with the system.

2. Given a group of process transformations (7). Find the Lagrange-function ℓ of a particular system, which for the given group is consistent with the invariance criterium (8). After that, find the observables ω_κ associated with the given group.

Both problems are practically important. The first problem is mainly related to the investigation of specific properties of a given particular system, i.e. it allows for defining specific observables or conservation quantities of a particular system. The second problem is particularly important for the unified definition of observables which are no more specific only for single systems, but which are associated with a whole class of systems or even with the whole nature. For the second problem I give three examples, which lead back to thermodynamics.

A. Homogeneity of time and the observable "energy".

Each process of each closed physical system can be reproduced at any time. This statement, being part of the fundamental causality principle, is associated with the one-parameter group of process transformations

$$\underline{T}_\varepsilon : \begin{cases} \bar{t} = t + \varepsilon \quad , \quad -\infty < \varepsilon < +\infty \\ \bar{x} = x \\ \bar{\psi} = \psi \end{cases} \tag{12}$$

by means of which each process is merely shifted in time. This transformation is the mathematical correlate of the homogeneity of time with respect to all closed systems.

Along the line of eqs.(7-11) time homogeneity (12) is once and for all and universily associated with an observable of the first kind, called energy. Eqs.(9-11) are specified as

energy balance
$$\partial_t u + \operatorname{div} \underset{(u)}{\vec{J}} = 0 \quad , \tag{13}$$

energy density
$$u = u(\psi_i, \partial\psi_i) = \sum_{i=1}^{N} \frac{\partial \ell}{\partial(\partial_t \psi_i)} \partial_t \psi_i - \ell \quad , \tag{14}$$

energy flux density
$$\underset{(u)}{\vec{J}} = \underset{(u)}{\vec{J}}(\psi_i, \partial\psi_i) = \sum_{i=1}^{N} \frac{\partial \ell}{\partial(\nabla\psi_i)} \partial_t \psi_i \quad . \tag{15}$$

Of course the particular Lagrangian ℓ has to be defined consistently with the invariance criterium (8) as applied to the process transformations (12).

Obviously eq.(13) can be identified with the first law of thermodynamics, provided thermal processes can be included into Lagrange-Formalism. But this can be done, as will be shown below. Furthermore Eqs. (14,15) define constitutive equations of the system as far as energy density and flux are concerned. As a result of Lagrange-Formalism these two quantities in general depend on the process variables ψ_i as well as on their spatial and time derivatives.

A further remark should be done concerning the methodical nature of the first law of thermodynamics: Within the framework of Lagrange-Formalism it is deduced on the basis of three fundamental statements:

I the existence of a Lagrangian,
II the universality of Hamilton's Principle,
III the homogeneity of time.

Especially the last point manifests the methodical nature of the whole affair in so far, as the experimentalist permanently has to take care of the reproducibility of experiments either by sufficiently screening of the system or by appropriately extending the system in order to get a closed one, i.e. he permanently produces homogeneity of time with respect to his actual experiments. On the other hand nobody doubts about the universality of energy conservation including thermodynamic systems. I cannot see a reason why the first two principles I and II should break down exclusively in thermodynamics. In my opinion it is rather natural to integrate thermodynamics into Lagrange-Formalism which already works quite well for non-thermal processes.

B. Homogeneity of space and the observable "linear momentum".

The procedure is quite analogous to that one of time homogeneity: Each process of each closed system can be reproduced at any site in space. Again we are dealing with a methodical and experimental postulate, namely with homogeneity of space with respect to all closed systems. The associated universal group of process transformations is the three parameter translation group in Euclidean space

$$\underline{T}_{\varepsilon^1\varepsilon^2\varepsilon^3} : \begin{cases} \bar{t} = t \\ \bar{x}^\alpha = x^\alpha + \varepsilon^\alpha , \quad \alpha = 1,2,3, \\ \bar{\psi} = \psi . \end{cases} \tag{16}$$

Once and for all this group is associated with linear momentum as a vector-valued observable of the first kind. Eqs.(9-11) now read as follows:

Balance of linear momentum:

$$\partial_t \vec{p} + \operatorname{div} \underline{\sigma} = 0 . \tag{17}$$

Density vector of linear momentum:

$$\vec{p} = \vec{p}(\psi_i, \partial\psi_i) = - \sum_{i=1}^{N} \frac{\partial \ell}{\partial(\partial_t \psi_i)} \nabla\psi_i . \tag{18}$$

Flux density tensor of linear momentum:

$$\underline{\sigma} = \underline{\sigma}(\psi_i, \partial\psi_i) = - \sum_{i=1}^{N} \frac{\partial \ell}{\partial(\nabla\psi_i)} \otimes \nabla\psi_i + \underline{1}\cdot\ell . \tag{19}$$

($\otimes$ indicates tensor product.) In a mechanical context $\underline{\sigma}$ is also called "stress tensor".

One should keep in mind, that from this point of view linear momentum is no quantity which exclusively is due to mechanical degrees of freedom. From eqs.(18,19) we see, that linear momentum is principally associated with each process variable. Light pressure is an example for momentum flux due to non-mechanical degrees of freedom. So we should not be surprised, if thermal processes are endowed with linear momentum, too.

C. Gauge transformations of complex process variables and mass-like observables.

From quantum mechanics it is wellknown, that complex-valued variables (matter fields) occur as process variables in a real-valued Lagrangian. If in such a case the <u>gauge transformations</u>

$$\underline{T}_\varepsilon : \begin{cases} \bar{t} = t \\ \bar{x} = x \\ \bar{\psi} = \psi \cdot e^{i\Lambda\varepsilon} \\ \bar{\psi}^* = \psi^* \cdot e^{-i\Lambda\varepsilon} \end{cases} \tag{20}$$

define a one-parameter group of process transformations in the sense of eqs.(7,8), we are lead from eqs. (9-11) to <u>mass-like observables</u> with <u>mass-like balance equation</u>

$$\partial_t w + \operatorname{div} \underset{(w)}{\vec{J}} = 0 \qquad , \tag{21}$$

<u>mass-like density function</u>

$$w = w(\psi,\psi^*,\partial\psi,\partial\psi^*) = i\Lambda\left(\frac{\partial \ell}{\partial(\partial_t\psi)}\psi - \frac{\partial \ell}{\partial(\partial_t\psi^*)}\psi^*\right) \tag{22}$$

and <u>mass-like flux density vector</u>

$$\underset{(w)}{J} = \underset{(w)}{J}(\psi,\psi^*,\partial\psi,\partial\psi^*) = i\Lambda\left(\frac{\partial \ell}{\partial(\nabla\psi)}\psi - \frac{\partial \ell}{\partial(\nabla\psi^*)}\psi^*\right) . \tag{23}$$

ψ^* is the complex conjugate of ψ and Λ a gauge factor. In the case of Schrödinger's matter field we get by means of eqs.(21-23) the mass balance or the electrical charge balance depending on the particular choise of the gauge factor Λ. Thus with respect to matter fields the gauge transformations (20) have to be regarded as universal process transformations in the same way as the more obvious time and space translations (12) and (16). The question arises, if there are further complex process variables in physics which have to be submitted to the gauge transformations (20), too.

A final remark is due to the question of how to get universal process transformations quite generally. Looking at the well-established examples they altogether show a principal experimental uncertainty as a common feature: The transformations (12) and (16) are the mathematical correlates of the physical facts, that there is no possibility to distinguish principally any particular instant or spatial point as compared with any others. In the case of complex matter fields there is no possibility to fix absolutely the phase, i.e. by means of interference experiments only relative phase differences can be determined. The gauge transfromations (20) are again the mathematical correlate of an experimental uncertainty.

Lagrange-Formalism takes account of such uncertainties by means of the invariance principle (8) as applied to an appropriate group of process transformations which are the result of analysing the real experimental situation.

3. THE FIELD OF THERMAL EXCITATION - HEAT CONDUCTION IN THE FRAMEWORK OF LAGRANGE-FORMALISM

Let me give a heuristic and a mathematical argument for the insufficiency of the temperature T as the only variable describing heat conduction by means of Lagrange-Formalism:

Stationary heat conduction ($\partial_t T = 0$) is associated with a non-vanishing energy flux $\vec{J}_{(u)}$. Let me assume the existence of a Lagrangian $\ell(T, \partial_t T, \nabla T)$. Then by eq.(15) the energy flux should be

$$\vec{J}_{(u)} = \frac{\partial \ell}{\partial(\nabla t)} \partial_t T \tag{24}$$

which in the stationary case leads to $\vec{J}_{(u)} = 0$ in contradiction to our experimental knowledge. Thus the set of fundamental process variables has to be extended.

Fourier's fundamental law of heat conduction

$$c\, \partial_t T - \lambda\, \Delta T = 0 \tag{25}$$

is a non-selfadjoint equation, which cannot be looked upon as the Euler-Lagrange-equation (5) of a particular Lagrangian /7/. This equation must be supplemented by at least one further equation in order to make the whole set of equations selfadjoint. But this again is necessarily associated with an extension of the fundamental set of process variables.

As the <u>fundamental process variable of thermal processes</u> I introduce a complex-valued <u>field of thermal excitation $\psi(x,t)$</u>, which together with its complex conjugate ψ^* enters into ℓ in such a way, as to define a real-valued Lagrangian /8/. The <u>absolute temperature</u> T is defined from ψ as a measurable but nevertheless as a secondary quantity from the Lagrangian point of view:

$$T(x,t) = \psi^*(x,t) \cdot \psi(x,t) > 0 \tag{26}$$

By this equation the <u>existence of the absolute zero-point of tempera-</u>

ture is included into the theory from the very beginning: Wherever there is thermal excitation ($\psi \neq 0$), there is a temperature $T > 0$, and vice versa, and wherever there is no thermal excitation ($\psi = 0$) the system is found at the zero-point $T = 0$, and vice versa. These statements give a first visualization of the thermal excitation field.

The Lagrangian for heat conduction in a non-deformable material being at rest is given by

$$\ell = - c\, \psi^*\psi$$
$$- \frac{c}{\omega} \left[\frac{1}{2i} (\psi^*\partial_t\psi - \psi\, \partial_t\psi^*) + \frac{1}{2} \frac{\ln \frac{\psi^*\psi}{T_0}}{\frac{\psi^*\psi}{T_0}} \partial_t(\psi^*\psi) \right]$$
$$+ \sum_{\alpha,\beta=1}^{3} \frac{\lambda^{\alpha\beta}}{\omega} \left[\frac{1}{\psi^*\psi} \frac{1}{2i} \left((\psi^*\partial_\alpha\psi)(\psi^*\partial_\beta\psi) - (\psi\partial_\alpha\psi^*)(\psi\partial_\beta\psi^*) \right) \right. \tag{27}$$
$$\left. + \frac{T_0}{2(\psi^*\psi)^2} \partial_\alpha(\psi^*\psi)\partial_\beta(\psi^*\psi) \right] .$$

It is of the first order in the spatial and time derivatives $\partial_\alpha = \frac{\partial}{\partial x^\alpha}$ and $\partial_t = \frac{\partial}{\partial t}$, of the first degree in the time derivatives and of the second degree in the spatial derivatives. The following physical meanings of the constants shall get apparent with subsequent considerations:

- c : specific heat,
- $\underline{\lambda} = \lambda^{\alpha\beta}$, $\alpha,\beta = 1,2,3$,: tensor of heat conductivity,
- T_0 : a reference temperature,
- ω : a frequency which cannot be specified a priori, but which is necessary because of dimension. (The Lagrangian has the dimension of an energy density.)

For simplicity the material moduli c and $\underline{\lambda}$ are assumed to be constant. It will now be shown, that the Lagrangian (27) completely reproduces Onsager's theory of heat conduction.

The Lagrangian being not invariant with respect to time reversal defines irreversible processes.

The last bracket [...] in eq.(27) is symmetric with respect to the indices α and β. Thus from the tensor $\underline{\lambda}$ only the symmetric part is relevant:

$$\lambda^{\alpha\beta} = \lambda^{\alpha\beta} \tag{28}$$

Obviously Onsager's reciprocity relations concerning the moduli of heat conductivity are reproduced.

The complex-valued field ψ comprises its real and imaginary parts as two independent functional degrees of freedom. Instead we may look on ψ and ψ^* as being functionally independent concerning the variation in Hamilton's Principle. From eqs.(27,5) we arrive at two highly non-linear fundamental field equations for heat conduction:

1.
$$c\psi + \frac{c}{\omega}\frac{1}{i}\partial_t\psi + \frac{1}{\omega}\underline{\lambda} \cdot\cdot \Big[\frac{T_0}{\psi^*\psi} \nabla\otimes\nabla\psi + \Big(\frac{T_0}{\psi^*\psi} - \frac{1}{i} \Big) \frac{\psi}{\psi^*} \nabla\otimes\nabla\psi^* - \frac{1}{i}\frac{1}{\psi^*} \nabla\psi^*\otimes\nabla\psi - \Big(\frac{T_0}{\psi^*\psi} + \frac{1}{2i}\Big) \frac{1}{\psi} \nabla\psi\otimes\nabla\psi - \Big(\frac{T_0}{\psi^*\psi} - \frac{1}{2i}\Big) \frac{\psi}{\psi^{*2}} \nabla\psi^*\otimes\nabla\psi^* \Big] = 0 \tag{29}$$

2. complex conjugate of eq.(29) (30)

They are supplemented by two boundary conditions (see below).

These equations look rather complicated. However,using eq.(26) I can equally well take the temperature T and the phase function ϕ as independent variables:

$$\psi(x,t) = \sqrt{T(x,t)}\; e^{i\phi(x,t)} \quad . \tag{31}$$

The Lagrangian then reads

$$\ell = - cT - \frac{c}{\omega} \Big[T\,\partial_t\phi + \frac{1}{2} \frac{\ln \frac{T}{T_0}}{\frac{T}{T_0}} \partial_t T \Big] + \frac{1}{\omega}\underline{\lambda} \cdot\cdot \Big[\nabla T\otimes\nabla\phi + \frac{T_0}{2T^2} \nabla T\otimes\nabla T \Big] \tag{32}$$

and the fundamental field equations of heat conduction take the alternative form

$$c\partial_t T - \underline{\lambda} \cdot\cdot \nabla\otimes\nabla T = 0 \tag{33}$$

and

$$c\partial_t\phi + \underline{\lambda} \cdot\cdot \nabla\otimes\nabla\phi = - c\omega - \underline{\lambda} \cdot\cdot \Big(\frac{T_0}{T^2} \nabla\otimes\nabla T - \frac{T_0}{T^3} \nabla T\otimes\nabla T\Big) \tag{34}$$

From Hamilton's Principle the boundary conditions of a closed system are found to be

$$\frac{1}{\omega}\vec{n}\cdot(\underline{\lambda}\cdot\nabla T) = 0 \tag{35}$$

and

$$\frac{1}{\omega}\vec{n}\cdot(\underline{\lambda}\cdot\nabla\phi) = 0 \tag{36}$$

by free variation of T and ϕ at the boundary. $\vec{n}$ is the outward unit normal vector of the boundary. Of course eqs.(33,34) can also be found by suitably combining eqs.(29,30).

Obviously eq.(33) is Fourier's law of heat conduction. However this is supplemented by the second field equation (34), which is due to the introduction of the complex fundamental variable ψ instead of T. The first equation is decoupled from the second one. This gives the possibility to solve the equations consecutively. Non-linearities are completely shifted into the second equation.

Given the temperature field T(x,t) of a real heat flow process, i.e. given a solution of the first field equation (33) and of the boundary condition (35). Then eq.(34) can be solved by the particular but nevertheless fundamental expression

$$\phi(x,t) = -\omega t + \frac{T_0}{2T(x,t)} \tag{37}$$

which on the basis of eq.(35) satisfies the boundary condition (36). The fundamental nature of this solution gets apparent by some deeper investigations which are related with the systematic determination of the Lagrangian and with generalizations of the theory towards temperature dependent material moduli c(T) and $\underline{\lambda}$(T). I shall not go into more details here.

Using eq.(31) the heat conduction process can be associated with a thermal excitation wave running down the slope of the temperature gradient:

$$\psi(x,t) = \sqrt{T(x,t)}\cdot e^{-i\omega t + i\frac{T_0}{2T(x,t)}} \tag{38}$$

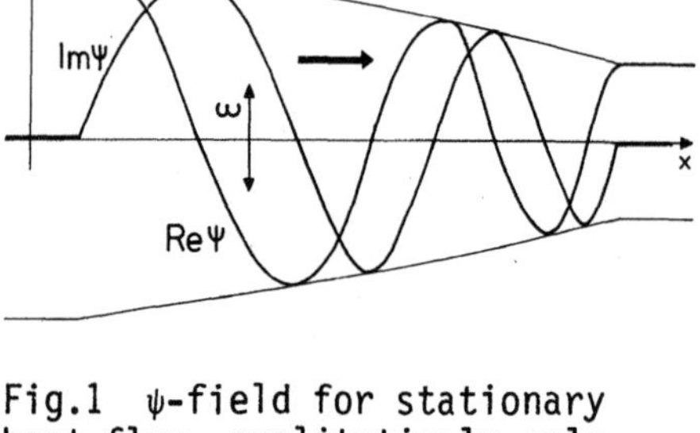

Fig.1 ψ-field for stationary heat flow, qualitatively only

Figure 1 visualizes the situation for the case of stationary heat flow between two heat reservoirs. With linearly decreasing temperature the wave length decreases from left to right. The frequency ω being the unspecified constant in the Lagrangian determines the temporal oscillations in the wave. In the case of non-stationary heat flow the field T(x,t) contributes to the temporal oscillations, too. Because of the non-linearity of the field equation (29,30) or (33,34) both the real and the imaginary part of ψ must be regarded simultaneously. In the case of constant temperature the excitation wave degenerates into a homogeneous oscillation with frequency ω.

The complex-valued excitation wave is a carrier of energy, thus filling the missing link which appeared in connection with eq.(24). From eqs.(14,15) I get

$$u = \frac{\partial \ell}{\partial(\partial_t T)} \partial_t T + \frac{\partial \ell}{\partial(\partial_t \phi)} \partial_t \phi - \ell$$
$$= cT - \frac{1}{\omega} \underline{\lambda} \cdot\cdot (\nabla T \otimes \nabla\phi + \frac{T_0}{2T^2} \nabla T \otimes \nabla\phi) \tag{39}$$

and

$$\vec{J}_{(u)} = \frac{\partial \ell}{\partial(\nabla T)} \partial_t T + \frac{\partial \ell}{\partial(\nabla\phi)} \partial_t \phi$$
$$= \frac{1}{\omega} \underline{\lambda} \cdot\cdot ((\nabla\phi + \frac{T_0}{2T^2} \nabla T) \partial_t T + \nabla T \partial_t \phi) \tag{40}$$

Eq.(40) shows that in the stationary case ($\partial_t T = 0$) due to $\partial_t \phi \neq 0$ there is a term left in the energy flux giving rise to $\vec{J}_{(u)} \neq 0$. Using (37) I finally arrive at the <u>density of thermal energy (internal energy)</u>

$$u = cT \tag{41}$$

and at the <u>flux density of thermal energy (heat flux)</u>

$$\vec{J}_{(u)} = - \underline{\lambda} \cdot \nabla T \ , \tag{42}$$

thus reproducing constitutive equations of Onsager's theory. - With (42) the meaning of the boundary condition (35) gets obvious: Due to our closed system there is no heat flux across the boundary.

With respect to linear momentum the Lagrange-Formalism of heat conduction goes beyond Onsager's theory. Evaluating eqs.(18,19) for the Lagrangian (32) and using (37) I get the <u>density of linear momentum</u>

$$\vec{p} = - \frac{1}{\omega} \frac{c}{2} \left(\frac{T_0}{T} - \frac{\ln \frac{T}{T_0}}{\frac{T}{T_0}} \right) \nabla T \tag{43}$$

and the <u>flux density of linear momentum</u>

$$\underline{\sigma} = \frac{1}{\omega} \frac{c}{2} \left(\frac{T_0}{T} - \frac{\ln \frac{T}{T_0}}{\frac{T}{T_0}} \right) \partial_t T \cdot \underline{1} \tag{44}$$

associated with heat conduction.

This result cannot be related immediately with experiment, because on the one hand I am dealing with a closed thermal system without any mechanical degrees of freedom, on the other hand linear momentum due to non-mechanical variables is always experimentally determined by means of linear momentum exchange between these variables and directly observable mechanical variables of a measuring instrument. To observe linear momentum associated with heat flux it is thus necessary to couple the thermal process variable ψ with mechanical variables. Investigations towards this problem are in progress. For the moment we should only keep in mind, that the unknown frequency ω (see (27,32)) enters into linear momentum of heat contuction, which might give an experimental means to determine this quantity later on. Furthermore the proportionalities $\vec{p} \sim - \nabla T$ and $\underline{\sigma} \sim \underline{1} \cdot \partial_t T$ are qualitatively quite satisfactory.

The Lagrangian (27) is obviously invariant with respect to the gauge transformations (20) as applied to the thermal excitation field ψ. The associated mass-like quantities w and $\underset{(w)}{\vec{J}}$ are provisionally called heat mass density and heat mass flux density. They are given by

$$w = \Lambda c T \tag{45}$$

and

$$\underset{(w)}{\vec{J}} = - \Lambda \underline{\lambda} \cdot \nabla T \quad . \tag{46}$$

It is remarkable, that this balance quantities are proportional to the energy quantities u and $\underset{(u)}{\vec{J}}$ (see eqs. (41,42)). However one should carefully distinguish between the observables "heat mass" and "energy". These observables are related with qualitatively different groups of process transformations, namely with gauge transformations and time translations respectively. The specification of the gauge factor Λ is still open. I guess, that the gauge group (20) is the mathematical correlate of a universal mass conservation according to eq.(21) and that the gauge factor Λ is related with a universal mass-energy-equivalence, which already is known from relativistic physics. (However one should keep in mind, that I am dealing so far within the scope of non-relativistic physics.) If these statements are true, then eqs.(21,45,46) are a manifestation of the universal mass conservation and of the universal mass-energy-equivalence in the special case of heat conduction. Investigations are in progress.

4. OBSERVABLES OF THE SECOND KIND AND THE SECOND LAW OF THERMODYNAMICS

The n-parameter group of process transformations (7) can be used to define another set of n observables which I call observables of the second kind and which are related with non-homogeneous balance equations. For simplicity the formal procedure is presented for a one-parameter group.

Let $\psi_i(x,t)$ be a real process and let $\underline{T}_\varepsilon$ be the process transformations, where $\varepsilon = 0$ indicates the identity. Then

$$\bar{\psi}_i(x,t,\varepsilon) = \underline{T}_\varepsilon \, \psi_i(x,t) \quad , \quad i = 1,\dots,N \, , \tag{47}$$

is a one-parameter class of real processes which satisfies the field equations (5). Within this class the functions

$$\eta_i(x,t) = \left. \frac{\partial \bar{\psi}_i(x,t,\varepsilon)}{\partial \varepsilon} \right|_{\varepsilon=0} , \quad i = 1,\dots,N \, , \tag{48}$$

define real variations of the process $\psi_i(x,t)$ with respect to the parameter ε ($\delta\psi = \eta\delta\varepsilon$). The set of functions $\eta_i(x,t)$ satisfies Jacobi's equations

$$\partial_t \frac{\partial \Omega}{\partial(\partial_t \eta_i)} + \nabla \cdot \frac{\partial \Omega}{\partial(\nabla \eta_i)} - \frac{\partial \Omega}{\partial \eta_i} = 0 \quad , \quad i = 1,\dots,N, \tag{49}$$

which are the variational equations of the field equations (5). The kernel Ω is completely defined by means of the Lagrangian. It depends on the fixed real reference process

$$\psi_i(x,t) = \bar{\psi}_i(x,t,\varepsilon = 0) \tag{50}$$

and on the associated real process variations $\eta_i(x,t)$:

$$2\Omega(\psi,\partial\psi,\eta,\partial\eta) = \sum_{i,j=1}^{N} \left[\frac{\partial^2 \ell}{\partial\psi_i \partial\psi_j} \eta_i \eta_j + 2 \sum_{\alpha=0}^{3} \frac{\partial^2 \ell}{\partial\psi_i \partial(\partial_\alpha \psi_j)} \eta_i \partial_\alpha \eta_j + \sum_{\alpha,\beta=0}^{3} \frac{\partial^2 \ell}{\partial(\partial_\alpha \psi_i)\partial(\partial_\beta \psi_j)} \partial_\alpha \eta_i \partial_\beta \eta_j \right] . \tag{51}$$

It is a homogeneous form of second degree with respect to η and $\partial\eta$. (To contract the formula ∂_t is denoted by ∂_0.) Applying Euler's formula to Ω and taking account of Jacobi's equations (49) it can be shown, that the class (47) of real processes, or equivalently the group of process transformations $\underline{T}_\varepsilon$, is associated in a natural way with a non-homogeneous balance equation which is fulfilled by each real process:

$$\partial_t s + \operatorname{div} \underset{(s)}{\vec{J}} = \underset{(s)}{\sigma} \quad . \tag{52}$$

The quantities s, $\underset{(s)}{\vec{J}}$ and $\underset{(s)}{\sigma}$ are uniquely defined by Ω and thus, via eq.(51), they are finally defined from the Lagrangian:

<u>Density function</u>:

$$s = \sum_{i=1}^{N} \frac{\partial\Omega}{\partial(\partial_t \eta_i)} \eta_i = s(\psi, \partial\psi_i \eta, \partial\eta) \quad . \tag{53}$$

<u>Flux density vector</u>:

$$\underset{(s)}{\vec{J}} = \sum_{i=1}^{N} \frac{\partial\Omega}{\partial(\nabla\eta_i)} \eta_i = \underset{(s)}{\vec{J}} (\psi, \partial\psi_i \eta, \partial\eta) \quad . \tag{54}$$

<u>Density of the production rate</u>:

$$\underset{(s)}{\sigma} = 2\Omega = \underset{(s)}{\sigma} (\psi, \partial\psi_i \eta, \partial\eta) \quad . \tag{55}$$

These quantities must be evaluated for the real process $\psi_i(x,t)$ (eq. (50)) and its associated real variation $\eta_i(x,t)$ (eq.(48)) which is due to the group of process transformations (47).

The set of eqs.(52-55) defines an <u>observable of the second kind</u> which is associated with the group of process transformations (47). If this group is a universal one, we arrive at a universal observable of the second kind which by its formal structure within Lagrange-Formalism is qualitatively different from the observable of the first kind associated with the same group.

The observable of the second kind associated with the universal group of gauge transformations (20) is called <u>entropy</u>. Then eq.(52) is the first part of the <u>second law of thermodynamics</u> and eqs.(53-55) define the <u>constitutive equations</u> for <u>entropy density</u>, <u>entropy flux</u> and <u>entropy production rate</u>. Applying this concept to the complex thermal excitation field ψ and to the Lagrangian (27) of heat conduction we arrive at the following definitions:

<u>Entropy density</u>:

$$s = c \ln \frac{T}{T_o} \quad . \tag{56}$$

<u>Entropy flux density</u>:

$$\underset{(s)}{\vec{J}} = - \frac{\underline{\lambda} \cdot \nabla T}{T} \quad . \tag{57}$$

<u>Density of entropy production rate</u>:

$$\underset{(s)}{\sigma} = \frac{\nabla T \cdot \underline{\lambda} \cdot \nabla T}{T^2} \quad . \tag{58}$$

Again fundamental equations of Onsager's theory are reproduced.

The relation

$$\sigma_{(s)} > 0 \qquad (59)$$

being the second part of the second law of thermodynamics, is no straightforward result of Lagrange-Formalism. It seems to distinguish between stable and unstable processes in a particular Ljapunov sense. In this context I refer also to Jacobi's equations (49) which play a fundamental part in stability theory. Obviously within Lagrange-Formalism the proposed concept of entropy is again a methodical one.

A final remark is due to Jacobi's equations (49). They have the form of Euler-Lagrange-equations of a particular variational principle which defines a generalized principle of least entropy production and which suffers no restrictions as compared with the analogous principle in Onsager's theory /5/. Investigations are in progress.

5. FINAL REMARKS

It has been shown, that heat conduction being a special case of an irreversible process can be included into Lagrange-Formalism. The price we have to pay is the introduction of a new fundamental variable, namely the thermal excitation field. The temperature gets a secondary quantity which is associated with energy balance (see eqs.(13,41,42)) and which therefore can be measured by means of energy exchange, as is always done when we use any thermometer.

By the thermal excitation field oscillations are brought into the play. In the case of pure heat conduction they are hidden from the phenomenological point of view, i.e. in the observables, which are the only measurable quantities (exchange processes!), the oscillations are cancelled. However, by a suitable coupling of the thermal excitation field with additional system variables it is not hard to produce oscillating observables. So the question arises, if the oscillations of the thermal excitation field can be regarded as a preformation of those temporal and spatial oscillations which are discussed in synergetics by means of other methods.

Looking at eqs.(37,38) the thermal excitation field gets singular at $T = 0$. It seems, that this behaviour is associated with the third law of thermodynamics, which asserts, that it is impossible to reach the absolute zero-point. Further investigation are in progress.

Having completely reproduced Onsager's theory I now can get rid of its restrictions, such as linearity of constitutive equations (e.g. eq.(42)) or the taking over of constitutive equations from thermostatics (e.g. eq.(41)). The general structure of Lagrange-Formalism and the excitation field ψ are retained. A particular material behaviour is then defined by a particular ansatz for the Lagrangian.

Thermostatics is included into the theory as the special class of those processes for which all fluxes vanish and all densities are stationary.

6. REFERENCES

1. E. Schmutzer: Symmetrien und Erhaltungsätze der Physik (Akademie-Verlag, Berlin; Pergamon Press, Oxford; Vieweg, Braunschweig, 1972, WTB Nr. 75).

2. E.M. Corson: Introduction to Tensors, Spinors and Relativistic Wave equations (Blackie a. Son, 1957).

3. H. Rund: The Hamilton-Jacobi Theory in the Calculus of Variation (D. van Nostrand Comp., London,Toronto,New York,Princeton, 1966).

4. I. Gyarmati: Non-Equilibrium Thermodynamics, Field Theory and Variational Principles (Springer, 1970).

5. S.R. De Groot, P. Mazur: Non-Equilibrium Thermodynamics (North-Holland Publ. Comp., 1969).

6. I. Müller: Thermodynamik. Die Grundlagen der Materialtheorie (Bertelsmann, Universitätsverlag, Düsseldorf, 1973).

7. R.M. Santilli: Foundations of Theoretical Mechanics I (Springer, 1978).

8. K.-H. Anthony: 'A New Approach Describing Irreversible Processes' in Continuum Models of Discrete Systems 4, ed. by O. Brulin, R.K.T. Hsieh (North-Holland Publ. Comp., 1981).

A BOUNDARY-LAYER THEORY FOR THE DYNAMICS AND THERMODYNAMICS OF PHASE-INTERFACES

Thorsten Alts
Institut für Theorie der Kondensierten Materie
Freie Universität Berlin
Arnimallee 14
D-1000 Berlin 33

ABSTRACT. Existing continuum physical theories of phase boundaries model these as mass, momentum, energy and entropy carrying non-material singular surfaces. The physical interpretation of these surface fields is not well understood and leads in fact to misleading results.

Real phase interfaces are thin boundary-layers across which all bulk fields experience smooth though rapid changes when crossing from one phase to the other. This is modelled in a new boundary-layer theory for curved phase boundaries. Comparison with the theory of singular surfaces allows physical interpretation of the surface fields in terms of mean values of bulk fields, but it also requires satisfaction of dynamical consistency conditions for tangential momentum and surface stress. These yield new results for the curvature dependence of surface tension, for phase-change processes and the dynamics of nucleation. A stability analysis proves the impossibility of certain nuclei.

Some descriptive applications to ice formation in nature and to the phenomena of undercooling and superheating will corroborate the new results.

1. INTRODUCTION

Nuclei formation and their growth and decay, formation of phase boundaries between a solid and its fluid or between a fluid and its steam are examples for dynamical self-organization of matter. Ordering effects caused by intermolecular forces are opposed by disordering effects due to thermal motion. The description of phase change is therefore basically a problem for statistical thermodynamics. Within this frame the equilibrium properties of phase boundaries have been studied intensively[1]. Following the work of Gibbs[2] surface tension, adsorption, surface structure[3,4], electrostatic double-layer[5,6] found theoretical and experimental attention[7,8]. It is now positive that the phase interface forms a layer between adjacent bulk phases, across which bulk properties change smoothly though rapidly within a thickness of some molecular distances.

C. W. Kilmister (ed.), Disequilibrium and Self-Organisation, 93–127.

The dynamics of phase boundary formation and the transport properties of mass and heat across moving phase interfaces are non-equilibrium processes. Statistical theories for the description of these processes are not available. However, in the last three decades progress in continuum physics has originated a few papers on the non-equilibrium properties of interfaces. One group of authors working in physical chemistry extended the ideas of Gibbs and modelled the phase boundary as a thin transition layer between the bulk phases[9,10,11]. Another group of authors coming from mathematics considered the interface as a mass, momentum and energy bearing singular surface[12,13,14,15,16]. Both groups of continuum theories allow the set up of balance laws for the motion of the interface. The transport properties of mass and heat across the interface, however, have not been considered at all or lead to results which cannot be accepted for phase interfaces.

I compare the boundary-layer theory of phase interfaces with the singular-surface theory and find out that both theories are kinematically and dynamically consistend then and only then, if certain consistency conditions on the tangential motion and the normal shear stress are satisfied. Besides a physical interpretation of surface fields this comparison leads to new results for the curvature dependence of surface tension, for phase-change processes and the dynamics of nucleation. A stability analysis of the boundary layer proves the impossibility of certain nuclei. Furthermore, the temperature dependences of surface tension and of boundary-layer thickness are determined explicitly and in accordance with observations.

Basic to the new boundary-layer theory is the concept of a curvature dependent excentricity between the mathematical surface of curvature and a physical surface of inertia within the layer. It emerges from the kinematic and dynamic consistency conditions and is quite general for all boundary layers of phase change of first order. It is a dynamical concept which cannot be obtained by equilibrium considerations, but it changes also equilibrium results. As a consequence, Gibbs' relation for surface entropy in equilibrium and some of its implications must be changed.

The results of this new boundary-layer theory of phase interfaces are a little more complicated than the classical ones. However, the corrections are all explicit and therefore ready to experimental investigation. Detailed results will be given for the ice/water interface. Some descriptive applications to ice formation in nature and the phenomena of undercooling and superheating will corroborate the new boundary-layer theory.

2. THE PHASE BOUNDARY AS A SINGULAR SURFACE

Mathematically the phase interface can be described as a non-material singular surface bearing surface mass, momentum, energy and entropy. It is in general a moving and curved surface dividing three-dimensional regions R^+ and R^-, that are occupied by adjacent bulk phases, see Fig. 1.

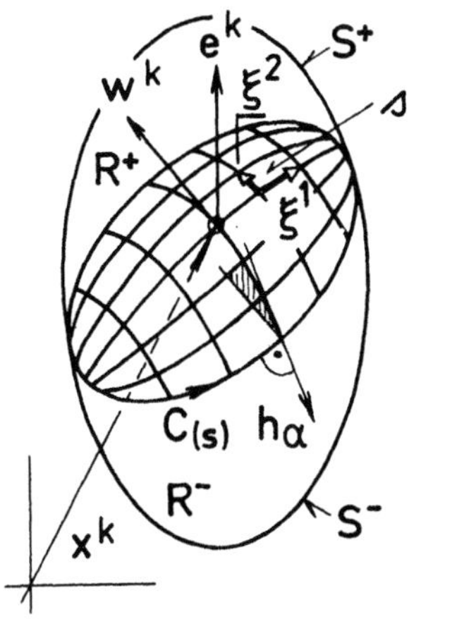

Fig. 1: Geometry and motion of a non-material singular surface dividing pure bulk phases.

2.1. Geometry and Motion

The motion of the surface $\mathcal{s}$ can be described by the parameter representations

$$x^k = {}_a\chi^k(\xi^\alpha,t) = {}_R\chi^k(\Xi^\Gamma,t) \; , \qquad \xi^\alpha = \xi^\alpha_{\mathcal{s}}(\Xi^\Gamma,t) \; . \tag{2.1}$$

x^k $(k=1,2,3)$ are the co-ordinates of surface points in a Cartesian co-ordinate system that is fixed to some inertial frame in Euklidean space. The ξ^α $(\alpha=1,2)$ denote surface co-ordinates in the local tangent planes of the actual surface at time t and Ξ^Γ $(\Gamma=1,2)$ are surface co-ordinates in some reference surface at time $t_R (\leqq t)$. The tangent vectors τ^k_α and the unit normal vector e_k

$$\tau^k_\alpha = \frac{\partial\, {}_a\chi^k(\xi^\beta,t)}{\partial \xi^\alpha} \; , \quad e_k = \frac{(\underline{\tau}_1 \times \underline{\tau}_2)_k}{|\underline{\tau}_1 \times \underline{\tau}_2|} \tag{2.2}$$

may form a right handed system. The cross denotes the exterior vector product. Henceforth, the summation convention will be applied throughout, according to which summation is performed over diagonally doubly repeated greek and latin indices. Metric and curvature of the surface are then given by

$$g_{\alpha\beta} = \tau^k_\alpha \tau_{\beta k} \; , \qquad b_{\alpha\beta} = e_k \frac{\partial \tau^k_\alpha}{\partial \xi^\beta} \; . \tag{2.3}$$

Both are symmetric surface tensors. The velocity of the surface is

$$\begin{aligned} w^k &= \frac{\partial\, {}_R\chi^k(\Xi^\Gamma,t)}{\partial t} = \frac{\partial\, {}_a\chi^k(\xi^\beta,t)}{\partial \xi^\alpha} \cdot \frac{\partial \xi^\alpha(\Xi^\Gamma,t)}{\partial t} + \frac{\partial\, {}_a\chi^k(\xi^\alpha,t)}{\partial t} \\ &= \tau^k_\alpha w^\alpha + e^k w_n \; , \end{aligned} \tag{2.4}$$

where w^α and w_n denote the tangential and normal components, for details see[17]. We shall denote the partial derivative of a surface field $\psi_{\dots}(\xi^\beta,t)$ with respect to ξ^α by $\psi_{\dots,\alpha}$ and the covariant derivative by $\psi_{\dots;\alpha}$. The inverse $g^{\alpha\beta}$ of the metric will be used in the usual way to raise surface indices.

Mean curvature K_M and Gaußian curvature K_G are defined by

$$K_M := \frac{1}{2} b^\alpha_\alpha = \frac{1}{2}(K_1 + K_2) \; , \qquad K_G := \det\|b^\alpha_\beta\| = K_1 \cdot K_2 \; . \tag{2.5}$$

Furthermore, the Hamilton-Cayley theorem yields

$$(\underline{\underline{b}}^2)^\alpha_\beta = 2\, K_M\, b^\alpha_\beta - K_G\, \delta^\alpha_\beta \; , \tag{2.6}$$

where δ^α_β is Kronecker's symbol in two dimensions.

2.2. Balance Laws

If the phase boundary is represented by a moving non-material singular orientable surface the general form of a balance law on that surface is

$$\frac{d}{dt}\int_{s(t)} \psi_s(\xi^\alpha,t)\,da = -\oint_{C(s)} \phi_s^\alpha h_\alpha\,ds - \int_{s(t)} [\![\phi^k + \psi_v(v^k-w^k)]\!]e_k\,da + \int_{s(t)} (\Pi_s+\sigma_s)\,da\,. \tag{2.7}$$

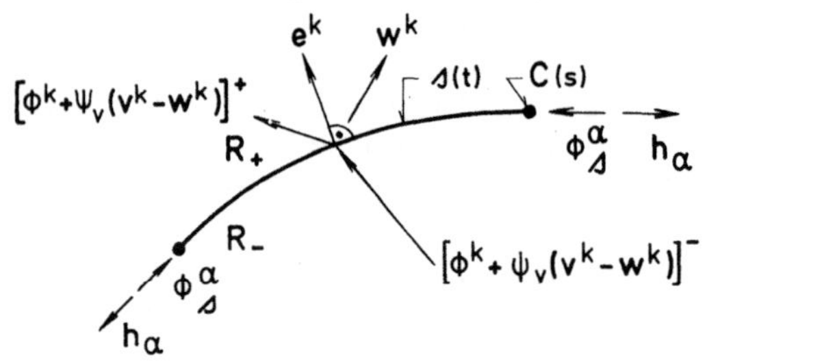

Fig. 2: Flux contributions at a non-material singular surface.

In this equation h_α is a unit vector in the tangent plane of the surface $s(t)$, perpendicular to the bounding curve $C(s)$ and orientated such, that it points out of the singular surface s, see Fig. 2. s denotes the arc length along $C(s)$; its orientation is counterclockwise around the surface normal e_k. This curve is chosen to be material in the sense that it possesses the same tangential velocity w^α as do material particles sitting instantaneously on the surface. The singular surface as a whole, however, is non-material, because matter may cross it when it represents a phase boundary. Moreover, da is the area element on $s(t)$, and d/dt denotes total time derivative following the advective velocity w^k. The field quantities have the meaning:

$\psi_s(\xi^\beta,t)$ = surface density of an additive quantity Ψ per unit area on $s(t)$.

$\phi_s^\alpha(\xi^\beta,t)$ = influx of Ψ along the tangent planes of $s(t)$ per unit time and unit length through the bounding curve $C(s)$ moving with velocity w^α.

π_s, σ_s = surface production and supply densities of Ψ per unit time and unit area of $s(t)$, respectively.

$\phi^k(\underline{x},t)$ = bulk influx of Ψ per unit time and unit area through a material surface in the bulk moving with material velocity v^k.

$\psi_v(\underline{x},t)$ = bulk density of Ψ per unit volume of the bulk material.

$$[\![\phi^k + \psi_v(v^k-w^k)]\!]e_k := [\phi_+^k + \psi_v^+(v_+^k - w^k)]e_k - [\phi_-^k + \psi_v^-(v_-^k - w^k)]e_k$$

= normal jump contribution of the bulk flux of Ψ per unit time and unit area through $s(t)$. It represents the influx to $s(t)$

from the adjacent bulk materials. $\phi^k_\pm$, $\psi^\pm_v$, $v^k_\pm$ are the limiting values of the bulk fields as the singular surface $s(t)$ is approached from R_+ and R_-, respectively.

With the identifications of Table I the special balance equations for mass, momentum, energy and entropy are obtained. The mass quantities in the first column are surface mean density ρ_s, surface momentum

	1	2	3	4	5	6
quantity	ψ_s	ϕ^α_s	π_s	σ_s	ψ_v	ϕ^m
mass	ρ_s	0	0	0	ρ	0
momentum	$\rho_s w^k$	$-t^{k\alpha}_s$	0	$\rho_s g^k$	ρv^k	$-t^{km}$
energy	$\rho_s(u_s+\frac{1}{2}\underline{w}^2)$	$q^\alpha_s - w_k t^{k\alpha}_s$	0	$\rho_s w_k g^k + \rho_s r_s$	$\rho(u+\frac{1}{2}\underline{v}^2)$	$q^m - v_k t^{km}$
entropy	$\rho_s s_s$	ϕ^α_s	$\pi_s \geqq 0$	σ_s	ρs	ϕ^m

Table I: Identification of fields for surface-balance equations.

density $\rho_s w^k$, surface energy density $\rho_s(u_s + \frac{1}{2} w^k w_k)$ and surface entropy density $\rho_s s_s$, all per unit area. u_s and s_s are the surface internal energy and surface entropy per unit mass, respectively. The second column contains the non-convective surface fluxes per unit time and unit length relative to particles moving with velocity w^α along moving material lines. $t^{k\alpha}_s$ is the surface stress, $q^\alpha_s - w_k t^{k\alpha}_s$ the energy flux, where q_s is the surface heat flux, and ϕ^α_s is the surface entropy flux. The corresponding mass flux vanishes, because C(s) is referred to moving material lines. The third column lists the production densities on the surface; those of mass, momentum and energy vanish since these quantities are conserved. The entropy production density satisfies the inequality $\pi_s \geqq 0$, hence entropy is produced within the phase boundary whenever phase-transition processes are irreversible. In the fourth column the supply densities due to external fields such as gravity and electromagnetism are collected. g^k is the gravitational force per unit mass acting on the surface particles, r_s is the heat supply due to radiative absorption per unit mass and unit time and σ_s is the entropy supply density, which will be specified later. The bulk densities in the fifth column are mass density ρ, momentum density ρv^k where v^k is the material velocity, energy density $\rho(u + \frac{1}{2} v^k v_k)$ and entropy density ρs in the adjacent bulk materials. u and s denote the specific internal energy and the specific entropy, respectively. Finally, the non-convective bulk fluxes in the last column have the following meaning: t^{km} is Cauchy's stress tensor, q^m the heat flux vector and ϕ^k the entropy flux vector.

The surface stress $t_s^{k\alpha}$ can be decomposed according to

$$t_s^{k\alpha} = \tau_\beta^k \, S^{\beta\alpha} + e^k \, S^\alpha \, . \tag{2.8}$$

$S^{\beta\alpha}$ is the tensor of surface tension, traction being positive; β indicates the direction of force and α the co-ordinate line to whose unit length the force is referred; S^α is a surface shear stress in the direction normal to the surface and α refers to the unit length of the co-ordinate lines. Assuming conservation of moment of momentum it can be shown, that

$$S^{\alpha\beta} = S^{\beta\alpha} \, , \qquad S^\alpha = 0 \, . \tag{2.9}$$

By routine methods of differential geometry, see[17], the local balance equations for any regular surface point is easily derived from (2.7). Using the convective time derivative for a surface density $\psi_s \, (\xi^\alpha, t)$

$$\frac{d\psi_s}{dt} = \frac{\partial \psi_s}{\partial t} + w^\alpha \, \psi_{s,\alpha} \, , \tag{2.10}$$

the identifications of Table I and the relations (2.3) - (2.5), (2.8), (2.9), the balance equations for surface mass, momentum, energy and entropy can be written as follows:

$$\frac{d\rho_s}{dt} = - \rho_s (w^\alpha_{;\alpha} - 2 K_M w_n) - \mathfrak{R} \, ,$$

$$\rho_s \frac{dw^\alpha}{dt} = - \rho_s \, {}_a\Gamma^\alpha_{\beta\gamma} w^\beta w^\gamma + \rho_s w_n (g^{\alpha\beta} w_{n,\beta} + 2 b^\alpha_\beta w^\beta) + S^{\alpha\beta}_{;\beta} + \rho_s \tau^\alpha_k g^k - \mathfrak{T}^\alpha ,$$

$$\rho_s \frac{dw_n}{dt} = - \rho_s w^\alpha (w_{n,\alpha} + b_{\alpha\beta} w^\beta) + S^{\alpha\beta} b_{\beta\alpha} + \rho_s e_k g^k - \mathfrak{T}_n \, ,$$

$$\rho_s \frac{du_s}{dt} = - q^\alpha_{s;\alpha} + S^{\alpha\beta} D_{\beta\alpha} + \rho_s r_s - \mathfrak{U} \, ,$$

$$\rho_s \frac{ds_s}{dt} = - \phi^\alpha_{s;\alpha} + \Pi_s + \sigma_s - \mathfrak{S} \, . \tag{2.11}$$

${}_a\Gamma^\alpha_{\beta\gamma} := g^{\mu\gamma} \tau_{\gamma k} \tau^k_{\alpha,\beta}$ is Christoffel's symbol of the actual surface and

$$D_{\alpha\beta} := w_{(\alpha;\beta)} - w_n \, b_{\alpha\beta} \tag{2.12}$$

denotes the tensor of deformation velocity of the surface; round brackets enclosing tensor indices indicating symmetrization with respect to these indices. Finally, $\mathfrak{R}$, $\mathfrak{T}^\alpha$, $\mathfrak{T}_n$, $\mathfrak{U}$ and $\mathfrak{S}$ are jump contributions across the interface:

$$\mathfrak{R}(\xi^\alpha,t) \;:=\; [\![\rho(v^m-w^m)]\!]e_m \;,$$

$$\mathfrak{T}^\alpha(\xi^\alpha,t) \;:=\; \tau^\alpha_k[\![-t^{km}+\rho(v^k-w^k)(v^m-w^m)]\!]e_m \;,$$

$$\mathfrak{T}_n(\xi^\alpha,t) \;:=\; e_k[\![-t^{km}+\rho(v^k-w^k)(v^m-w^m)]\!]e_m \;,$$

$$\mathfrak{U}(\xi^\alpha,t) \;:=\; [\![q^m-(v_k-w_k)t^{km}+\rho\big[(u-u_s)+\tfrac{1}{2}(\underline{v}-\underline{w})^2\big](v^m-w^m)]\!]e_m \;,$$

$$\mathfrak{S}(\xi^\alpha,t) \;:=\; [\![\phi^m+\rho(s-s_s)(v^m-w^m)]\!]e_m \;. \tag{2.13}$$

The balance laws $(2.11)_{1-4}$ govern development and motion of the phase interface and mass and energy transport across it. For irreversible phase change, the entropy production density $\Pi_s \geq 0$ is non-negative.

3. THE PHASE INTERFACE AS A BOUNDARY LAYER

Physically the phase interface is a thin layer (surrounding the singular surface s), across which all bulk fields change smoothly, though they may experience considerable changes over the thickness dimension of only a few molecular distances, see Fig. 3. The molecular structure and physical behaviour of matter inside this transition layer differ from those of the pure bulk phases.

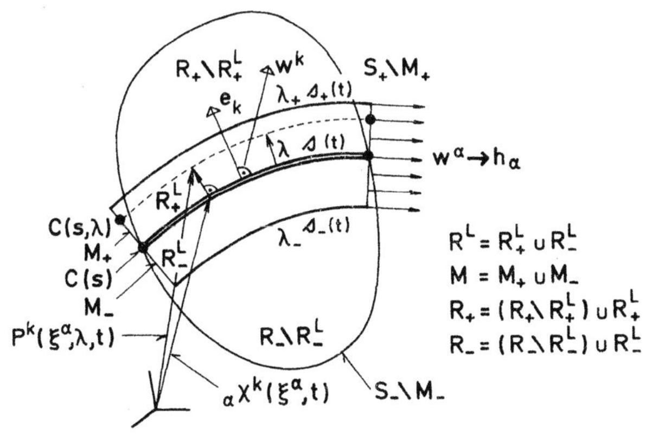

Fig. 3: Two-dimensional picture of a thin boundary layer between two bulk phases (for notation see also Fig. 1).

The surface fields of the previous chapter can be identified with mean values of bulk fields over the boundary layer. Thus the boundary can be reduced to a surface model. I follow closely the arguments of Gibbs[2], Buff[9] and Deemer & Slattery[11], but amend their procedure which is partly defective for curved interfaces.

The boundary layer is divided in a family

$$P^k(\xi^\alpha,\lambda,t) \;=\; {}_aX^k(\xi^\alpha,t) \;+\; \lambda e^k(\xi^\alpha,t) \tag{3.1}$$

of surfaces that are parallel to the singular surface $s(t)$. In this way integrations over the boundary-layer volume R^L can be separated into integrations over the surface area of $s(t)$ and over the thickness parameter λ and the volume balance equation can be formulated. We may extend the adjacent bulk fields $\psi_v^+ := \psi_v(\xi^\alpha,\lambda^+,t)$, $\phi_+^k := \phi^k(\xi^\alpha,\lambda^+,t)$ etc. into the part R_+^L of the layer up to the positive edge of the singular surface $s(t)$ and $\psi_v^- = \psi_v(\xi^\alpha,\lambda^-,t)$, $\phi_-^k = \phi^k(\xi^\alpha,\lambda^-,t)$ etc. into R_-^L of the transition layer up to the negative edge of $s(t)$ and formulate the volume balance equations over R_+^L and R_-^L, respectively. Subtraction of these balance equations from the balance equation over the true boundary layer in R^L yields then the balance equation for the surface fields

$$\frac{d}{dt}\int_{s(t)} da \int_{\lambda-}^{\lambda+}\left[\psi_v(\lambda)-\psi_v^{\pm}\right]\sqrt{\frac{g(\lambda)}{g}}\,d\lambda$$

$$= -\oint_{C(s)} ds\, h_\alpha \int_{\lambda-}^{\lambda+} \tau_k^\alpha(\lambda)\left[\left(\psi_v(\lambda)v^k(\lambda)-\psi_v^{\pm}v_{\pm}^k\right) - w^k\left(\psi_v(\lambda)-\psi_v^{\pm}\right) + \right.$$

$$\left. + \left(\phi^k(\lambda)-\phi_{\pm}^k\right)\right]\sqrt{\frac{g(\lambda)}{g}}\,d\lambda -$$

$$- \int_{s(t)} [\![\psi_v(v^k-w^k)+\phi^k]\!]e_k\,da +$$

$$+ \int_{s(t)} da \int_{\lambda-}^{\lambda+}\left[\left(\Pi_v(\lambda)-\Pi_v^{\pm}\right)+\left(\sigma_v(\lambda)-\sigma_v^{\pm}\right)\right]\sqrt{\frac{g(\lambda)}{g}}\,d\lambda\,, \qquad (3.2)$$

in which

$$\psi_v^{\pm} = \begin{cases}\psi_v^+ & \text{for } \lambda>0\,,\\ \psi_v^- & \text{for } \lambda<0\,,\end{cases} \quad \text{etc.} \qquad (3.3)$$

and

$$\sqrt{\frac{g(\lambda)}{g}} = (1-2\lambda K_M+\lambda^2K_G)\,, \qquad (3.4)$$

$$\tau_k^\alpha(\lambda) = \frac{1}{1-2\lambda K_M+\lambda^2K_G}\left[\delta_\beta^\alpha-\lambda(2K_M\delta_\beta^\alpha-b_\beta^\alpha)\right]\tau_k^\beta\,, \qquad (3.5)$$

$$h_\alpha = \varepsilon_{klm}\,\tau_\alpha^k\,\tau_\beta^l\,\frac{d\xi^\beta}{ds}\,e^m\,, \qquad g := \det\|g_{\alpha\beta}\|\,. \qquad (3.6)$$

Π_v and σ_v denote the production and the supply densities within the bulk and the other bulk fields have the interpretations of Table I. The factor $\sqrt{g(\lambda)/g}$ accounts for the change in surface area parallel to s and expresses it in terms of λ and the mean and Gaußian curvature of s. $\tau_k^\alpha(\lambda) = g^{\alpha\beta}(\lambda)\tau_{\beta k}(\lambda)$ is the dual of the tangent vector $\tau_{\alpha k}(\lambda)$ of a parallel surface at the distance λ from the dividing surface s. Finally, h_α is the outer unit normal vector to the bounding curve $C(s)$ in the local plane of the surface $\mathcal{s}$, where s is the arc length along C(s) (see Fig. 2, 3). The derivation of (3.2) is more or less technical using differential geometry. Details are given in[18].

Comparison of (3.2) with the surface balance (2.7) yields the following one-to-one correspondence

$$\psi_{\mathcal{s}}(\xi^{\alpha},t) := \int_{\lambda-}^{\lambda+} (\psi_v - \psi_v^{\pm}) \sqrt{\frac{g(\lambda)}{g}}\, d\lambda ,$$

$$\phi_{\mathcal{s}}^{\alpha}(\xi^{\alpha},t) := \int_{\lambda-}^{\lambda+} \left[(\psi_v v^k - \psi_v^{\pm} v_{\pm}^k) - w^k(\psi_v - \dot{\psi}_v^{\pm}) + (\phi^k - \phi_{\pm}^k)\right] \tau_k^{\alpha}(\lambda) \sqrt{\frac{g(\lambda)}{g}}\, d\lambda ,$$

$$\Pi_{\mathcal{s}}(\xi^{\alpha},t) := \int_{\lambda-}^{\lambda+} (\Pi_v - \Pi_v^{\pm}) \sqrt{\frac{g(\lambda)}{g}}\, d\lambda ,$$

$$\sigma_{\mathcal{s}}(\xi^{\alpha},t) := \int_{\lambda-}^{\lambda+} (\sigma_v - \sigma_v^{\pm}) \sqrt{\frac{g(\lambda)}{g}}\, d\lambda . \tag{3.7}$$

Accordingly, the surface fields are weighted mean values of bulk excess fields over the boundary layer of thickness $\lambda_+ - \lambda_-$. For flat surfaces with $g_{\alpha\beta}(\lambda) = g_{\alpha\beta}$ and $\tau_k^{\alpha}(\lambda) = \tau_k^{\alpha}$ the curvature depending weighting disappears.

We decompose now the bulk velocity, heat flux, entropy flux and Cauchy stress within the layer into tangential and normal components with respect to the singular surface:

$$v^k(\lambda) = v^{\alpha}(\lambda)\tau_{\alpha}^k + v_n(\lambda)e^k ,$$

$$q^k(\lambda) = q^{\alpha}(\lambda)\tau_{\alpha}^k + q_n(\lambda)e^k ,$$

$$\phi^k(\lambda) = \phi^{\alpha}(\lambda)\tau_{\alpha}^k + \phi_n(\lambda)e^k , \tag{3.8}$$

$$t^{kl}(\lambda) = t^{\alpha\beta}(\lambda)\tau_{\alpha}^k\tau_{\beta}^l + t^{\alpha}_{\bullet n}(\lambda)\tau_{\alpha}^k e^l + t_n^{\bullet\alpha}(\lambda)e^k\tau_{\alpha}^l + t_n(\lambda)e^k e^l .$$

Insertion of the identifications of Table I in (3.7) and use of (2.8), (2.9), (3.4), (3.5) and (3.8) yields then

i. <u>for surface mass density and surface mass flux:</u>

$$\rho_{\mathcal{s}} = \int_{\lambda-}^{\lambda+} (\rho - \rho^{\pm})(1 - 2\lambda K_M + \lambda^2 K_G)\, d\lambda , \tag{3.9}$$

$$0 = \int_{\lambda-}^{\lambda+} (\rho v^{\beta} - \rho^{\pm} v_{\pm}^{\beta}) \left[\delta_{\beta}^{\alpha} - \lambda(2K_M\delta_{\beta}^{\alpha} - b_{\beta}^{\alpha})\right] d\lambda - w^{\beta} \int_{\lambda-}^{\lambda+} (\rho - \rho^{\pm}) \left[\delta_{\beta}^{\alpha} - \lambda(2K_M\delta_{\beta}^{\alpha} - b_{\beta}^{\alpha})\right] d\lambda ; \tag{3.10}$$

ii. for surface momentum density and surface stress:

$$\rho_s w^\alpha = \int_{\lambda-}^{\lambda+} (\rho v^\alpha - \overset{\pm}{\rho} v_\pm^\alpha)(1 - 2\lambda K_M + \lambda^2 K_G)\,d\lambda \;,$$

$$\rho_s w_n = \int_{\lambda-}^{\lambda+} (\rho v_n - \overset{\pm}{\rho} \overset{\pm}{v}_n)(1 - 2K_M + \lambda^2 K_G)\,d\lambda \;; \tag{3.11}$$

$$s^{\alpha\beta} = \int_{\lambda-}^{\lambda+} \Big\{ (t^{\alpha\gamma} - t_\pm^{\alpha\gamma}) - \Big[\rho(v^\alpha - w^\alpha)(v^\gamma - w^\gamma) - \overset{\pm}{\rho}(v_\pm^\alpha - w^\alpha)(v_\pm^\gamma - w^\gamma)\Big]\Big\}\Big[\delta_\gamma^\beta - \lambda(2K_M\delta_\gamma^\beta - b_\gamma^\beta)\Big]d\lambda \;,$$

$$s^\alpha = 0 = \int_{\lambda-}^{\lambda+} \Big\{ (t_n^{\bullet\gamma} - t_{n\pm}^{\bullet\gamma}) - \Big[\rho(v_n - w_n)(v^\gamma - w^\gamma) - \overset{\pm}{\rho}(\overset{\pm}{v}_n - w_n)(v_\pm^\gamma - w^\gamma)\Big]\Big\}\Big[\delta_\gamma^\alpha - \lambda(2K_M\delta_\gamma^\alpha - b_\gamma^\alpha)\Big]d\lambda \;; \tag{3.12}$$

iii. for surface (internal) energy density and heat flux:

$$\rho_s u_s = \int_{\lambda-}^{\lambda+} \Big\{\rho\Big[u + \tfrac{1}{2}(\underline{v}-\underline{w})^2\Big] - \overset{\pm}{\rho}\Big[\overset{\pm}{u} + \tfrac{1}{2}(\overset{\pm}{\underline{v}}-\underline{w})^2\Big]\Big\}(1 - 2\lambda K_M + \lambda^2 K_G)\,d\lambda \;, \tag{3.13}$$

$$q_s^\alpha = \int_{\lambda-}^{\lambda+} \Big\{ (q^\gamma - q_\pm^\gamma) - \Big[(v_\beta - w_\beta)t^{\beta\gamma} - (\overset{\pm}{v}_\beta - w_\beta)t_\pm^{\beta\gamma}\Big] - \Big[(v_n - w_n)t_n^{\bullet\gamma} - (\overset{\pm}{v}_n - w_n)t_{n\pm}^{\bullet\gamma}\Big] + \Big[\rho\Big(u + \tfrac{1}{2}(\underline{v}-\underline{w})^2\Big)(v^\gamma - w^\gamma) - \overset{\pm}{\rho}\Big(\overset{\pm}{u} + \tfrac{1}{2}(\overset{\pm}{\underline{v}} - \underline{w})^2\Big)(v_\pm^\gamma - w^\gamma)\Big]\Big\}\Big[\delta_\gamma^\alpha - \lambda(2K_M\delta_\gamma^\alpha - b_\gamma^\alpha)\Big]d\lambda \tag{3.14}$$

iv. for surface entropy density and entropy flux:

$$\rho_s s_s = \int_{\lambda-}^{\lambda+} (\rho s - \overset{\pm}{\rho}\overset{\pm}{s})(1 - 2\lambda K_M + \lambda^2 K_G)\,d\lambda \;, \tag{3.15}$$

$$\phi_s^\alpha = \int_{\lambda-}^{\lambda+} \Big\{ (\phi^\beta - \phi_\pm^\beta) + \Big[\rho s(v^\beta - w^\beta) - \overset{\pm}{\rho}\overset{\pm}{s}(v_\pm^\beta - w^\beta)\Big]\Big\}\Big[\delta_\beta^\alpha - \lambda(2K_M\delta_\beta^\alpha - b_\beta^\alpha)\Big]d\lambda \tag{3.16}$$

Corresponding relations for surface supply and production densities are easily derived, but we do not need them in the sequal. Especially, the surface stress, internal energy and heat flux in non-equilibrium are rather complicated mean values over the non-material phase-boundary layer due to the appearance of convective contributions.

The requirement of vanishing normal shear $S^\alpha = 0$ is called the membrane approximation. Physically, it is motivated by the very small thickness of the boundary layer and conservation of moment of momentum, which in turn requires the symmetry $S^{\alpha\beta} = S^{\beta\alpha}$ of surface tension. Accordingly, the relations (3.12) contain three constraints on the possible profiles of Cauchy's stress within the phase boundary layer. These we call the *dynamical consistency conditions*.

Another set of constraints is included in (3.10) and $(3.11)_1$. Both must give the same tangential velocity w^α of the singular surface. This is a consequence of the definition of moving material lines (with velocity w^α) within the singular surface $\mathcal{s}(t)$, whose introduction is necessary in order that the concept of surface tension is physically significant. Elimination of w^α between (3.10) and $(3.11)_1$ yields thus the constraints

$$b^\beta_\alpha I^\alpha - K_G \delta^\beta_\alpha II^\alpha + K_G(2K_M\delta^\beta_\alpha - b^\beta_\alpha) III^\alpha = 0 \quad (\beta = 1,2) \;, \tag{3.17}$$

where I^α, II^α and III^α are abbreviations for

$$I^\alpha := \left(\int_{\lambda-}^{\lambda+} \varphi^\alpha d\lambda\right)\left(\int_{\lambda-}^{\lambda+} R\lambda d\lambda\right) - \left(\int_{\lambda-}^{\lambda+} \varphi^\alpha \lambda d\lambda\right)\left(\int_{\lambda-}^{\lambda+} R d\lambda\right) , \tag{3.18}$$

$$II^\alpha := \left(\int_{\lambda-}^{\lambda+} \varphi^\alpha d\lambda\right)\left(\int_{\lambda-}^{\lambda+} R\lambda^2 d\lambda\right) - \left(\int_{\lambda-}^{\lambda+} \varphi^\alpha \lambda^2 d\lambda\right)\left(\int_{\lambda-}^{\lambda+} R d\lambda\right) ,$$

$$III^\alpha := \left(\int_{\lambda-}^{\lambda+} \varphi^\alpha \lambda d\lambda\right)\left(\int_{\lambda-}^{\lambda+} R\lambda^2 d\lambda\right) - \left(\int_{\lambda-}^{\lambda+} \varphi^\alpha \lambda^2 d\lambda\right)\left(\int_{\lambda-}^{\lambda+} R\lambda d\lambda\right) ,$$

with $\quad \varphi^\alpha(\lambda) := \rho(\lambda)v^\alpha(\lambda) - \rho^\pm v^\alpha_\pm \;, \quad R(\lambda) := \rho(\lambda) - \rho^\pm \;.$

The relations (3.17) are constraints on the possible profiles $\rho(\lambda)$ and $v^\alpha(\lambda)$ of bulk density and bulk velocity within the layer; we call them the *kinematic consistency conditions*.

For flat phase boundaries relation (3.17) is trivially satisfied. For curved phase boundaries, however, they form two conditions on λ_+ and λ_- and fix the positions of $\mathcal{s}_+$ and $\mathcal{s}_-$ relative to the singular surface $\mathcal{s}(t)$ at $\lambda = 0$. Equivalent parameters are the thickness $d_\mathcal{s} := \lambda_+ - \lambda_-$ of the boundary layer and $a_\mathcal{s} := \lambda_+ + \lambda_-$, a measure of "excentricity". Given the position of the singular surface $\mathcal{s}$ (at $\lambda = 0$), its curvature, the distributions $\rho(\lambda)$ and $v^\alpha(\lambda)$ of density and tangential velocity within the layer and their values $\rho^\pm$ and $v^\alpha_\pm$ in the adjacent bulk phases, the thickness $d_\mathcal{s}$ and the excentricity $a_\mathcal{s}$ emerge as functions of the curvature and functionals of the distributions $\rho(\lambda)$ and $v^\alpha(\lambda)$. Because the actual distributions $\rho(\lambda)$ and $v^\alpha(\lambda)$ across the phase boundary layer are not known, they can be chosen such as to make $d_\mathcal{s}$ independent of curvature and $a_\mathcal{s}$ depending on the curvature only. This guarantees then that the tangential velocity w^α of the singular surface satisfies (3.10) *and* $(3.11)_1$. Under these conditions only does the singular surface represent a valid continuum physical image of the layer of phase transition.

4. CONSTITUTIVE ASSUMPTIONS FOR SURFACE FIELDS AND THEIR RELATION TO THE BOUNDARY-LAYER THEORY OF PHASE INTERFACES

The basic objective of thermodynamics on non-material singular surfaces, that represent phase change surfaces, is the determination of the fields

$$\rho_s(\xi^\alpha,t)\ ,\qquad {}_a\chi^k(\xi^\alpha,t)\ ,\qquad T_s(\xi^\alpha,t) \tag{4.1}$$

of density, motion and surface temperature. These can be determined from the balance equations $(2.11)_{1-4}$ for mass, momentum and energy, provided (for given supplies $g^k(\underline{x},t)$, $r_s(\underline{x},t)$) the surface fields

$$s^{\alpha\beta}(\xi^\gamma,t)\ ,\qquad q_s^\alpha(\xi^\gamma,t)\ ,\qquad u_s(\xi^\gamma,t) \tag{4.2}$$

and the jumps

$$\mathfrak{R}(\xi^\gamma,t)\ ,\qquad \mathfrak{T}^\alpha(\xi^\gamma,t)\ ,\qquad \mathfrak{T}_n(\xi^\gamma,t)\ ,\qquad \mathfrak{U}(\xi^\gamma,t) \tag{4.3}$$

are related in a "materially dependent manner" to the basic fields (4.1). Such relations are called "constitutive equations" for phase boundaries. They must reflect the peculiarities of the phase-change interface in non-equilibrium, which embrace dynamics and thermodynamics of freezing and melting or evaporation and condensation, effects of surface tension, internal friction etc. Moreover, the layer between a crystal and its melt, for instance, must match between the anisotropic structure on the solid side and the isotropic structure on the fluid side with a continuous loss of anisotropy over the thickness of the layer. The transition layer between a solid and its fluid is neither a solid nor a fluid, it is something in between. Corresponding statements hold for the fluid-gas transition layer. Furthermore, the transition layer contains extremely large density gradients, velocity gradients and (possibly) temperature gradients in the normal direction. Thermodynamic constitutive theories or non-equilibrium statistical theories, that would embrace all these peculiarities of the phase-change layer do presently not exist. This means, that the actual profiles of density $\rho(\lambda)$, velocity $v^\alpha(\lambda)$, $v_n(\lambda)$, Cauchy stress $t^{\alpha\beta}(\lambda)$, $t_n^{\cdot\alpha}(\lambda)$, $t^\alpha_{\cdot n}(\lambda)$, $t_n(\lambda)$, of internal energy $u(\lambda)$, heat flux $q^\alpha(\lambda)$, $q_n(\lambda)$ [and entropy $s(\lambda)$ and entropy flux $\phi^\alpha(\lambda)$, $\phi_n(\lambda)$] are not known. This implies through (3.9) - (3.16) also, that the constitutive equations for the surface fields (4.2) and the jump fields (4.3) are by no means evident.

4.1. Calculation of surface fields from distributions of boundary-layer fields

To get a first impression on the form of constitutive equations for the surface and jump fields we calculate them from the excess integrals (3.9) - (3.16) for assumed profiles for the bulk fields within the phase-transition layer. The simplest profiles which satisfy the kinematic and the dynamic consistency conditions are cubic profiles $f(\lambda)$ for all bulk fields, see Fig. 4:

$$f(\xi^\alpha,\lambda,t) = f_0(\xi^\gamma,t) + f_1(\xi^\alpha,t)\cdot\lambda + f_2(\xi^\alpha,t)\cdot\lambda^3\ . \tag{4.4}$$

$$f_1 = \frac{\partial f}{\partial \lambda}_{\lambda=0} \tag{4.5}$$

is the normal derivative at the location of $\mathscr{s}(t)$ and f_0 and f_2 are determined from the boundary conditions $f(\xi^\alpha, \lambda_\pm, t) = f^\pm(\xi^\alpha, t)$ at the sites of the adjacent bulk phases:

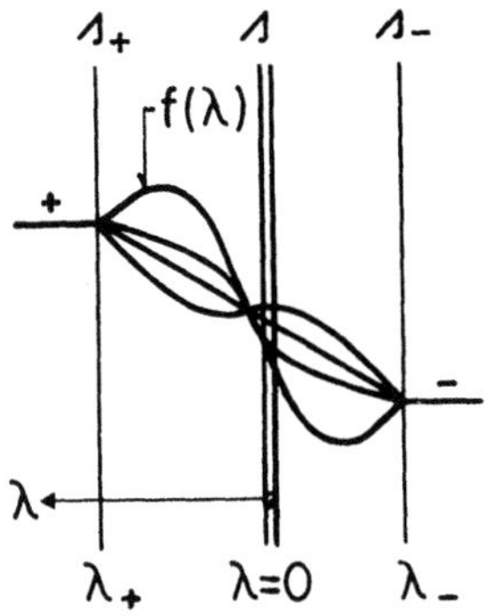

Fig. 4: Family of cubic profiles for the bulk fields within the boundary layer.

$$f_0 = \frac{1}{2}(f^+ + f^-) - \frac{a_\mathscr{s}}{2d_\mathscr{s}(3a_\mathscr{s}^2 + d_\mathscr{s}^2)} \times$$
$$\times \left\{ [\![f]\!] (a_\mathscr{s}^2 + 3d_\mathscr{s}^2) + f_1 2d_\mathscr{s} (a_\mathscr{s}^2 - d_\mathscr{s}^2) \right\} ,$$

$$f_2 = \frac{8}{2d_\mathscr{s}(3a_\mathscr{s}^2 + d_\mathscr{s}^2)} ([\![f]\!] - f_1 d_\mathscr{s}) \tag{4.6}$$

with

$$d_\mathscr{s} := \lambda_+ - \lambda_- , \quad a_\mathscr{s} := \lambda_+ + \lambda_- . \tag{4.7}$$

These families of profiles with adjustable normal slopes f_1 at $\lambda = 0$ contain sufficient freedom to satisfy the constraints (3.17) and to allow a membrane approximation $[s^{\alpha\beta} = s^{\beta\alpha},\ s^\alpha = 0]$ for the phase interface. Insertion of the cubic profiles into (3.9) - (3.16) and calculation of the surface fields is tedious routine work, details of which may be found in[18]. Here we discuss the main results.

The constraints (3.17) can be satisfied, if the slopes ρ_1 and v_1^α of the density and the tangential velocity profiles are proportional to the corresponding jumps

$$\rho_1 = \frac{X}{d_\mathscr{s}} [\![\rho]\!] , \quad v_1^\alpha = \frac{X}{d_\mathscr{s}} [\![v^\alpha]\!] \tag{4.8}$$

with the same factor of proportionality. The identities (3.17) reduce then to two non-linear equations for the dimensionless factor X and the dimensionless excentricity $\alpha := a_\mathscr{s}/d_\mathscr{s}$ which require that

$$\alpha = \alpha(k_M, k_G) , \quad X = X(k_M, k_G) \tag{4.9}$$

are functions of the dimensionless curvatures $k_M := K_M d_\mathscr{s}$ and $k_G := K_G d_\mathscr{s}^2$. Numerical solutions show, that within the physically permissible interval $-1 \leqq \alpha \leqq 1$ exactly two families, $\alpha_1(k_M, k_G) < 0$ and $\alpha_2(k_M, k_G) > 0$, exist; see Fig. 5.

Using these results, the surface density and tangential velocity follow from (3.9) and $(3.11)_1$ and are given by

$$\rho_\mathscr{s} = [\![\rho]\!] \frac{d_\mathscr{s}}{2} F_1(k_M, k_G) , \tag{4.10}$$

$$\rho_\mathscr{s} w^\alpha = [\![\rho v^\alpha]\!] \frac{d_\mathscr{s}}{2} F_1(k_M, k_G) + [\![\rho]\!] \cdot [\![v^\alpha]\!] \frac{d_\mathscr{s}}{6} F_2(k_M, k_G) . \tag{4.11}$$

The dimensionless functions F_1 and F_2 are, strictly, functions of the curvatures, of $\alpha(k_M, k_G)$ and $X(k_M, k_G)$. Plots for spherical geometry, $k_M^2 = k_G$, are given in Fig. 5.

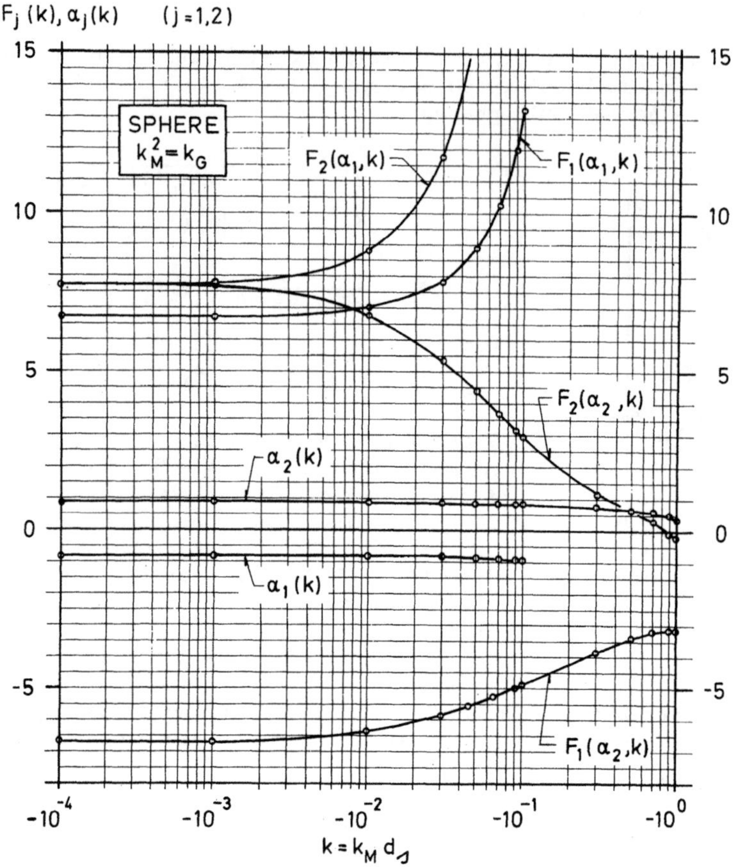

Fig. 5:
Solutions of equation (3.17) for spherical geometry, $K_M^2 = K_G$. Plotted on semi-logarithmic scale is $\alpha_1 < 0$ and $\alpha_2 > 0$ as a function of $k = K_M d_\delta$. For $|k| > 0.1$ one has $|\alpha_1| > 1$ which is unphysical and therefore not shown. Displayed are also the values of the functions F_1 and F_2 arising in (4.10) and (4.11). Dependence on curvature is significant only for $|k| > 10^{-3}$.

Let us interpret (4.10) first. F_1 can be positive or negative. Its sign must be chosen such that $\rho_\delta > 0$. Hence, two cases may be distinguished, which, for spherical geometries, are explained in Fig. 6. [Recall that the +side of the interface is always that side into which the surface normal vector points.] For an inclusion of the denser in the less dense phase (Case A: water in ice) the positive family

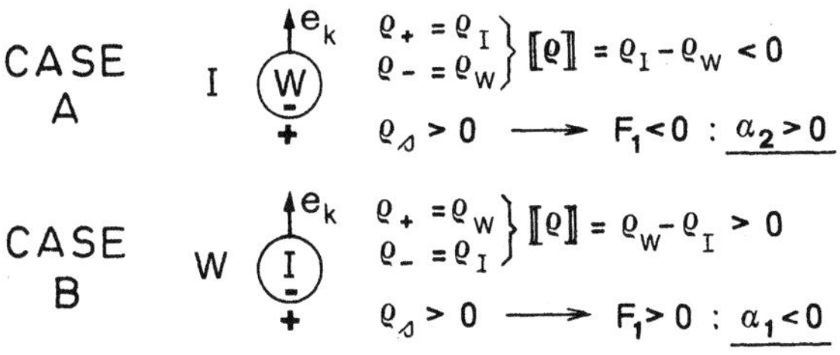

Fig. 6: Relative positions of the denser phase (water = W) and the less dense phase (ice = I) and choice of sign of the dimensionless excentricity α and function F_1.

$1 \geqq \alpha_2(k_M,k_G) > 0$ of excentricities must be chosen. Since for this family (3.17) possess solutions *for all* curvatures, inclusion of the denser phase in the less dense one (water in ice) do exist down to bubble sizes of nucleation dimensions. The situation is different, when inclusions of the less dense phase in the dense one are considered (Case B: ice in water). In this case the negative family $-1 \leqq \alpha_1(k_M,k_G) < 0$ must be selected. Since for this family (3.17) possess no physically meaningful solution ($|\alpha_1| < 1$) when $|k_M| \geqq 0.1$, we conclude that very small inclusions of the less dense phase in the dense one (ice in water) cannot exist at nucleation dimensions. Ice nucleation in steam, however, is possible (case A is applicable!). From this we infer, that nucleation of ice from pure water must start from the water surface. This, in fact, is observed in nature. The non-existence of ice-nucleation in water may also explain the phenomenon of undercooling in clean water. Ice formation in water, however, is possible, when nucleation kernels are present with dimensions that are larger than ten times the boundary layer thickness, or $|k_M| \leqq 0.1$.

Let us shortly discuss the order of magnitude of the surface mass density ρ_s. Assume a thickness $d_s = 100$ Å $= 10^{-6}$ cm for the boundary layer between ice and water at the normal freezing point, Golecki & Jaccard[8]. For a flat phase boundary we have thus

$$\rho_s = (\rho_W - \rho_I)\frac{d_s}{2}|F_1(0,0)| \cong 2{,}81\cdot 10^{-7}\ \frac{\text{g}}{\text{cm}^2} \qquad \text{(Ice/Water)}.$$

For a flat phase boundary between water and its vapour at normal pressure and room temperature we have correspondingly, assuming a thickness of again $d_s = 100$ Å,

$$\rho_s = (\rho_W - \rho_V)\frac{d_s}{2}|F_1(0,0)| \cong 3{,}35\cdot 10^{-6}\ \frac{\text{g}}{\text{cm}^2} \qquad \text{(Water/Vapour)}.$$

These are small numbers, consequently, if there is only a single phase boundary between large masses of ice and water and its vapour, its mass contribution to the total mass is negligible. However, in two phase mixtures as in the mushy zone of temperate glaciers, or in dense fogs and clouds, the area of internal phase boundaries per unit volume is considerable (up to 10^5 cm²/cm³), and therefore the mass contribution of the phase boundaries to the total mass of the mixture cannot be ignored.

The boundary layer model is also valid for the fluid-vapour phase transitions. In this case the critical point can be reached, at which the phase boundary must disappear, since the phases are no longer distinguishable. This property is included in the foregoing results, for with $\|\rho\| \to 0$ also $\rho_s \to 0$ for all curvatures: the surface field ρ_s

vanishes when the phase boundary layer disappears. All other surface fields must have similar behaviour at the critical point.

A brief discussion of formula (4.11) for surface momentum may be added. For adherence of the bulk material, viz $v_+^\alpha = v_-^\alpha$, we deduce $w^\alpha = v_+^\alpha = v_-^\alpha$: the tangential velocity of the interface equals the material velocities of the adjacent bulk materials. For $v_+^\alpha \neq v_-^\alpha$ the surface velocity w^α differs from both.

Next, we consider the normal motion of the interface. Inserting the cubic profile for $v_n(\lambda)$ into $(3.11)_2$ and choosing the normal slope $v_n' = X/d_s [\![v_n]\!]$ we obtain accordingly

$$\rho_s w_n = [\![\rho v_n]\!]\frac{d_s}{2} F_1\ (k_M,k_G) + [\![\rho]\!]\cdot[\![v_n]\!]\frac{d_s}{6} F_2\ (k_M,k_G)\ . \tag{4.12}$$

For a non-material phase interface w_n must be different from the material velocity components v_n^+ and v_n^- of the adjacent bulk materials. This requires $v_n^+ \neq v_n^-$. The material interface with $v_n^+ = v_n^-$ and $w_n = v_n^+ = v_n^-$ is included as a special case and so is the critical point for which $[\![\rho]\!] = 0$.

More complex are the results for surface tension, heat- and entropy flux and internal energy and entropy. Neglecting all convective contributions in (3.12) - (3.16) and satisfying the membrane approximation we obtain for the cubic profile, c.p.[18]:

$$\begin{aligned}
s^{\alpha\beta} - s^{\alpha\beta}\Big|_E &= -[\![t^{\gamma\delta} + p_E g^{\gamma\delta}]\!]\frac{d_s}{2}\left\{G_1\delta^\alpha_\gamma\delta^\beta_\delta + \frac{d_s}{4} G_2\delta_\gamma^{(\alpha}b_\delta^{\beta)} + \frac{d_s^2}{16} G_3 b^\alpha_\gamma b^\beta_\delta\right\},\\
q_s^\alpha &= -[\![q^\beta]\!]\frac{d_s}{2}\left\{J_1\delta^\alpha_\beta + \frac{d_s}{4} J_2 b^\alpha_\beta\right\},\\
\phi_s^\alpha &= -[\![\phi^\beta]\!]\frac{d_s}{2}\left\{J_1\delta^\alpha_\beta + \frac{d_s}{4} J_2 b^\alpha_\beta\right\},\\
\rho_s u_s &= [\![\rho u]\!]\frac{d_s}{2} F_1 + [\![\rho]\!]\cdot[\![u]\!]\frac{d_s}{6} F_2\ ,\\
\rho_s s_s &= [\![\rho s]\!]\frac{d_s}{2} F_1 + [\![\rho]\!]\cdot[\![s]\!]\frac{d_s}{6} F_2\ .
\end{aligned} \tag{4.13}$$

G_1, G_2, G_3, J_1, J_2 are dimensionless functions of the dimensionless curvatures k_M, k_G. Equilibrium surface tension $s^{\alpha\beta}\big|_E = \sigma_E g^{\alpha\beta}$ satisfies the relations

$$\begin{aligned}
\sigma_E &= \left[[\![p_E]\!] - (\rho_s e^k)\big|_E\, g_k\right]\cdot\frac{d_s}{2}\frac{1}{k_M}\ ,\\
\sigma_{E,\alpha} &= -(\rho_s \tau^k_\alpha)\big|_E\, g_k\ .
\end{aligned} \tag{4.14}$$

$(4.14)_1$ is Laplace's formula in the gravitational field.

Common structure in all of these equations is, that the surface fields depend linearily on jumps of corresponding bulk fields, on the thickness of the boundary layer and on mean and Gaußian curvature. This structure will be the guide for the set up of more general constitutive equations.

4.2. Constitutive equations for surface fields

Constitutive equations for the surface and the jump fields will be treated separately.

The surface fields

$$c_s(\xi^\gamma,t): \begin{cases} s^{\alpha\beta}(\xi^\gamma,t) , & q_s^\alpha(\xi^\gamma,t) , \quad u_s(\xi^\gamma,t) \\ & \phi_s^\alpha(\xi^\gamma,t) , \quad s_s(\xi^\gamma,t) \end{cases} \tag{4.15}$$

must be related to the independent fields (4.1) by constitutive equations. For this we assume the interface between the bulk material as a fluid-like, viscous and heat conducting (two-dimensional) continuum, which is mathematically defined by the constitutive equations

$$c_s(\xi^\gamma,t) = \mathcal{L}_s(\rho_s, T_s, w^\alpha, w_n, T_{s,\alpha}, w^\alpha_{;\beta}, w_{n,\beta}; \tau^k_\alpha, e^k, \tau^k_{\alpha;\beta}, e^k_{,\beta}) \tag{4.16}$$

Commas $(\ )_{,\alpha}$ and semi-colons $(\ )_{;\alpha}$ denote partial derivatives or covariant differentiations with respect to ξ^α. T_s is the surface temperature which may be different from the temperatures T_+ and T_- in the adjacent bulk materials. The surface gradients $T_{s,\alpha}$, $w^\alpha_{;\beta}$, $w_{n,\beta}$ account for heat conduction along and internal viscous friction within the phase boundary, respectively. The tangent vectors and the normal vector and their derivatives allow that constitutive quantities also depend on the geometry of the surface.

4.3. Constitutive equations for jump fields

For vanishing surface fields the jump contributions (2.13) $\mathfrak{R}$, $\mathfrak{T}^\alpha$, $\mathfrak{T}_n$, $\mathfrak{U}$ and $\mathfrak{S}$ in the balance laws (2.11) vanish; and (2.13) degenerate to the classical boundary conditions at immaterial moving surfaces. Mass, momentum etc. bearing real phase boundaries, however, require constitutive assumptions for these jump quantities. If we split off the equilibrium pressure $p_E^\pm$ from the Cauchy stress of the adjacent bulk materials, $t_\pm^{kl} = -p_E^\pm \delta^{kl} + \hat{t}_\pm^{kl}$, and use the well known result $\phi_\pm^k = 1/T_\pm \, q_\pm^k$ for the entropy flux within the bulk materials (T = absolute temperature), inspection of (2.13) shows that they are explicitly functions of the limiting fields

$$\begin{aligned} W_+^\alpha &:= v_+^\alpha - w^\alpha , & \hat{t}^\alpha_{n+} &:= \tau^\alpha_k \, \hat{t}_+^{km} \, e_m , \\ W_n^+ &:= v_n^+ - w_n , & \hat{t}_n^+ &:= e_k \, \hat{t}_+^{km} \, e_m , \\ \theta_+ &:= T_+ - T_s , & q_n^+ &:= q_+^k \, e_k \end{aligned} \tag{4.17}$$

of relative tangential velocity W_+^α *and* shear stress $\hat{t}^\alpha_{n+}$, relative normal velocity W_n^+ *and* normal stress $\hat{t}_n^+$ and temperature difference ϑ_+ *and* normal heat flux q_n^+. The jump contributions (2.13) contain then explicitly $[\![\rho]\!]$, $[\![u]\!]$, $[\![p_E]\!]$, $[\![s]\!]$ and the following jump quantities

$$\mathfrak{K}(\xi^\alpha,t) : \begin{cases} \mathfrak{v}^\alpha := \tau^\alpha_k[\![v^k]\!], & \mathfrak{C}^\alpha := \tau^\alpha_k[\![\hat{t}^{km}]\!]e_m \;; \\ \mathfrak{v}_n := e_k[\![v^k]\!], & \mathfrak{C}_n := e_k[\![\hat{t}^{km}]\!]e_m \;; \\ \mathfrak{K} := \left[\!\!\left[\frac{1}{T}\right]\!\!\right], & \mathfrak{Q} := e_k[\![q^k]\!] \;; \end{cases} \tag{4.18}$$

and can be rewritten:

$$\begin{aligned} \mathfrak{R} &= [\![\rho]\!]w^+_n + (\rho^+ - [\![\rho]\!])\mathfrak{v}_n \,, \\ \mathfrak{T}^\alpha &= -\mathfrak{C}^\alpha + [\![\rho]\!]w^\alpha_+ w^+_n + (\rho^+ - [\![\rho]\!])\left[\mathfrak{v}_n w^\alpha_+ + \mathfrak{v}^\alpha w^+_n - \mathfrak{v}^\alpha\mathfrak{v}_n\right] \,, \\ \mathfrak{T}_n &= -\mathfrak{C}_n + [\![p_E]\!] + [\![\rho]\!](w^+_n)^2 + (\rho^+ - [\![\rho]\!])\left[2\mathfrak{v}_n w^+_n - \mathfrak{v}^2_n\right] \,, \\ \mathfrak{U}_n &= \mathfrak{Q} + \left\{\left[\!\!\left[\rho\left(u - u_s + \frac{p_E}{\rho} + \frac{1}{2}(\underline{v} - \underline{w})^2\right)\right]\!\!\right] - \mathfrak{C}_n\right\}(w^+_n - \mathfrak{v}_n) + \\ &\quad + \left\{\rho^+\left[u^+ - u_s + \frac{p^+_E}{\rho^+} + \frac{1}{2}(\underline{v}^+ - \underline{w})^2\right] - \hat{t}^+_n\right\}(w^+_n - \mathfrak{v}_n) - \\ &\quad - \mathfrak{C}_\alpha(w^\alpha_+ - \mathfrak{v}^\alpha) - \mathfrak{v}_\alpha \hat{t}^\alpha_{n+} \,, \\ \mathfrak{S} &= \mathfrak{K}q^+_n + \left(\frac{1}{T_+} - \mathfrak{K}\right)\mathfrak{Q} + [\![\rho(s - s_s)]\!](w^+_n - \mathfrak{v}_n) + \mathfrak{v}_n\rho^+(s^+ - s_s) \,. \end{aligned} \tag{4.19}$$

The fields (4.17) may be independently prescribed on one side of the phase boundary. The outcome on the other side, however, depends on this input *and* the constitutive properties of the phase interface. Hence, the jump contributions (4.18) may depend on the variables listed in (4.16) and (4.17):

$$\begin{aligned} \mathfrak{K}_s(\xi^\alpha,t) = \mathfrak{K}_s(&\rho_s, T_s, w^\alpha, w_n, T_{s,\alpha}, w^\alpha_{;\beta}, w_{n,\beta}; \tau^k_\alpha, e^k, \tau^k_{\alpha;\beta}, e^k_{,\beta}; \\ &T_+, \rho^+; w^\alpha_+, w^+_n, \hat{t}^\alpha_{n+}, \hat{t}^+_n, \theta^+, q^+_n) \,. \end{aligned} \tag{4.20}$$

The same holds for the jump contributions (4.19), since they are combinations of (4.18), (4.17) and (4.15).

Inserting the constitutive equations for the surface fields and the jump contributions into $(2.11)_{1-4}$ yields a set of field equations for the determination of $\rho_s(\xi^\alpha,t)$, ${}_a\chi^k(\xi^\alpha,t)$ and $T_s(\xi^\alpha,t)$. Every solution for given supplies $g^k(\xi^\alpha,t)$ and $r_s(\xi^\alpha,t)$ and given bulk fields $\rho^+(\xi^\beta,t)$, $T_+(\xi^\beta,t)$, $v^\alpha_+(\xi^\beta,t)$, $v^+_n(\xi^\beta,t)$, $t^\alpha_{n+}(\xi^\beta,t)$, $t^+_n(\xi^\beta,t)$ and $q^+_n(\xi^\beta,t)$ on one side of the phase boundary is called a *thermodynamic process* upon phase interfaces.

5. RESTRICTIONS ON THE CONSTITUTIVE EQUATIONS

The constitutive equations are restricted by three general requirements: The rule of material frame indifference, the principle of material symmetry and the entropy principle. The evaluation of these requirements is the content of the thermodynamic constitutive theory of phase boundaries. We cannot go into the details, we merely state the requirements and the results.

5.1. Rule of material frame indifference

This principle requires invariance of the constitutive equations under Euklidean-transformations of observer frames. The surface and jump fields (4.15) and (4.20) are scalar invariants under these transformations. Hence their constitutive equations can only depend on scalar invariants; they reduce to

$$c_s(\xi^\gamma,t) = \mathcal{L}_s(\rho_s, T_s, T_{s,\alpha}, D_{\alpha\beta}; g_{\alpha\beta}, b_{\alpha\beta}) \ , \tag{5.1}$$

$$\begin{aligned} K_s(\xi^\gamma,t) = \mathcal{K}_s(&\rho_s, T_s, T_{s,\alpha}, D_{\alpha\beta}; g_{\alpha\beta}, b_{\alpha\beta}; \\ &T_+, \rho^+; W^\alpha_+, W^+_n, \hat{t}^\alpha_{n+}, \hat{t}^+_n, \theta_+, q^+_n) \ . \end{aligned} \tag{5.2}$$

5.2. Principle of material symmetry

The reduced constitutive equations (5.1) and (5.2) hold quite generally for fluid-like anisotropic or isotropic phase interfaces. Considering, for instance, the ice/water transition, the boundary layer must match between the hexagonal symmetry of ice on one side and the isotropy of water on the other side. The two-dimensional continuum onto which we map the boundary layer must therefore be anisotropic, where the corresponding symmetry group is some compromise between the hexagonal crystal of ice and the isotropic water. However, we do not know this compromise group. For the water/vapour transition the phase boundary layer matches two isotropic bulk materials. In this case the phase interface is transversely isotropic with respect to rotations around the *local* normal vector e_k and *local* in-plane inflections. Less accurately we assume this also for the ice/water interface and hope that this approximation is not too bad.

The constitutive equations (5.1) and (5.2) are then isotropic functions of their arguments. The *irreducible isotropic invariants* of the basic surface fields are then

$$\begin{aligned}
K_M &= \frac{1}{2} b^\alpha_\alpha \ , \quad K_G = \frac{1}{2}\left[(\mathrm{tr}\underline{\underline{b}})^2 - \mathrm{tr}\underline{\underline{b}}^2\right] \ ; \\
I_D &= D^\alpha_\alpha \ , \quad II_D = \mathrm{tr}(\underline{\underline{b}}\underline{\underline{D}}) \ , \quad III_D = \frac{1}{2}\left[(\mathrm{tr}\underline{\underline{D}})^2 - \mathrm{tr}\underline{\underline{D}}^2\right] \ ; \\
H_1 &= \underline{T}_s \cdot \underline{T}_s = g^{\alpha\beta}\, T_{s,\alpha}\, T_{s,\beta} \ , \\
H_2 &= \underline{T}_s \cdot \underline{\underline{b}}\, \underline{T}_s = b^{\alpha\beta}\, T_{s,\alpha}\, T_{s,\beta} \ , \\
H_3 &= \underline{T}_s \cdot \underline{\underline{D}}\, \underline{T}_s = D^{\alpha\beta}\, T_{s,\alpha}\, T_{s,\beta} \ , \\
H_4 &= \underline{T}_s \cdot (\underline{\underline{b}}\,\underline{\underline{D}} - \underline{\underline{D}}\,\underline{\underline{b}})\underline{T}_s = (\underline{\underline{b}}\,\underline{\underline{D}} - \underline{\underline{D}}\,\underline{\underline{b}})^{\alpha\beta}\, T_{s,\alpha}\, T_{s,\beta} \ .
\end{aligned} \tag{5.3}$$

The *linearized irreducible representations* for the *surface fields* are thus

$$
\begin{aligned}
u_s &= u_s(\rho_s, T_s, K_M, K_G; I_D, II_D) , \\
s_s &= s_s(\text{———————————}) , \\
q_s^\alpha &= -\left[\kappa g^{\alpha\beta} + \kappa_2(b^{\alpha\beta} - K_M g^{\alpha\beta})\right] T_{s,\beta} , \\
\phi_s^\alpha &= -\left[\varphi g^{\alpha\beta} + \varphi_2(b^{\alpha\beta} - K_M g^{\alpha\beta})\right] T_{s,\beta} , \\
S^{\alpha\beta} &= \sigma g^{\alpha\beta} + \sigma_2(b^{\alpha\beta} - K_M g^{\alpha\beta}) + (\mu_1 I_D + \mu_2 II_D) g^{\alpha\beta} + \\
&\quad + (\mu_3 I_D + \mu_4 II_D)(b^{\alpha\beta} - K_M g^{\alpha\beta}) + \mu_5 D^{\alpha\beta} + \mu_6 (\underline{\underline{b}}\,\underline{\underline{D}} - \underline{\underline{D}}\,\underline{\underline{b}})^{(\alpha\beta)} ;
\end{aligned}
\tag{5.4}
$$

and the corresponding ones for the *jump fields* are

$$
\begin{aligned}
\mathfrak{V}_n &= \mathfrak{V}_1 I_D + \mathfrak{V}_2 II_D + \mathfrak{V}_3 \overset{+}{w}_n + \mathfrak{V}_4 \hat{\overset{+}{t}}_n + \mathfrak{V}_5 \overset{+}{q}_n + \mathfrak{V}_6 \theta_+ , \\
\mathfrak{G}_n &= \mathfrak{G}_1 I_D + \mathfrak{G}_2 II_D + \mathfrak{G}_3 \overset{+}{w}_n + \mathfrak{G}_4 \hat{\overset{+}{t}}_n + \mathfrak{G}_5 \overset{+}{q}_n + \mathfrak{G}_6 \theta_+ , \\
\mathfrak{Q} &= \mathfrak{Q}_1 I_D + \mathfrak{Q}_2 II_D + \mathfrak{Q}_3 \overset{+}{w}_n + \mathfrak{Q}_4 \hat{\overset{+}{t}}_n + \mathfrak{Q}_5 \overset{+}{q}_n + \mathfrak{Q}_6 \theta_+ , \\
\mathfrak{K} &= \mathfrak{K}_1 I_D + \mathfrak{K}_2 II_D + \mathfrak{K}_3 \overset{+}{w}_n + \mathfrak{K}_4 \hat{\overset{+}{t}}_n + \mathfrak{K}_5 \overset{+}{q}_n + \mathfrak{K}_6 \theta_+ ; \\
\mathfrak{V}^\alpha &= \left[\mathfrak{D}_1 g^{\alpha\beta} + \mathfrak{D}_2 (b^{\alpha\beta} - K_M g^{\alpha\beta})\right] T_{s,\beta} + \left[\mathfrak{D}_3 g^{\alpha\beta} + \mathfrak{D}_4 (b^{\alpha\beta} - K_M g^{\alpha\beta})\right] \overset{+}{w}_\beta + \\
&\quad + \left[\mathfrak{D}_5 g^{\alpha\beta} + \mathfrak{D}_6 (b^{\alpha\beta} - K_M g^{\alpha\beta})\right] \hat{t}^n_{\beta+} , \\
\mathfrak{G}^\alpha &= \left[\mathfrak{C}_1 g^{\alpha\beta} + \mathfrak{C}_2 (b^{\alpha\beta} - K_M g^{\alpha\beta})\right] T_{s,\beta} + \\
&\quad + \left[\mathfrak{C}_3 g^{\alpha\beta} + \mathfrak{C}_4 (b^{\alpha\beta} - K_M g^{\alpha\beta})\right] \overset{+}{w}_\beta + \\
&\quad + \left[\mathfrak{C}_5 g^{\alpha\beta} + \mathfrak{C}_6 (b^{\alpha\beta} - K_M g^{\alpha\beta})\right] \hat{t}^n_{\beta+} .
\end{aligned}
\tag{5.5}
$$

u_s, s_s, σ, σ_2 may be linear functions of I_D and II_D and depend on ρ_s, T_s, K_M, K_G, of which also the transport coefficients $\kappa,\dots,\mu_6$ in (5.4) can depend. The transport coefficients $\mathfrak{V}_1,\dots,\mathfrak{C}_6$ may be functions of ρ_s, T_s, K_M, K_G and of ρ^+, T^+. The linear representations (5.5) for non-spherical geometry (for a sphere holds $b^{\alpha\beta} - K_M g^{\alpha\beta} = 0$!) contain 36 unknown transport coefficients. Their number will be reduced later by physically motivated additional requirements.

5.3. Entropy principle

The entropy principle is the most restrictive among the general requirements on the constitutive functions. It guarantees dissipation and is the expression of the second law of thermodynamics. We use the following generalization of Müller's[19] postulates:

(i) There exists an additive surface entropy. It satisfies the balance law $(2.11)_5$ for every surface point and all times.

(ii) The specific surface entropy s_s, its non-convective flux ϕ_s^α and its input density $\mathcal{S}$ from the adjacent bulk materials are given by constitutive equations, which satisfy the rules of material frame indifference and material symmetry. For the heat conducting, viscous interface between ice and water in the membrane approximation they are given by (5.1) and (5.2) together with (4.15), (4.18) and $(4.19)_5$ and if the phase boundary is isotrop, by $(5.4)_{2,3}$, (5.5) and $(4.19)_5$.

(iii) The entropy-supply density σ is a linear combination of the densities of the supplies of momentum and internal energy:

$$\sigma_s = \lambda^s \cdot \rho_s r_s + \lambda^s_\alpha \cdot \rho_s \tau^\alpha_k g^k + \lambda^s_n \rho_s e_k g^k \,, \tag{5.6}$$

where λ^s, λ^s_α and λ^s_n are independent of r_s and g^k.

(iv) There exist ideal heat conducting singular quasi-material *lines* ℓ, which move with velocity w^α and divide different surface materials (Fig. 7). Let h_α be the unit normal vector to the singular line inside the local tangential planes of the phase interface. Assume that the surface temperature T_s is continuous across this line (definition of a dia-thermal line). A dia-thermal line is called *ideal*, if the normal jump of the entropy flux vanishes across the line whenever the normal component of the heat flux is continuous. Thus,

$$\ell \text{ is ideal} \iff [\![T_s]\!] = 0 \,,$$

$$[\![q_s^\alpha h_\alpha]\!] = 0 \implies [\![\phi_s^\alpha h_\alpha]\!] = 0 \,. \tag{5.7}$$

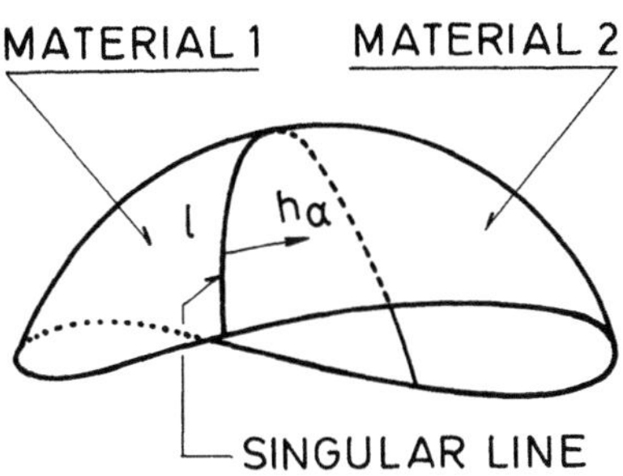

Fig. 7 a:
Singular line ℓ dividing two materials on the singular surface $s(t)$, which represents the phase boundary.

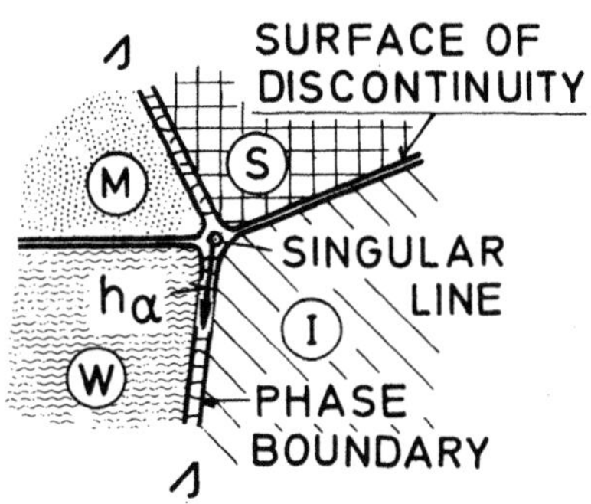

Fig. 7 b:
Intersection of four bulk materials S, M, I, W and singular line. Visualization, how a singular line on a singular surface may be understood.

(v) The entropy production density Π_s is non-negative for every thermodynamic process:

$$\Pi_s \geqq 0 \;, \qquad \forall \text{ thermodynamic processes.} \tag{5.8}$$

The entropy inequality (5.8) is the key for the evaluation of further restrictions on the constitutive equations. It is especially the requirement that (5.8) holds for thermodynamic processes, which connects the constitutive quantities of the entropy balance with those in the balance laws for mass, momentum and energy. For it means, that the inequality must be valid for all physically admissible processes which are solutions of the conservation laws of mass, momentum and energy. The balance laws $(2.11)_{1-4}$ are therefore *constraints* of the possible thermodynamic processes and restrict them. One can free the formulation from these constraints by introducing Lagrange-parameters and then obtains an equivalent inequality, Liu[20],

$$\begin{aligned}
\Pi_s = {} & \rho_s \frac{ds_s}{dt} + \phi^{\alpha}_{s;\alpha} - \rho_s(\lambda^{s}_{\alpha}\tau^{\alpha}_{k}g^{k} + \lambda^{s}_{n}e_{k}g^{k} + \lambda^{s}r_s) + \mathfrak{S} - \\
& - \Lambda^{s}_{\rho}\left[\frac{d\rho_s}{dt} + \rho_s D^{\alpha}_{\alpha} + \mathfrak{R}\right] - \Lambda^{s}_{\alpha}\Big[\rho_s \frac{dw^{\alpha}}{dt} + \rho_{s\,a}\Gamma^{\alpha}_{\beta\gamma}w^{\beta}w^{\gamma} - \\
& - \rho_s w_n(g^{\alpha\beta}w_{n,\beta} + 2b^{\alpha}_{\beta}w^{\beta}) - S^{\alpha\beta}_{\;;\beta} - \rho_s\tau^{\alpha}_{k}g^{k} + \mathfrak{T}^{\alpha}\Big] - \\
& - \Lambda^{s}_{n}\left[\rho_s \frac{dw_n}{dt} + \rho_s w^{\alpha}(w_{n,\alpha} + b_{\alpha\beta}w^{\beta}) - S^{\alpha\beta}b_{\beta\alpha} - \rho_s e_k g^k + \mathfrak{T}_n\right] - \\
& - \Lambda^{s}_{\varepsilon}\left[\rho_s \frac{du_s}{dt} + q^{\alpha}_{s;\alpha} - S^{\alpha\beta}D_{\beta\alpha} - \rho_s r_s + \mathfrak{U}\right] \geqq 0 \;,
\end{aligned} \tag{5.9}$$

which now holds *for arbitrary surface fields* $\rho_s(\xi^{\alpha},t)$, ${}_a\chi^{k}(\xi^{\alpha},t)$, $T_s(\xi^{\alpha},t)$ and *for unrestricted bulk fields* $W^{\alpha}_{+}(\xi^{\beta},t)$, $W^{+}_{n}(\xi^{\beta},t)$, $\hat{t}^{\alpha}_{n+}(\xi^{\beta},t)$, $\hat{t}^{+}_{n}(\xi^{\beta},t)$, $\theta_{+}(\xi^{\beta},t)$ and $q^{+}_{n}(\xi^{\beta},t)$ on one side of the phase boundary.

Exploring inequality (5.9) is routine work; we shall not go into details but describe the main deductions and results. Inserting the constitutive equations (5.4) for the surface fields yields an inequality, that is linear in the derivatives

$$\frac{d\rho_s}{dt}\,,\quad \frac{dT_s}{dt}\,,\quad \frac{dT_{s,\alpha}}{dt}\,,\quad \frac{dD_{\alpha\beta}}{dt}\,,\quad \frac{dg_{\alpha\beta}}{dt}\,,\quad \frac{db_{\alpha\beta}}{dt}\,,\quad \frac{dw^{\alpha}}{dt}\,,$$
$$\frac{dw_n}{dt}\,,\quad \rho_{s,\alpha}\,,\quad w_{n,\alpha}\,,\quad T_{s;\alpha\beta}\,,\quad D_{\alpha\beta;\gamma}\,,\quad b_{\alpha\beta;\gamma}\,. \tag{5.10}$$

The inequality could be violated unless the corresponding factors vanish. This yields a multitude of identities relating entropy and entropy flux to internal energy, stress, heat flux and the Lagrange-parameters. From these, together with the requirements (iii) and (iv) of the entropy principle the following results may be deduced, details of which are given in[18].

5.3.1. Lagrange parameters.

$$\Lambda^{s}_{\alpha} \equiv 0, \quad \Lambda^{s}_{n} \equiv 0; \quad \Lambda^{s}_{\varepsilon}(T_s) = \frac{1}{T_s}, \quad \Lambda^{s}_{\rho}(T_s,\rho_s) = \frac{\sigma(T_s,\rho_s)}{\rho_s T_s}\,. \tag{5.11}$$

The Lagrange-parameters of momentum vanish, the Lagrange-parameter for energy is a universal function of (absolute) surface temperature and the Lagrange-parameter of mass is only a function of temperature and density and given by the scalar surface tension σ.

5.3.2. Entropy, internal energy and surface tension. Surface entropy, internal energy and surface tension are only functions of T_s and ρ_s:

$$s_s = s_s(T_s,\rho_s) \ , \qquad u_s = u_s(T_s,\rho_s) \ ,$$

$$\sigma = \sigma(T_s,\rho_s) \quad , \quad \sigma_2 \equiv 0 \ ; \tag{5.12}$$

and are related by Gibbs' equation

$$ds_s = \frac{1}{T_s}\,(du_s + \frac{\sigma}{\rho_s^2}\,d\rho_s) \ . \tag{5.13}$$

The second coefficient in the representation $(5.4)_5$ of surface tension vanishes.

Remark: In classical theories of surface tension (5.13) is not used in this form. For the special case that the temperature T_s and the density $\rho_s = m_s/A_s$ are uniform over a phase boundary with mass m_s and area A_s, the total surface entropy $S_s = m_s s_s$, internal energy $U_s = m_s u_s$ and surface tension are connected by the relation

$$dS_s = \frac{1}{T_s}\,(dU_s - \sigma dA_s) - \frac{1}{T_s}\,G_s\,\frac{1}{m_s}\,dm_s \ , \tag{5.14}$$

where

$$G_s := U_s - T_s S_s - \sigma A_s = m_s\left(u_s - T_s s_s - \frac{\sigma}{\rho_s}\right) := m_s g_s \tag{5.15}$$

is the total free enthalpy. The differential (5.14) is an immediate consequence of (5.13). For surfaces with constant mass ($dm_s = 0$) it implies the well known result for the surface entropy. However, growing phase boundaries do not preserve their mass, and since $G_s \neq 0$ (as we shall prove later) the application of (5.14) without the last term to problems of nucleation growth (as is done) is questionable.

5.3.3. Entropy and heat flux. Entropy flux and heat flux are proportional and are related as follows

$$\phi_s^\alpha = \frac{1}{T_s}\,q_s^\alpha \ . \tag{5.16}$$

5.3.4. Entropy supply. The coefficients in (5.6) are

$$\lambda^s = \Lambda_\varepsilon^s \ , \quad \lambda_\alpha^s \equiv 0 \ , \quad \lambda_n^s \equiv 0 \ . \tag{5.17}$$

Hence entropy supply and energy supply are proportional

$$\sigma_s = \frac{1}{T_s}\,\rho_s\,r_s \ . \tag{5.18}$$

5.3.5. Entropy production density. We may introduce the viscous contribution to surface tension by

$$S_V^{\alpha\beta} := S^{\alpha\beta} - \sigma\, g^{\alpha\beta} \; . \tag{5.19}$$

Together with the foregoing results the entropy production density (5.9) reduces to

$$T_s \Pi_s = S_V^{\alpha\beta} D_{\beta\alpha} - \frac{1}{T_s} q_s^{\alpha} T_{s,\alpha} + \mathfrak{P} \geqq 0 \tag{5.20}$$

where

$$\mathfrak{P} := T_s \mathfrak{S} - (\mathfrak{U} + \frac{\sigma}{\rho_s} \mathfrak{R}) \; . \tag{5.21}$$

5.3.6. Thermostatic equilibrium. In view of (5.5), (5.4) and (4.19) the entropy production density Π_s is at least a quadratic function of the variables

$$X_A = \left\{ D_{\alpha\beta}, T_{s,\alpha}; W_+^{\alpha}, W_n^{+}, \hat{t}_{n+}^{\alpha}, \hat{t}_n^{+}, q_n^{+}, \theta_+ \right\} \tag{5.22}$$

with the property that Π_s vanishes, whenever $X_A = 0$. A process with $X_A = 0$ is called thermostatic equilibrium and will be denoted by an index $|_E$. From (5.20) then follows that Π_s assumes its minimum value zero in equilibrium. Necessary conditions for equilibrium are thus:

$$\left.\frac{\partial \Pi_s}{\partial X_A}\right|_E = 0 \; , \qquad \left\| \left.\frac{\partial^2 \Pi_s}{\partial X_A \partial X_B}\right|_E \right\| = \text{positive-semidefinit} \tag{5.23}$$

In performing the indicated differentiations it is suitable to introduce the specific free enthalpies of the adjacent bulk materials and the phase boundary:

$$g^{\pm} := u^{\pm} - T_{\pm} s^{\pm} + \frac{p_E^{\pm}}{\rho^{\pm}} \; , \qquad g_s := u_s - T_s s_s - \frac{\sigma}{\rho_s} \; . \tag{5.24}$$

Condition (5.23) yields then a single result:

$$g^+|_E = g^-|_E = g_s|_E \; . \tag{5.25}$$

Thus the specific free enthalpies of the adjacent bulk materials and the specific enthalpy of the phase boundary have the same values in equilibrium.

Remark 1: The first part of (5.25), $g^+|_E = g^-|_E$, is a well known result of thermostatics of phase equilibria and has been derived in various ways, see i.e. Baehr[21], p. 167. Investigating the behaviour of jump quantities at flat immaterial phase boundaries leads to the same result (Müller[19], p. 98). However, the second part of (5.25), $g_s|_E = g^+|_E = g^-|_E$, is new and has considerable consequences for the description and understanding of nucleation phenomena. It contradicts the basic assumption, made in classical nucleation theories, that surface tension equals the free energy of the surface per unit area, see i.e. Becker[22], p. 59. In our notation this assumption would read $\sigma_E = \rho_s^E(u_s^E - T_s s_s^E)$ or $g_s^E \equiv 0$. We conclude that the classical nucleation theory is thermodynamically inconsistent. Note that this conclusion is independent of the fact whether $\rho_s = 0$ or $\rho_s \neq 0$.

Remark 2: To draw inferences from (5.24) explicit expressions for the Gibbs-relations of bulk materials must be known. For viscoelastic solids and viscous fluids these can be written in a common form, and are in non-equilibrium, see[23, 24]:

$$ds = \frac{1}{T}\left[du - \frac{p_E}{\rho^2}\, d\rho - \frac{1}{\rho}\, \hat{t}^k_{\ell}\Big|_E\, dG^{\ell}_{\cdot K} (\underline{\underline{G}}^{-1})^K_{\cdot k}\right] , \tag{5.26}$$

where p_E is the hydrostatic pressure, $\hat{t}^k_{\ell}|_E = t^k_{\ell}|_E + p_E \delta^k_{\ell}$ is the trace-free Cauchy-stress in equilibrium and

$$G^k_{\cdot K} := \frac{F^k_{\cdot K}}{(\det \underline{\underline{F}})^{1/3}} , \quad F^k_{\cdot K} := \frac{\partial x^k}{\partial X^K} , \quad \det \underline{\underline{G}} = 1 \tag{5.27}$$

is a generalized shear and is constructed with the deformation gradient $F^k_{\cdot K}$ and $\det \underline{\underline{F}} = \rho_R/\rho$ (ρ_R = reference density in an undistorted state). Specific entropy s, internal energy u, pressure p_E and Cauchy-stress $\hat{t}^{k\ell}|_E$ in the solid are thus functions of T, ρ and $G^k_{\cdot K}$, whereas in the fluid they only depend on T and ρ because $\hat{t}^{k\ell}|_E \equiv 0$. Legendre-transformations of (5.26) for the bulk materials and of (5.13) for the surface yield

$$dg = - s\, dT + \frac{1}{\rho}\, dp_E + \frac{1}{\rho}\, \hat{t}^k_{\ell}\Big|_E\, dG^{\ell}_{\cdot K} (\underline{\underline{G}}^{-1})^K_{\cdot k} ,$$
$$dg_s = - s_s\, dT_s - \frac{1}{\rho_s}\, d\sigma , \tag{5.28}$$

demonstrating that the free enthalpy $g = g(T, p_E, G^k_{\cdot K})$ for the solid depends on temperature, pressure and shear in non-equilibrium, that $g = g(T, p_E)$ for the fluid is a function of temperature and pressure and that $g_s = g_s(T_s, \sigma)$ of the interface depends on temperature and surface tension.

In equilibrium, the bulk temperatures $T_\pm$ of the adjacent materials are both equal to the surface temperature T_s. This follows from (5.22) and $(4.19)_6$: $T_+ = T_- = T_s$. Moreover, because the fluidlike boundary layer between the solid and its fluid cannot transmit equilibrium shear stresses, the adjacent solid bulk material must equally be free of shear stresses, explicitly $\hat{t}^{k\ell}_+|_E = 0$ or $\hat{t}^{k\ell}_-|_E = 0$, depending on at which side the solid is situated. Consequently, in equilibrium, relations (5.28) reduce to

$$dg^{\pm}_E = - s^{\pm}_E dT_s + \frac{1}{\rho^{\pm}_E} dp^{\pm}_E , \quad dg^E_s = - s^E_s dT_s - \frac{1}{\rho^E_s} d\sigma_E . \tag{5.29}$$

These relations will frequently be used in the sequel.

Evaluation of the second equilibrium condition $(5.23)_2$ remains. It will not be discussed, except for a special case.

6. SURFACE TENSION AND MODIFIED CLAUSIUS-CLAPEYRON EQUATIONS

In this chapter only equilibrium is considered. This is characterized by the thermostatic equilibrium conditions (5.25) and the mechanical equilibrium results (4.14), which follow from the momentum balance equations $(2.11)_{2,3}$. A dimensional analysis shows that the gravitational contributions are negligible for all curved phase boundaries, except for almost flat ones with dimensionless curvatures $|k_M| \leqq 10^{-10}$. We shall henceforth neglect these. Surface temperature and the temperatures in the adjacent bulk materials are equal in equilibrium, $T_s = T_+ = T_-$. Phase equilibrium at phase change interfaces is therefore characterized by the following conditions:

$$g^+(T_s, \overset{+}{p}_E) = g_s(T_s, \sigma_E) \ , \qquad g^-(T_s, \overset{-}{p}_E) = g_s(T_s, \sigma_E) \ ,$$
$$\overset{+}{p}_E - \overset{-}{p}_E = \sigma_E \cdot 2K_M \ . \tag{6.1}$$

Because in equilibrium the phase boundary is fixed in space, K_M is fixed and (6.1) are three equations for the determination of the four fields $\overset{+}{p}_E$, $\overset{-}{p}_E$, σ_E and T_s. Assuming T_s to be prescribed and supposing that (6.1) can be inverted, it follows that

$$\overset{+}{p}_E = p^+(T_s, K_M), \quad \overset{-}{p}_E = p^-(T_s, K_M), \quad \sigma_E = \sigma(T_s, K_M) \ . \tag{6.2}$$

For $K_M \neq 0$ the equilibrium pressures of ice and water differ from each other. With (6.2) the equilibrium densities, entropies and internal energies of the bulk materials and the phase boundary become functions of the temperature (and curvature) alone, viz.

$$\overset{\pm}{\rho}_E(T_s, K_M) = \rho^\pm\left[T_s, \overset{\pm}{p}_E(T_s, K_M)\right] \ ,$$
$$\overset{E}{\rho}_s(T_s, K_M) = \rho_s\left[T_s, \sigma_E(T_s, K_M)\right] \ , \qquad \text{etc.} \tag{6.3}$$

Phase equilibrium between a solid and its liquid is possible at all temperatures $T_s \geqq T_{tr}$ above the triple point, it is possible between the solid and its vapour for temperatures $T_s \leqq T_{tr}$ below the triple point and it exists between the liquid and its vapour in a temperature interval $T_{tr} \leqq T_s \leqq T_{cr}$ between the triple point and the critical point. Phase equilibria at curved phase boundaries are possible at any curvature for temperatures $T_s \leqq T_{cr}$.

Change of curvature at constant temperature shifts the phase equilibrium as does change of temperature at constant curvature. The inferences, however, differ from each other and will be discussed in due course.

6.1. Change of the curvature at fixed temperature

Differentiating the equilibrium conditions (6.1) with respect to K_M and using (5.29) yields

$$\frac{\partial \overset{+}{p}_E}{\partial K_M} = -\frac{\overset{+}{\rho}_E}{\overset{E}{\rho}_s}\frac{\partial \sigma_E}{\partial K_M}, \quad \frac{\partial \overset{-}{p}_E}{\partial K_M} = -\frac{\overset{-}{\rho}_E}{\overset{E}{\rho}_s}\frac{\partial \sigma_E}{\partial K_M}, \quad \frac{\partial \overset{+}{p}_E}{\partial K_M} - \frac{\partial \overset{-}{p}_E}{\partial K_M} = 2K_M\frac{\partial \sigma_E}{\partial K_M} + 2\sigma_E. \tag{6.4}$$

Eliminating the pressure derivatives gives

$$\frac{1}{\sigma_E}\frac{\partial\sigma_E}{\partial K_M} = -\frac{1}{K_M + \frac{[\![\rho_E]\!]}{2\rho_\delta^E}} , \tag{6.5}$$

a differential equation for the determination of the curvature dependence of the surface tension. Neglecting gravity implies that the mean curvature is constant along the equilibrium phase boundary. The only geometrical surfaces with constant mean curvature are spheres ($K_G = K_M^2$) and circular cylinders ($K_G = 0$). In other words, in the absence of gravity, the equilibrium shapes of all isotropic phase boundaries that separate isotropic bulk materials are either spheres or circular cylinders.

Insert now the result (4.10) of the boundary-layer theory into the right hand side of (6.5), pass to dimensionless curvature $k_M = K_M d_\delta$ and integrate the resulting expression. What obtains reads

$$\frac{\sigma_E}{\sigma_E^0} = \exp\left[-\int_0^{k_M}\frac{F_1(k)}{1+kF_1(k)}\,dk\right] , \tag{6.6}$$

where $\sigma_E^0 = \sigma_E(T_\delta, k_M{=}0)$ is the surface tension for the flat phase boundary and where $F_1(k_M)$ equals either $F_1(k_M, k_G = k_M^2)$ for the sphere or $F_1(k_M, k_G = 0)$ for the circular cylinder.

Equation (6.6) exhibits a distinct and very interesting curvature dependence of the surface tension. For spherical geometry the result of numerical integration with the values for $F_1(k_M)$ from Fig. 5 is drawn in Fig. 8. Due to the different signs of $F_1(k_M)$ the surface tension for a water*) inclusion in ice*) increases with increasing curvature $|k_M|$ (case A); for an ice inclusion in water (case B), however, it decreases with increasing curvature and vanishes when $|k_M| = 0.09$. Vanishing surface tension, however, corresponds physically to disappearance of the phase boundary. We conclude, that ice inclusions in water of nucleation dimensions $|k_M| \gtrsim 0.09$ do not exist. This confirms our earlier deductions in paragraph 4.1 about the unlikelyhood of ice nucleation in pure water.

As far as we know, no similar result on the curvature dependence of the surface tension has so far been published. [Experiments on surface tension are performed at too small dimensionless curvatures as to cover the regime where the surface tension splits up into the two branches.] So, (6.6) is a new result with important consequences for the physical understanding and the mechanism of nucleation. It also describes a rather general phenomenon of isotropic phase boundaries. [Anisotropic phase boundaries possess in equilibrium a tensor of surface tension and correspondingly more curvature measures!] We feel, therefore, justified to repeat the assumptions on which (6.6) is based upon. Besides isotropy, the boundary-layer theory required (i) that the thickness d_δ of the layer

---

*) The terms "water, ice or vapour" stand for the dense and the less dense phases. They are used to shorten the subsequent description in the following text and to combine it with a visualization. The results of this chapter, however, are independent of special choices of bulk materials.

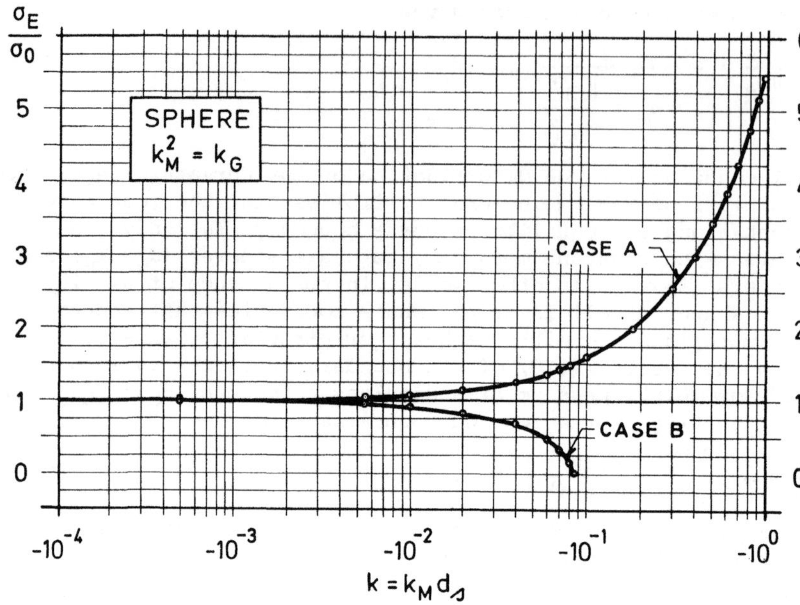

Fig. 8:
Normalized surface tension σ_E/σ_E^0 as function of curvature according to (6.6) for spherical geometry. Case A: Inclusion of the denser in the less dense phase. Case B: Inclusion of the less dense in the denser phase.

Whereas embedding of the denser phase in the less dense phase (water in ice or water in vapour) is possible down to nucleation dimensions, inclusion of the less dense in the denser phase (ice in water or vapour in water) is not possible at very small inclusion size. This may explain the phenomenae of undercooling and overheating observed in water.

be independent of the curvature and (ii) that the density profile is a cubic function of the boundary layer coordinate. Because the layer is extremely thin (of the order of a few molecular distances) and the cubic profile guarantees a smooth connection of the bulk phases, these are fairly realistic assumptions. Minor changes are possible, they certainly will not change the qualitative behaviour, as discussed in Fig. 8 and will result in only small quantitative changes. Especially, the splitting of the $\sigma_E(k_M)$-relation into two curves must be preserved under any reasonable boundary-layer assumptions. This is so because the kinematic compatibility condition (3.17) requires two families of solutions for the excentricity parameter $\alpha = a_\Delta/d_\Delta$. And the compatibility condition, in turn, resulted from the necessity to define material lines for the motion within the surface, a concept without which a surface tension could not have been introduced. So the chain of argumentations closes. Basic to (6.6) is thus kinematic consistency of the boundary-layer theory of singular surfaces. By purely static reasoning a result like (6.6) for the curvature dependence of the static surface tension can *never* be achieved.

It may be stated that using (6.6) the adjacent bulk values of pressure $p_E^{\pm}$, of internal energy $u_E^{\pm}$ and entropy $s_E^{\pm}$ can be calculated as functions of curvature K_M and temperature. These calculations, however, require the constitutive equation $p = p(T,\rho)$ for the pressure in the adjacent bulk materials and yield different results for different materials. We do not go into further details.

6.2. Change of temperature at fixed curvature

Let us now turn to the consequences of the equilibrium conditions (6.1), when the temperature is shifted at fixed curvature. Differentiating (6.1) with respect to T_s and using (5.29) yields:

$$
\begin{aligned}
- s_E^{+} + \frac{1}{\rho_E^{+}} \left.\frac{\partial p_E^{+}}{\partial T_s}\right|_{K_M} &= - s_s^{E} - \frac{1}{\rho_s^{E}} \left.\frac{\partial \sigma_E}{\partial T_s}\right|_{K_M} ,\\
- s_E^{-} + \frac{1}{\rho_E^{-}} \left.\frac{\partial p_E^{-}}{\partial T_s}\right|_{K_M} &= - s_s^{E} - \frac{1}{\rho^{E}} \left.\frac{\partial \sigma_E}{\partial T_s}\right|_{K_M} ,\\
\left.\frac{\partial p_E^{+}}{\partial T_s}\right|_{K_M} - \left.\frac{\partial p_E^{-}}{\partial T_s}\right|_{K_M} &= 2K_M \left.\frac{\partial \sigma_E}{\partial T_s}\right|_{K_M} ,
\end{aligned}
\tag{6.7}
$$

from which, on eliminating the right hand sides of $(6.7)_{1,2}$ we deduce

$$
\begin{aligned}
\left[\!\left[\frac{1}{\rho_E}\right]\!\right] \left.\frac{\partial p_E^{+}}{\partial T_s}\right|_{K_M} &= [\![s_E]\!] - \frac{1}{\rho_E^{-}} \left.\frac{\partial \sigma_E}{\partial T_s}\right|_{K_M} \cdot 2K_M ,\\
\left[\!\left[\frac{1}{\rho_E}\right]\!\right] \left.\frac{\partial p_E^{-}}{\partial T_s}\right|_{K_M} &= [\![s_E]\!] - \frac{1}{\rho_E^{+}} \left.\frac{\partial \sigma_E}{\partial T_s}\right|_{K_M} \cdot 2K_M .
\end{aligned}
\tag{6.8}
$$

In the limit as $K_M \to 0$, $s_E^{\pm} \to s_0^{\pm}$, $\rho_E^{\pm} \to \rho_0^{\pm}$, $p_E^{\pm} \to p_E^{0}$, and the relations (6.8) reduce to the well known *Clausius-Clapeyron equation*

$$
\left[\!\left[\frac{1}{\rho_E^{0}}\right]\!\right] \frac{dp_E^{0}}{dT} = [\![s_E^{0}]\!]
\tag{6.9}
$$

for phase equilibrium pressure at flat phase boundaries. For this reason (6.8) will be called the *generalized Clausius-Clapeyron equations*.

The generalized Clausius-Clapeyron equations (6.8) are rather complex and depend on the materials and the kind of phase equilibrium (solid/fluid, solid/vapour or fluid/vapour) and curvature, c.p. (6.2), (6.3), (6.6).

A considerable simplification can be achieved for phase equilibrium between ice and water. Assuming both bulk materials to be approximately incompressible, such that $\rho^{\pm}(T_s, p^{\pm}) \cong \rho^{\pm}(T_s) = \rho_0^{\pm}(T_s)$, the bulk densities are independent of curvature and (6.8) simplify to

$$
\begin{aligned}
\left[\!\left[\frac{1}{\rho_0}\right]\!\right] \left(\left.\frac{\partial p_E^{+}}{\partial T_s}\right|_{K_M} - \frac{dp_E^{0}}{dT_s} \right) &= - \frac{2K_M}{[\![\rho_0]\!]} \left([\![\beta_0]\!] \sigma_E + \frac{[\![\rho_0]\!]}{\rho_0^{-}} \left.\frac{\partial \sigma_E}{\partial T_s}\right|_{K_M} \right) ,\\
\left[\!\left[\frac{1}{\rho_0}\right]\!\right] \left(\left.\frac{\partial p_E^{-}}{\partial T_s}\right|_{K_M} - \frac{dp_E^{0}}{dT_s} \right) &= - \frac{2K_M}{[\![\rho_0]\!]} \left([\![\beta_0]\!] \sigma_E + \frac{[\![\rho_0]\!]}{\rho_0^{+}} \left.\frac{\partial \sigma_E}{\partial T_s}\right|_{K_M} \right) ,
\end{aligned}
\tag{6.10}
$$

where $\beta_0^{\pm}(T_s)$ is the thermal volume expansion coefficient of bulk ice or water.

These relations describe the departures of the equilibrium pressures in the adjacent bulk materials at curved phase boundaries from the usual pressure melting curve. These departures can be considerable when curvatures are large and cannot be neglected for inclusions of nucleation dimensions.

We may state, that (6.8) and (6.9) may be used together with (6.6) to calculate also surface entropy s_s and surface internal energy u_s in terms of $p_E^0(T_s)$ at flat interfaces, temperature T_s and curvature K_M. Those calculations have been performed for the ice/water interface and may be found in[18].

A remark may be added concerning latent heat of melting or evaporation. At flat phase boundaries $T_s[\![s_E^0]\!]$ has this meaning. On curved phase boundaries, however, the quantity $T_s[\![s_E]\!]$ cannot be interpreted as the latent heat of phase change.

7. THERMOSTATIC STABILITY CONDITIONS FOR PHASE-BOUNDARY LAYERS

Let us consider a fixed equilibrium state

$$T_+ = T_- = T_s := T_s^E, \quad \sigma = \sigma(T_s^E, K_M) := \sigma_E, \quad p^{\pm} = p^{\pm}(T_s^E, K_M) := p_E^{\pm}, \tag{7.1}$$

where the surface tension σ_E satisfies the mechanical equilibrium conditions

$$\begin{aligned} \sigma_E \cdot 2K_M - [\![p_E]\!] + \rho_s (e^k g_k)|_E &= 0 , \\ \sigma_{E,\alpha} \qquad\qquad + \rho_s (\tau_\alpha^k g_k)|_E &= 0 \end{aligned} \tag{7.2}$$

for every point ξ^α on a *motionless* phase boundary in equilibrium. Such an equilibrium state is called *thermostatically stable*, provided that the phase boundary under any virtual departure from (7.1) and (7.2) is forced to return back to this equilibrium state.

From the entropy inequality, the balance laws for mass, momentum and energy and the equilibrium conditions it can then be proven that the *local thermostatic stability condition for phase boundaries* takes on the form

$$(g_s - g_s^E) + s_s(T_s - T_s^E) + \frac{1}{\rho_s}(\sigma - \sigma_E) \geq 0 , \tag{7.3}$$

where $g_s = g_s(T_s, \sigma)$, $s_s = s_s(T_s, \sigma)$, $\rho_s = \rho_s(T_s, \sigma)$ are the free surface enthalpy, entropy and density in any virtual non-equilibrium state T_s, σ of a motionless phase boundary. The proof of (7.3) is akin to corresponding derivations for simple bulk materials, see[25], and may be found in[18].

Taylor series expansion of (7.3) around the equilibrium state T_s^E, σ_E yields a positive-semi definite quadratic form in $\delta T_s = T_s - T_s^E$, $\delta\sigma = \sigma - \sigma_E$ and implies the inequalities

$$\left(\frac{\partial s_s}{\partial T_s}\right)_{\sigma_E} \geq 0 , \qquad - \left(\frac{\partial \rho_s}{\partial \sigma}\right)_{T_s^E} \geq 0 ,$$

$$- \left(\frac{\partial s_s}{\partial T_s}\right)_{\sigma_E} \left(\frac{\partial \rho_s}{\partial \sigma}\right)_{T_s^E} - \left[\frac{1}{\rho_s^E}\left(\frac{\partial \rho_s}{\partial T_s}\right)_{\sigma_E}\right]^2 \geqq 0 \; . \tag{7.4}$$

Evaluation of the consequences of (7.4) is complicated by the fact, that we do not know s_s and ρ_s in terms of T_s and σ_E, but only in terms of T_s and K_M. With the relation $\sigma_E = \sigma_E(T_s, K_M)$, however, we may introduce T_s and K_M as independent variables. What emerges for the *curved ice/water interphase* are the following stability conditions:

(i) Conditions on geometry:
Case A: a spherical water inclusion in ice is *stable* for
$0 \leqq |k_M| \leqq 1$.
Case B: a spherical ice inclusion in water is *stable* for
$0 \leqq |k_M| \leqq 0{,}086$, unstable for $|k_M| \geqq 0{,}086$. (7.5)
This proves the earlier statements on existence or non-existence of ice or water nuclei in the respective other phase.

(ii) Condition on boundary-layer thickness:
The phase interface is stable then and only then, if the temperature dependent thickness d_s satisfies

$$d_s(T_s) = d_s(T_R) \cdot \frac{|[\![\rho_0(T_R)]\!]|}{|[\![\rho_0(T_s)]\!]|} \; , \tag{7.6}$$

where T_R (= 273 K) is some reference temperature. The boundary-layer thickness is inversely proportional to the absolute value of the density jump of the flat surface. Measurements of Golecki & Jaccard[8] indicate, that $d_s(T_s)$ is decreasing with decreasing temperature below the normal freezing point. This means that $|[\![\rho_0(T_R)]\!]|/|[\![\rho_0(T_s)]\!]| \leqq 1$ whenever $T_s \leqq T_R = 273$ K. This conclusion can be checked via density measurements in undercooled water, however, those measurements are not known to us.

(iii) Condition on surface tension:
Stability of the phase interface requires

$$\frac{d^2\sigma_0}{dT_s^2} = 0 \; . \tag{7.7}$$

Hence, surface tension of flat interfaces between ice and water must be a linear function of absolute temperature. This is confirmed by observations, see Hobbs[7].

The conclusion (7.5) about stable and unstable inclusions of ice in water provides a new understanding of nucleation phenomena.

8. LINEAR TRANSPORT EQUATIONS ON AND ACROSS PHASE-BOUNDARY LAYERS

The thickness of the boundary layer is in the order of a few molecular distances, say $d_s = 100$ Å. This is much smaller than the dimensions of the smallest available thermometer. We therefore assume

(i) The temperature is continuous across the phase boundary and equal to the surface temperature

$$T_+ = T_- = T_s(\xi^\alpha, t) \ , \quad \forall\, \xi^\alpha \in s(t) \ , \quad \forall\, t \ . \tag{8.1}$$

Moreover, the processes of freezing and melting or evaporation and condensation may be slow, such that they take place near to equilibrium, in which the entropy production vanishes. We assume thus

(ii) The phase-change process is reversible, implying

$$\Pi_s \equiv 0 \quad \forall \text{ thermodynamic processes} \ . \tag{8.2}$$

These two requirements are consistent with the entropy principle for phase-boundary layers and allow a considerable simplification of the transport equations (5.4) and (5.5) without reducing them to triviality.

Due to continuity of temperature we have from (5.5) and (5.25):

$$\begin{aligned} &\mathfrak{K}_j = 0 \quad (j = 1,2,\dots,6) \ , \ \mathfrak{V}_6 = \mathfrak{G}_6 = \mathfrak{Q}_6 = \mathfrak{K}_6 = 0 \ , \\ &g^\pm = g_E^\pm = g_s^E = g_s \ . \end{aligned} \tag{8.3}$$

For reversible phase change (5.20) reduces to

$$\begin{aligned} T_s \Pi_s = {} & S_V^{\alpha\beta} D_{\beta\alpha} - \frac{1}{T_s} q_s^\alpha T_{s,\alpha} + \mathfrak{G}_\alpha (W_+^\alpha - \mathfrak{v}^\alpha) + \\ & + \mathfrak{v}_\alpha \hat{t}^\alpha_{n+} + \mathfrak{G}_n (W_n^+ - \mathfrak{v}_n) + \mathfrak{v}_n \hat{t}^+_n \equiv 0 \ . \end{aligned} \tag{8.4}$$

As a matter of simplification we use the boundary layer result

$$\mathfrak{v}^\alpha = \frac{1}{\mathcal{F}} W_+^\alpha \quad \text{with} \quad \mathcal{F} := 1 - \frac{1}{3}\frac{F_2}{F_1} - \frac{\rho^+}{\rho_s}\frac{d_s}{2} F_1 \ , \tag{8.5}$$

which is just another way of writing (4.10) and (4.11). Comparison with $(5.5)_5$ yields

$$\mathfrak{D}_1 = \mathfrak{D}_2 = 0 \ , \quad \mathfrak{D}_3 = \frac{1}{\mathcal{F}} \ , \quad \mathfrak{D}_4 = \mathfrak{D}_5 = \mathfrak{D}_6 = 0 \ . \tag{8.6}$$

Inserting $(5.4)_{2-4}$, (5.5) into (8.4) yields a lengthly expression which is identically zero for every thermodynamic process, if and only if (details are omitted):

$$\begin{aligned} &\kappa = \kappa_2 = 0 \ ; \\ &\mu_1 = \mu_3 K_M \ , \quad \mu_2 = -\mu_3 \ , \quad \mu_4 = \mu_5 = 0 \ , \quad \mu_6 \neq 0 \ ; \\ &\mathfrak{C}_1 = \mathfrak{C}_2 = \mathfrak{C}_3 = \mathfrak{C}_4 = 0 \ , \quad \mathfrak{C}_5 = -\frac{\mathfrak{D}_3}{1-\mathfrak{D}_3} \ , \quad \mathfrak{C}_6 = 0 \ ; \\ &\mathfrak{v}_1 = \mathfrak{v}_2 = 0 \ , \quad \mathfrak{v}_3 \neq 0 \ , \quad \mathfrak{v}_4 = \mathfrak{v}_5 = 0 \ ; \\ &\mathfrak{G}_1 = \mathfrak{G}_2 = \mathfrak{G}_3 = 0 \ , \quad \mathfrak{G}_4 = -\frac{\mathfrak{v}_3}{1-\mathfrak{v}_3} \ , \quad \mathfrak{G}_5 = 0 \ . \end{aligned} \tag{8.7}$$

The following linear transport equations are therefore necessary and sufficient in order that reversible phase transition arise

$$q_{s}^{\alpha} = 0 \, , \quad \phi_{s}^{\alpha} = 0 \, ,$$
$$S^{\alpha\beta} = \sigma_E g^{\alpha\beta} + \mu_3 (I_D b^{\alpha\beta} - II_D g^{\alpha\beta}) + \mu_6 (\underline{\underline{b}}\underline{\underline{D}} - \underline{\underline{D}}\underline{\underline{b}})^{(\alpha\beta)} \tag{8.8}$$

and

$$\mathfrak{V}_n = \mathfrak{V}_3 \, W_n^+ \, , \quad \mathfrak{G}_n = - \frac{\mathfrak{V}_3}{1-\mathfrak{V}_3} \, \hat{t}_n^+$$
$$\mathfrak{v}^{\alpha} = \mathfrak{D}_3 \, W_+^{\alpha} \, , \quad \mathfrak{G}^{\alpha} = - \frac{\mathfrak{D}_3}{1-\mathfrak{D}_3} \, \hat{t}_{n+}^{\alpha} \, . \tag{8.9}$$

Using the boundary layer result (4.12), which may be written as $\mathfrak{V}_n = \frac{1}{\mathcal{F}} W_n^+$ we conclude finally:

$$\mathfrak{V}_n = \frac{1}{\mathcal{F}} W_n^+ \, , \quad \mathfrak{G}_n = \frac{1}{1-\mathcal{F}} \hat{t}_n^+ \, ,$$
$$\mathfrak{v}^{\alpha} = \frac{1}{\mathcal{F}} W_+^{\alpha} \, , \quad \mathfrak{G}^{\alpha} = \frac{1}{1-\mathcal{F}} \hat{t}_{n+}^{\alpha} \, . \tag{8.10}$$

By the boundary-layer result $(4.13)_2$ it can be shown that the heat conductivity coefficients for the interface between ice and water are very small, $\kappa = O(10^{-8})\mathrm{WK}^{-1}$, $\kappa_2 = O(10^{-17})\mathrm{WmK}^{-1}$. Compared to the heat conductivities of the adjacent bulk materials, these can be neglected. Hence a reversible phase-change interface seems to be reasonable. It allows two viscosity coefficients within the reversible phase-change surface. The transport coefficient $\mathcal{F}$ in the transverse transport equations (8.10) is a known function of curvature and temperature. (8.10) constitute *generalized boundary conditions* for tangential and normal velocities and shear and normal stress. Corresponding boundary conditions for in-plane-stress can be derived from the boundary-layer result $(4.13)_1$. The only thing left is the jump condition for the normal component of the heat flux. This can be obtained from the balance equation of internal energy, which in the special case of continuous temperatures degenerates to a constraint. The result is

$$\mathfrak{Q} = \mathfrak{Q}_1 \, I_D + \mathfrak{Q}_3 \, W_n^+ \tag{8.11}$$

with

$$\mathfrak{Q}_1 = T_{s} \left(\frac{\partial \sigma}{\partial T_{s}} \right)_{\rho_{s}}^{E} \, , \tag{8.12}$$
$$\mathfrak{Q}_3 = - \rho_E^+ \left[h_E^+ - h_{s}^E - T_{s} \frac{1}{\rho_{s}^E} \left(\frac{\partial \sigma}{\partial T_{s}} \right)_{\rho_{s}}^{E} \right] + \rho_E^- \left[h_E^- - h_{s}^E - T_{s} \frac{1}{\rho_{s}^E} \left(\frac{\partial \sigma}{\partial T_{s}} \right)_{\rho_{s}}^{E} \right] \left(1 - \frac{1}{\mathcal{F}} \right)$$

where

$$h_E^{\pm} = u_E^{\pm} + \frac{p_E^{\pm}}{\rho_E^{\pm}} \, , \quad h_{s}^E = u_{s}^E - \frac{\sigma_E}{\rho_{s}^E} \tag{8.13}$$

are the specific enthalpies of the adjacent bulk materials and the surface in equilibrium.

The *classical result* would be obtained for an "incompressible" phase boundary with $d\rho_{s}/dt = 0$, $I_D = 0$ and $\mathfrak{R} = \rho^+ W_n^+ - \rho^- W_n^- = 0$; namely

$$\mathfrak{Q} = - [\![h_E]\!] \rho^+ \, W_n^+ \, , \tag{8.14}$$

where $[\![h_E]\!]$ is the latent heat of phase change. However, in view of the boundary-layer theory incompressibility is not acceptable.

We thus have the remarkable result that *all transport coefficients* for reversible phase transitions across inviscid phase boundaries *are known functions of the geometry and temperature*. The boundary-layer theory for the phase transition between ice and water contains thus only *one single constant*, namely the thickness $d_\delta(T_R)$ at some reference temperature, as a phenomenological parameter.

9. CONCLUSIONS

There remains the question whether the boundary-layer theory for phase boundaries is mathematically well-posed. This means, is the set of constitutive relations for the surface and the set of generalized boundary conditions sufficient, that the motion of the phase boundary can be calculated. For simple geometries this question can be answered positively. So we expect it to be the case also for arbitrary geometry.

REFERENCES

1. S. Ono & S. Kondo: 'Molecular Theory of Surface Tension in Liquids', Handbuch der Physik, Bd. 10, p. 134 (Ed. S. Flügge), Berlin, Göttingen, Heidelberg 1960

2. J. W. Gibbs: 'The Collected Work of J. Willard Gibbs', Vol. 1, Yale University Press, New Haven 1928

3. C. A. Croxton: 'Statistical Mechanics of the Liquid Surface', John Wiley & Sons, Chichester, New York, Brisbane, Toronto 1980

4. N. H. Fletscher: Philos. Magazine, Eigth Ser., 18, 1287 (1966)

5. W. A. Weyl: J. Colloid Sci., 6, 389 (1951)

6. S. L. Carnie & G. M. Torrie: 'The Statistical Mechanics of the Electrical Double Layer', Advances in Chemical Physics, Vol. LVI, 141 (1984)

7. P. V. Hobbs: 'Ice Physics', Clarendon Press, Oxford 1974

8. I. Golecki & C. Jaccard: Physics Letters 63A, 374 (1977)

9. F. P. Buff: J. Chem. Phys., 25, 146 (1956)

10. F. P. Buff & H. Saltsburg: J. Chem. Phys., 26, 23, 1526 (1957)

11. A. D. Deemer & J. C. Slattery: Int. J. Multiphase Flow, 4, 171 (1978)

12. L. E. Scriven: Chem. Engng. Sci., 12, 98 (1960)

13. G. P. Moeckel: Arch. Rat. Mech. Anal., 57, 255 (1974)

14. M. E. Gurtin & A. I. Murdoch: Arch. Rat. Mech. Anal., 57, 291 (1975)

15. K. A. Lindsay & B. Straughan: Arch. Rat. Mech. Anal., 71, 307 (1979)

16. A. Grauel: Physica 103 A, 468 (1980)

17. R. Aris: 'Vectors, Tensors and the Basic Equations of Fluid Mechanics', Prentice-Hall, Inc., Englewood Cliffs, N.Y., 1962

18. T. Alts & K. Hutter: 'Towards a Theory of Temperate Glaciers: Dynamics and Thermodynamics of Phase Boundaries between Ice and Water', Mitteilungen der Versuchsanstalt für Wasserbau, Hydrologie und Glaziologie, Nr. 82 (Ed. D. Vischer), ETH Zürich 1986

19. I. Müller: 'Thermodynamik - Die Grundlagen der Materialtheorie' Bertelsmann Universitätsverlag, Düsseldorf 1973

20. I-Shih Liu: Arch. Rat. Mech. Anal., 46, 131 (1972)

21. H. D. Baehr: 'Thermodynamik. Eine Einführung in die Grundlagen und ihre technischen Anwendungen', Springer-Verlag, Berlin, Heidelberg, New York (3rd ed.), 1973

22. R. Becker: 'Theorie der Wärme', Springer-Verlag Berlin, Göttingen, Heidelberg 1961

23. T. Alts: 'Zur Thermodynamik der Vorgänge in festen und fluiden Mischungen', Ph.D. Dissertation, Rheinisch-Westfälische Technische Hochschule Aachen, 1971

24. T. Alts: 'Thermodynamik elastischer Körper mit thermo-kinematischen Zwangsbedingungen - fadenverstärkte Materialien', Habilitation Dissertation, Technische Universität Berlin, Universitätsbibliothek der TU Berlin, Abteilung Publikationen 1979

25. M. E. Gurtin: Arch. Rat. Mech. Anal., 59, 63 (1975)

LARGE SCALE MAGNETIC FIELDS IN FLUIDS - AN EXAMPLE OF A DISSIPATIVE STRUCTURE

Konrad Bajer
Department of Applied Mathematics and Theoretical Physics
University of Cambridge
Cambridge CB3 9EW, U.K.
(on leave from the Institute of Geophysics, Univ. of Warsaw, Poland)

ABSTRACT. Large scale magnetic fields in fluids are remarkable examples of dissipative structures. In this paper an elementary description of their formation in laminar and turbulent flows is given. As an example of such a process, a simple qualitative model of the generation of the Earth's magnetic field is sketched.

1. INTRODUCTION

The magnetic field surrounding the Earth, which, for centuries has made compass needles point North seems to be a highly ordered large scale structure over the whole surface of our globe. With the development of geophysical sciences it has turned out however that it is not invariable (at least on the geological time scale). Present models of the generation and evolution of the Earth's magnetic field show that it is a dissipative structure i.e. an ordered large scale structure formed through energy flow in a system which is open in the thermodynamical sense. The essence of the concept of a dissipative structure is that the scale of created order (in time or in space) is much bigger than the scale of any phenomenon associated with the creating energy flow itself. Examples are the macroscopic spatial scales of the structures appearing in the diffusion of chemically reacting compounds or the size of living organisms.

For a long time people thought of the Earth's magnetic field as being as invariable as the terrestrial globe itself or changing very little. It seemed to be a "fossil" relic of the period of the Earth's formation. There was therefore no need to consider its generation. The situation changed when the techniques of determining the magnitude of the field in distant geological eras were developed. In 1955 S. K. Runcorn discovered that the Earth's magnetic field is recorded in ancient rocks and may be determined over almost the whole of the Earth's history. The investigations showed that this field does not decay, which would be the case if it were a fossil relic. Moreover it undergoes a complicated evolution. The field is predominantly poloidal with a dipole axis coinciding with the Earth's rotation axis (as it does at present).

C. W. Kilmister (ed.), Disequilibrium and Self-Organisation, 129–140.

Randomly in time however there occur reversals of the dipole orientation, North and South magnetic poles are interchanged [Moffatt 1978 and the references there] (fig.1).

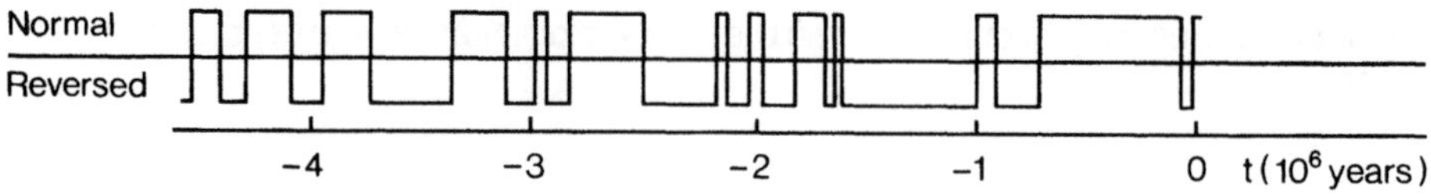

Figure 1. Record of reversals of the Earth's dipole polarity over the past 4×10^6 years (derived from Cox, 1969). The figure indicates only the direction of the dipole vector not its intensity (after Moffatt 1978).

The typical period between the reversals is of the order of 10^5 years. The reversals take place on a time scale of the order of 10^4 years [Cox 1972]. The occurrence of the reversals and the fact that the field recovers its original magnitude after each reversal (during the reversal the field intensity decreases to about one quarter of its usual value) proves the existence of a dynamical process generating the field in the Earth's core. Our globe consists of a solid mantle (0.45 of Earth radius) (fig.2), liquid outer core (0.36 of the radius) and small solid inner core (0.19 of the radius). The liquid core is a highly electrically conducting fluid and that is where the dynamo generating the magnetic field operates. Convective motions occur in the core and the magnetic field induces current in the moving conductor (Faraday's law).

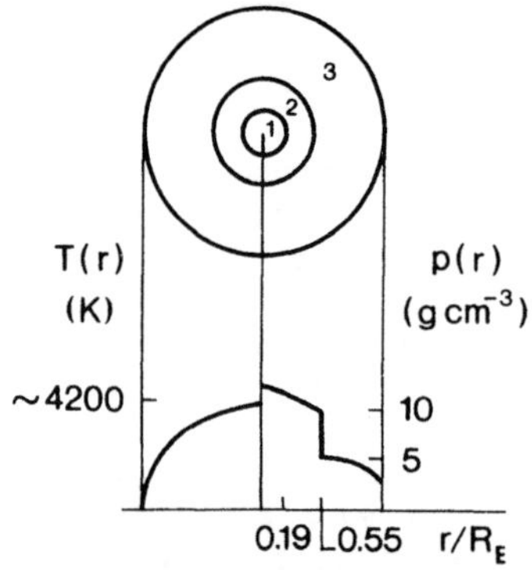

Figure 2. Interior structure of the Earth (R_E=6380 km): 1. solid inner core, iron/nickel alloy; 2. liquid outer core, iron and some lighter elements; 3. solid mantle; (after Moffatt 1978)

Current on the other hand is a source of magnetic field (Biot and Savart law) which gives the feedback necessary for dynamo action. If we treat the motion as given and consider only these interactions we have the so-called "kinematic dynamo problem". Taking into account the Lorentz force acting on a conductor moving in a magnetic field and thus influencing the initial motion we deal with the dynamically consistent dynamo. Let us explain some details of the kinematic dynamo.

2. BASIC EQUATIONS

The electric field $\vec{E}$ in an electrically conducting medium drives the electric current $\vec{J}$. If the medium is at rest in the frame of reference in which $\vec{E}$ and $\vec{J}$ are measured they are related by Ohm's law:

$$\vec{J} = \sigma\vec{E} \tag{2.1}$$

where σ is electrical conductivity of the medium which is a material constant if the field $\vec{E}$ is not too strong (for very strong $\vec{E}$, σ becomes dependent on $\vec{E}$). If the medium is in motion in the observer's frame of reference and there exists a magnetic field $\vec{B}$ the above formula becomes:

$$\vec{J} = \sigma(\vec{E}+\vec{u}\times\vec{B}) \tag{2.2}$$

where $\vec{u}$ is the velocity of the medium. Thus $\vec{E} = \frac{1}{\sigma}\vec{J} - \vec{u}\times\vec{B}$ which together with Faraday's law

$$\frac{\partial\vec{B}}{\partial t} = -\nabla\times\vec{E} \tag{2.3}$$

gives us:

$$\frac{\partial\vec{B}}{\partial t} = \nabla\times(\vec{u}\times\vec{B}) - \frac{1}{\sigma}\nabla\times\vec{J} \tag{2.4}$$

where $\vec{J}$ is given by Maxwell's law:

$$\mu_0\vec{J} = \nabla\times\vec{B} - \frac{1}{c^2}\frac{\partial\vec{E}}{\partial t} \tag{2.5}$$

(c being a speed of light) and hence:

$$\frac{\partial\vec{B}}{\partial t} = \nabla\times(\vec{u}\times\vec{B}) + \lambda\nabla^2\vec{B} + \lambda\nabla\times\frac{1}{c^2}\frac{\partial\vec{E}}{\partial t} \tag{2.6}$$

Assuming that the velocity of the medium is small compared to the speed of light we can neglect the last term of the above equation as being very small compared to the advection term $\nabla\times(\vec{u}\times\vec{B})$. Finaly we get:

$$\frac{\partial\vec{B}}{\partial t} = \nabla\times(\vec{u}\times\vec{B}) + \lambda\nabla^2\vec{B} \tag{2.7}$$

which is the so-called induction equation of magnetohydrodynamics (MHD) and governs the evolution of the magnetic field when the velocity $\vec{u}$ of the medium is given.

If our medium is an incompressible viscous fluid then the second equation giving the evolution of the velocity $\vec{u}$ under the influence of a magnetic field is just the Navier-Stokes equation of hydrodynamics with the Lorentz force as a body force:

$$\frac{\partial\vec{u}}{\partial t} + (\vec{u}\cdot\nabla)\vec{u} = -\frac{1}{\rho}\nabla P + \vec{J}\times\vec{B} + \nu\nabla^2\vec{u} \tag{2.8}$$

$$\nabla\cdot\vec{u} = 0$$

where ν is a kinematic viscosity of the fluid, ρ is its density and P is the pressure. Equations (2.7) and (2.8) form a complete set of MHD equations.

In this paper we are concerned with the kinematic part of the problem thinking of the velocity field $\vec{u}$ as being given and hence the induction equation (2.7) is the centre of our interest.

3. ADVECTION AND DIFFUSION OF THE MAGNETIC FIELD IN THE FLUID

When the conductivity of the fluid is very high the magnetic diffusivity $\lambda=1/\mu_0\sigma$ becomes very small and we can neglect the diffusive term in the induction equation. In this case the latter takes the form:

$$\frac{\partial\vec{B}}{\partial t} = \nabla(\vec{u}\times\vec{B}) \tag{3.1}$$

which is the induction equation for a perfectly conducting medium ($\sigma=\infty$ or in other words $\lambda=0$). In this case it is easy to imagine the evolution of the magnetic field in a geometrical way by introducing the concept of field lines. This enables us to understand the behaviour of the field without refering to the equation itself. A field line is a curve in space such that at every point of the line the vector $\vec{B}$ is tangential to it. They have two important properties:

i) The field lines in a perfectly conducting fluid are frozen which means that they are carried with it like lines marked with a non-diffusing dye (Alfven theorem, [Alfven 1942, Moffatt 1978])

ii) They can be stretched and stretching implies an increase of the field along the line.

Suppose that we have a perfectly conducting fluid flowing between two horizontal parallel planes and the velocity has a parabolic profile all the time i.e. it is maximal half way between the plates and decreases to zero on them (fig. 3a). Suppose also that at some initial instant t=0 we have a single tube of vertical magnetic field across the slab. The field lines are vertical straight lines filling the tube and there is no field outside the tube (fig. 3a). The field lines are frozen i.e. they are carried with the fluid so after some time the tube will look like the figure 3b. The field lines are stretched, that is, the field in the tube increases with time. The kinetic energy of the motion is transformed into magnetic field energy (since we consider the motion as being given, we don't take into account any loss of kinetic energy of the motion and the velocity profile remains the same all the time). We need energy to stretch magnetic field lines just as we need energy to stretch a rubber band. However we must remember that this simple picture is correct only in the limit of infinite conductivity of the fluid when the diffusive term in equation (2.6) may be neglected. We say that this is the limit of large magnetic Reynolds number $R_m = ul_o/\lambda$, l_o and u being respectively characteristic length and velocity of the problem.

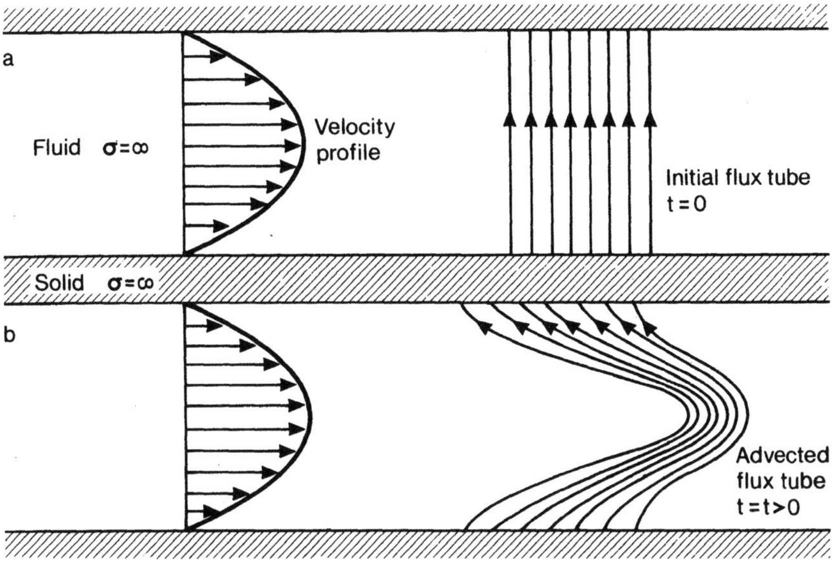

Figure 3. Advection of a magnetic flux tube by a two-dimensional flow with parabolic velocity profile in a slab: a) initial situation (t=0), b) distorted tube after some time ($t=\tau>0$)

The other extreme case takes place when the diffusive term dominates and the advection term may be neglected compared to it (this is small R_m limit). For example in the solid conductor with finite conductivity we have $\vec{u}=0$ and the advection term strictly vanishes ($R_m=0$). Equation (2.6) becomes:

$$\frac{\partial \vec{B}}{\partial t} = \lambda \nabla^2 \vec{B} \tag{3.2}$$

which is a diffusion equation for a vector field analogous to the diffusion equation for a scalar field like the temperature in the solid, for example. A diffusing field decays (the total energy of the field decreases with time and is transformed into heat). Consider for example a single Fourier mode:

$$\vec{B}(\vec{x},t) = \vec{b}(t)\, e^{i\vec{k}\cdot\vec{x}} \tag{3.3}$$

we obtain

$$\frac{\partial \vec{b}}{\partial t} = -\lambda k^2 \vec{b}(t) \tag{3.4}$$

and hence

$$\vec{b}(t) = \vec{B}_0 \, e^{-k^2 \lambda t} \tag{3.5}$$

$$\vec{B}(\vec{x},t) = \vec{B}_0 \, e^{i\vec{k}\cdot\vec{x}} \, e^{-\lambda k^2 t} = \vec{B}_0 \, e^{i\vec{k}\cdot\vec{x}} \, e^{-t/\tau} \tag{3.6}$$

where $\tau=(k^2\lambda)^{-1}$ is a diffusion time. The field lines are not carried by the fluid any more and stretching them doesn't imply any increase of the field.

Suppose now that we have an infinite horizontal slab of a solid conductor with $\sigma<\infty$ surrounded by a perfect solid conductor on both sides and a vertical magnetic tube at the initial instant $t=0$ (fig. 4a). The tube will diffuse and after a long time will be as shown in (fig. 4b). Geometrically the decrease of the field may be thought of as the increase of the volume occupied by a given set of field lines (in a limit $R_m \to 0$).

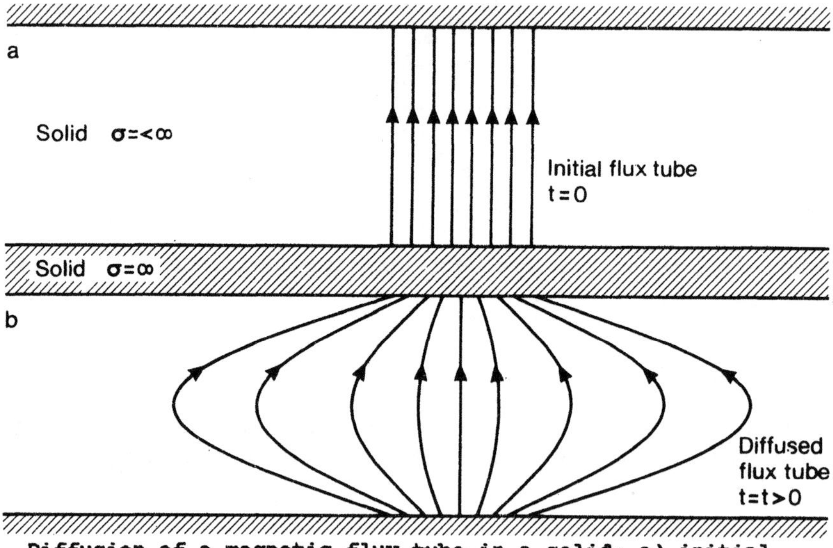

Figure 4. Diffusion of a magnetic flux tube in a solid: a) initial situation, b) diffused tube after some time ($t=\tau>0$)

These two aspects of magnetic field evolution in the fluid described above (advection and diffusion) dominate respectively the two parts of the Earth's dynamo cycle which we are going to describe qualitatively in the next section.

4. QUALITATIVE DESCRIPTION OF THE EARTH'S DYNAMO [Parker 1983]

Imagine a sphere of fluid rotating around its axis with a nonuniform angular velocity increasing towards the axis. Such a motion we call "differential rotation". Suppose that the fluid is very highly conducting and that initially there is a poloidal magnetic field symmetric

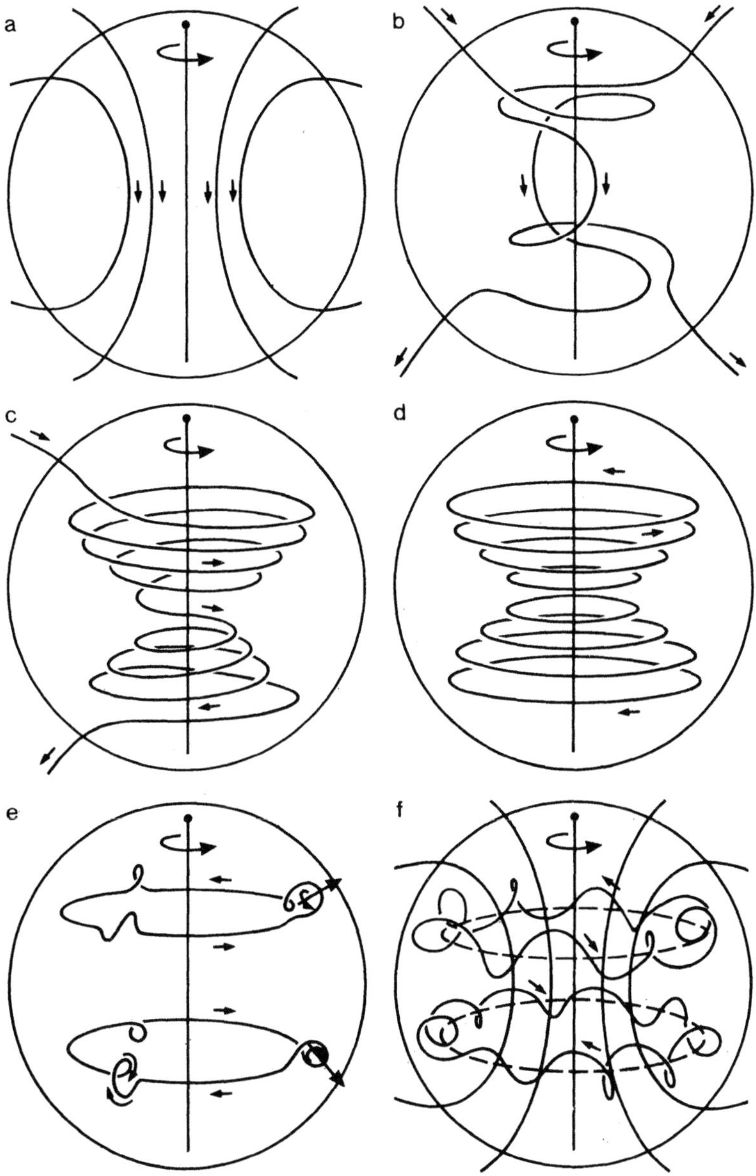

Figure 5. Differential rotation (DR) and helical convection (HC) in a sphere (after Parker 1983): a) initial poloidal field, b),c) wrapping of the poloidal field around the axis by DR, d) toroidal field produced by DR, e) formation and twisting of the poloidal field loops by HC, f) production and diffusion of the poloidal field

about the axis of differential rotation (fig. 5a). The field lines are almost frozen so the differential rotation will wrap them around the axis many times before any influence of the small diffusion may be noticed (fig. 5b),c)). Lines are stretched, the field is increased and its direction changed. Differential rotation produces a strong toroidal field from initial poloidal one (fig 5d). There is however a small but finite diffusion so this toroidal field will diffuse slowly towards the surface of the sphere. Suppose that in the outermost layer of the sphere convective motion occurs i.e. there are cells of fluid with upwelling in the middle of the cell and downwelling at the boundaries. In fact, because of the Coriolis force, fluid will move towards the surface along helical paths. Such a motion is called "helical convection" and produces twisted loops from toroidal field lines. Figure 6 shows such an isolated "cyclonic event" while figure 5e shows the situation in our sphere. This rather complicated flow pattern was necessary to give

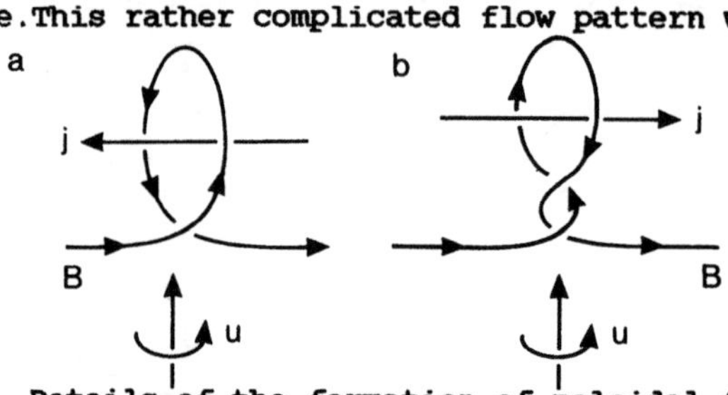

Figure 6. Details of the formation of poloidal field loops by helical convection. Mechanism of the production of current parallel (or antiparallel) to the field

the positive feedback to the dynamo i.e. to produce poloidal magnetic field again from the toroidal one. Small loops of poloidal field generated in the convection zone diffuse again becoming bigger and form the large scale poloidal field inside and outside the sphere (fig. 5f). This completes the cycle:

$$\text{poloidal field} \qquad \vec{B}_P \rightleftarrows \vec{B}_T \qquad \text{toroidal field}$$

which is a basis of all the theories of fluid dynamos. The second part of the cycle i.e. $\vec{B}_T \to \vec{B}_P$ is the most difficult to achieve especially in the models with laminar flow. As we can see it requires for example a fairly complex flow pattern like helical convection. There is however a well developed formalism to handle this problem when the flow is turbulent.

5. THE α - EFFECT

The major difficulty encountered in the previous section was to produce the loops of poloidal field lines (lying in meridional planes). The current associated with such a loop is a vector perpendicular to the plane of the loop i.e. tangent to the parallels of latitude in geographical terms (and also tangent to the toroidal field lines). Reformulating the problem we can say that the point is to produce a current parallel to the field. The advective part of the current $\vec{J}_A = \sigma(\vec{u}\times\vec{B})$ is trivially always perpendicular to the field (if both are measured at the same point of space, of course they don't have to be perpendicular at distant points):

$$\vec{J}_A \cdot \vec{B} = \sigma(\vec{u}\times\vec{B})\cdot\vec{B} = 0 \tag{5.1}$$

so we cannot get a parallel component of the current. This apparently trivial formula is in fact a strong geometrical constraint. In the model of the last section we avoided this difficulty by producing parallel current in a distant region (the convective zone) and then diffusing the associated poloidal field back towards the centre.

If however the flow is turbulent and we are interested only in the mean values of the current and field (averaged in time or over a region of space small compared with the scale of the mean field but much larger than the characteristic length of the turbulence) we can have a component of the advective part of the current parallel to the field:

$$0 = \langle\vec{J}_A\cdot\vec{B}\rangle \neq \langle\vec{J}_A\rangle\cdot\langle\vec{B}\rangle \neq 0 \tag{5.2}$$

where the angular brackets denote the avarage mentioned above. A simple illustration of the fact that the mean scalar product of two vectors may not be equal to the product of their mean values is as follows (fig. 7). Consider two points moving around an ellipse. Let the origin of the frame of reference be at the focus of the ellipse and suppose that motion of the points is such that their position vectors are always perpendicular (these vectors are denoted by $\vec{a}$ and $\vec{b}$). $\vec{a}$ and $\vec{b}$ are perpendicular but their time avarages are parallel being both parallel to the longer axis of the ellipse. Thus we have $0 = \langle\vec{a}\cdot\vec{b}\rangle \neq \langle\vec{a}\rangle\cdot\langle\vec{b}\rangle$.

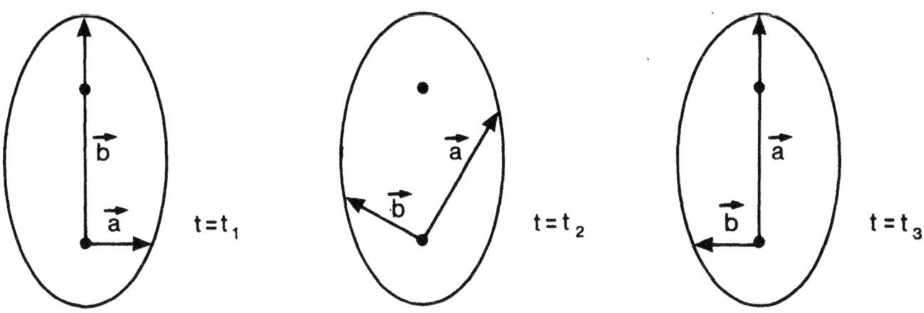

Figure 7. Rotating position vectors $\vec{a},\vec{b}$. They are perpendicular for all times but nevertheless their time avarages are parallel being both vertical

Suppose that both the magnetic and velocity fields are stochastic and decompose them into mean and fluctuating parts:

$$\vec{B}(\vec{x},t) = \vec{B}_0(\vec{x},t) + \vec{b}(\vec{x},t) \;;\quad \langle\vec{b}\rangle = 0 \;;\quad \langle\vec{B}\rangle = \vec{B}_0 \tag{5.3}$$

$$\vec{u} = \vec{u}_0 + \vec{v} \;;\quad \langle\vec{v}\rangle = 0 \;;\quad \langle\vec{u}\rangle = \vec{u}_0 \tag{5.4}$$

and substitute them into the induction equation:

$$\frac{\partial \vec{B}_0}{\partial t} + \frac{\partial \vec{b}}{\partial t} = \nabla\times(\vec{u}_0\times\vec{B}_0) + \nabla\times(\vec{u}_0\times\vec{b}) + \nabla\times(\vec{v}\times\vec{B}_0) + + \nabla(\vec{v}\times\vec{b}) + \lambda\nabla^2\vec{B}_0 + \lambda\nabla^2\vec{b} \tag{5.5}$$

Then taking the avarage of both sides we obtain:

$$\frac{\partial \vec{B}_0}{\partial t} = \nabla\times(\vec{u}_0\times\vec{B}_0) + \nabla\times\langle\vec{v}\times\vec{b}\rangle + \lambda\nabla^2\vec{B}_0 \tag{5.6}$$

Now subtract (5.6) from (5.5):

$$\frac{\partial \vec{b}}{\partial t} = \nabla\times(\vec{u}_0\times\vec{b}) + \nabla\times(\vec{v}\times\vec{B}_0) + \nabla\times(\vec{v}\times\vec{b}-\langle\vec{v}\times\vec{b}\rangle) + \lambda\nabla^2\vec{b} \tag{5.7}$$

Equations (5.6) and (5.7) govern the evolution of the mean and fluctuating parts of the magnetic field in a turbulent fluid [Moffatt 1978]. The mean field equation (5.6) now has an extra source term $\nabla\times\langle\vec{v}\times\vec{b}\rangle$ which couples it to (5.7) and then (5.6) cannot be solved separately. Equation (5.7) is however linear in $\vec{b}$ so if we assume homogeneous initial conditions like $\vec{b}(\vec{x},t)=0$ we will have that $\vec{b}$ satisfying (5.7) is a linear homogeneous functional of $\vec{B}_0$ and then $\langle\vec{v}\times\vec{b}\rangle$ must be such a functional too. The functional may be expanded in a series:

$$\langle\vec{u}\times\vec{b}\rangle_i = \alpha_{ij}B_{0_j} + \beta_{ijk}\frac{\partial B_{0_j}}{\partial x_k} + \gamma_{ijkl}\frac{\partial^2 B_{0_j}}{\partial x_k \partial x_l} + \ldots \tag{5.8}$$

In the first approximation we can write

$$\langle\vec{v}\times\vec{b}\rangle_i = \alpha_{ij}B_{0_j} \tag{5.9}$$

which provides a closure of (5.7), so we now have a closed equation for $\vec{B}_0$. If we assume that the turbulence is isotropic it follows that all the coefficients of the series must be isotropic tensors and in particular

$$\alpha_{ij} = \alpha\delta_{ij} \tag{5.10}$$

where α=constant and hence

$$\langle\vec{v}\times\vec{b}\rangle = \alpha\vec{B}_0 \quad ; \qquad \alpha \text{ - constant} \tag{5.11}$$

Finally the closed equation for $\vec{B}_0$ looks like:

$$\frac{\partial \vec{B}_0}{\partial t} = \nabla\times(\vec{u}_0\times\vec{B}_0) + \alpha\nabla\times\vec{B}_0 + \lambda\nabla^2\vec{B}_0 \tag{5.12}$$

$\nabla\times\vec{B}_0$ is the mean current ($\langle\vec{J}\rangle = \nabla\times\vec{B}_0\rangle = \nabla\times\vec{B}_0$) so the isolated α term gives us:

$$\frac{\partial \vec{B}_0}{\partial t} = \alpha\langle\vec{J}\rangle \qquad (5.13)$$

which means that during the evolution there appears a field parallel to the current which is exactly what we need to complete $\vec{B}_T \rightarrow \vec{B}_P$. We can see that in the first approximation the effect of turbulence on the mean magnetic field evolution may be incorporated by including an additional term $\alpha\nabla\times\vec{B}_0$ in the induction equation. This is the so-called α - effect. It can operate in both parts of the dynamo cycle: $\vec{B}_P \rightarrow \vec{B}_T$ and $\vec{B}_T \rightarrow \vec{B}_P$. If this is the case we say that the dynamo is of α^2 type. If $\vec{B}_P \rightarrow \vec{B}_T$ is done by differential rotation and $\vec{B}_T \rightarrow \vec{B}_P$ by the α - effect then dynamo is of $\alpha\omega$ type.

The second term of the expansion is interesting too. In the homogeneous case it is simply an additional diffusion term:

$$\frac{\partial \vec{B}_0}{\partial t} = \nabla\times(\vec{u}_0\times\vec{B}_0) + \alpha\nabla\times\vec{B}_0 + (\lambda+\beta)\nabla^2\vec{B}_0 \qquad (5.14)$$

β is a constant which in principle may be negative and it means that turbulence may reduce the diffusion of the magnetic field in contrast to the diffusion of a scalar field which is always increased by turbulence. There is the interesting possibility of having negative diffusion when $\lambda+\beta<0$. However this possibility has not yet been confirmed by numerical experiments.

6. CONCLUDING REMARKS

From the previous section we can see how the large scale magnetic field is produced by a dynamo process in the fluid. The laminar model of section 4 is plausible and likely to operate in the Earth, but it involves large-scale motions like differential rotation and helical convection. The scale of these motions is comparable to the scale of the resulting poloidal field. Also, the axial symmetry of the field is built into the motion through the axial symmetry of differential rotation. The situation is quite different in the homogeneous isotropic turbulence of section 5, especially with zero mean velocity. Such turbulence has, of course, no preferred direction (however, it must be reflectionally non-symmetric if the α - effect is to occur), also the typical length scale may be arbitrarily small. This highly symmetric motion may exhibit something which we can call spontaneous symmetry-breaking. Once the symmetry is broken by the apearance of a small mean magnetic field (which introduces a preferred direction) this fluctuation is increased and the mean field becomes stronger. Thus, the system prefers a non-symmetric state rather than a symmetric one even if the energy flow (turbulence) still retains its isotropy. The possibility of the creation of large scale mean magnetic fields through this mechanism has been confirmed by numerical experiments.

REFERENCES

1. H. Alfven, 'On the existence of electromagnetic-hydrodynamic waves', *Arkiv. f. Mat. Astron. Fysik* **29B**, no. 2 (1942)
2. A. Cox, 'Geomagnetic reversals: characteristic time constants and stochastic processes', in *Report on Int. Cong. on Core-Mantle Interface. Trans. Am. Geophys. Un.* **53**, 613 (1972)
3. H. K. Moffatt, *Magnetic field generation in electrically conducting fluids*, Cambridge University Press (1978)
4. E. N. Parker, 'Magnetic fields in cosmos', *Sci. Am.* **249**, no. 2 (1983)

FACING SOME PROBLEMS WITH ENTROPY

C. Blöss
GERENA Gesellschaft zur technischen Nutzung von Abwärme mbH
Prinzessinnenstraße 20
D - 1000 Berlin 61
GFR

ABSTRACT. This article is a short summary of the problems resulting with respect to the working fluid used in building low temperature thermal engines. It has been shown by experience that the relationship or interrelationship of the properties of materials in thermostatics discloses erroneous results in particular as far as heat capacity is concerned when exploiting low thermal heat in the supercritical state, ie in the range beyond the critical point. The discussion of these problems leads to a critical assessment of these interrelationships of the properties of materials. A new method of analysis is being proposed by which one arrives at a presentation of the complete profile of the working fluid starting from thermal measurements.

Considerable efforts have been invested for some time to find out how the large amounts of energy due to waste heat can be exploited technically by machines in converting said energy into an electrical current or in generating turning moment. The attempts at finding a solution comprise the use of low-boiling liquids such as Frigen or Freon in multistage turbines; other attempts take recourse to pure fluids and to the design of cyclic processes using high pressure without phase transition. GERENA Co. with seat in Berlin is developing a piston engine containing a working fluid with which the cyclic process is entirely run in the supercritical range.

The following ideas have been determinative in this decision:

- All fluids have in this state their relative highest coefficient of expansion with the consequence that considerable work in volume can be performed even at small differences in temperature.

- Moreover, the thermal capacities C_p and C_v vary in this state as a rule by at least one order of magnitude. This has determinative consequences for the design of the cyclic process which can be run, in principle, in a manner so that along those sections of the process where heat is taken up, the thermal capacity is on the average

C. W. Kilmister (ed.), Disequilibrium and Self-Organisation, 141–153.

large, while at those sections where heat has to be given off, such thermal capacity is small. The greater the difference in thermal capacity and the more divergent the temperature differences in heating and cooling, the more advantageous is its effect on efficiency.

These arguments are rendered relative on account of the supposition that the maximum efficiency η attainable at given temperatures T_1 and T_2 is delimited and defined by the formula

$$\eta \leq \frac{T_1 - T_2}{T_1} \tag{1}$$

The meaning of this supposition would be very sobering if it were not based on an assumption made with respect to entropy which results, in certain cases, in considerable discrepancies when giving an account of material properties. The discrepancies arise for instance when, in the vicinity of the critical point, the thermal capacity C_v is provided from thermal data such as the real gas factor Z(T,V). It should be remembered in this connection that C_v has, in general, a maximum along the isothermal line at variable volumes, which maximum increases the more, the more the critical point is approached. It is almost commonplace when Abdulgatov and Alibekov point out that the equation of state calculated from P-V-T data for C_v "cannot describe quantitatively the sharp anomalous growth of C_v and other thermodynamic functions near the critical point" (Abdulagatov et al., 1982, p.135) (cf. also figures (I) and (II)).

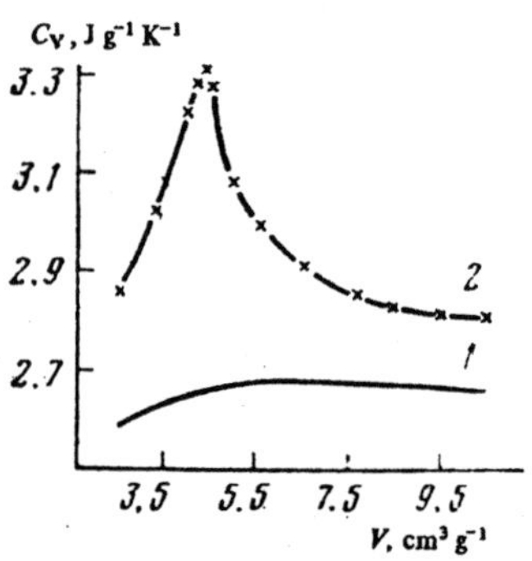

Figure (I). Comparison of the specific heats C_v of n-heptane at 543 K calculated from the thermal Benedict-Webb-Rubin equation (curve 1) with experimental data (curve 2).

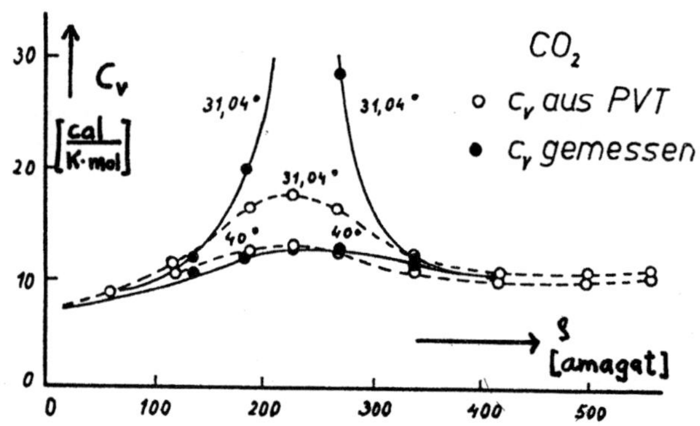

Figure (II). C_v for carbondioxide in proximity to the critical point.

References
Fig.(I): cf.Abdulagatov et al., 1982, p.804
Fig.(II): cf.Benzler, 1954,p.60

To be sure, the problem of giving a due presentation of C_p and C_v from P-V-T data is in no way trivial. Material properties relations as we know them are expressed by the formulae:

$$C_v = C_v^0 + \int \left\{ \frac{\partial^2 p}{\partial T^2} \bigg|_v \right\} dv \tag{2}$$

$$C_p = C_v + T \left\{ \frac{\partial p}{\partial T} \bigg|_v \right\} \left\{ \frac{\partial V}{\partial T} \bigg|_p \right\} \tag{3}$$

In order to be able to represent C_v it is necessary to take recourse to an equation of state P(V,T) which is sufficiently precise so that the requirements for precise presentation may be satisfied even after carrying out two partial derivations and one integration. Either the thermal equations of state used cannot live up to this requirement in certain cases or, however, the material properties relations (2) and (3) are (among others) principally not adequate.

Prior to entering into a discussion of the consequences to be derived from both of these possibilities, we would like that close attention be paid to two figures both of which relate to the rare gas argon. Figure (III) illustrates the field of state in which investigations carried out by Shah and Thodos give a "very good" representation of the thermal behaviour of argon by way of Van der Waals equation of state (cf. Landolt-Börnstein, II.1, [6]1967, p.298). This range of state corresponds by the relation T > 150° Kelvin and p < 48 atm. The reason why we refer to this figure is the following: if the P-V-T data can be approximated "very well" by the equation of state mentioned above, then - with equal certitude - the formulation for thermal capacity C_v holds: C_v is in the field of state a pure function of temperature, ie it is, in particular, independent of pressure. This follows directly from equation (2), for in this instance.

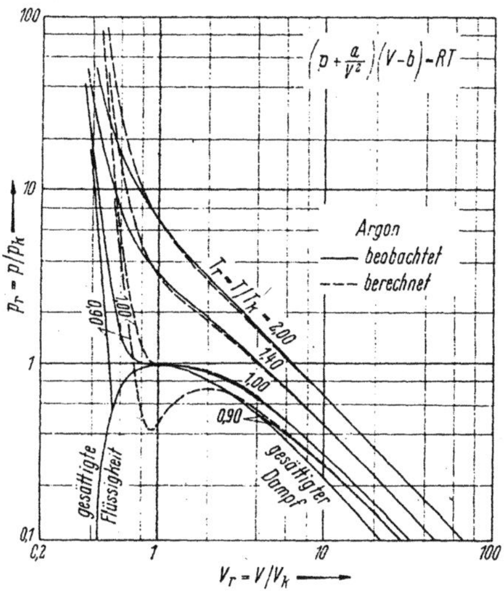

Figure (III). Comparison of the P-V Data for argon with those calculated by way of the Van der Waals equation.

$$\left.\frac{\partial^2 p}{\partial T^2}\right|_v \equiv 0 \tag{4}$$

and hence C_v is a pure function of temperature.

For this field of state Michels et al. (cf. Landolt-Börnstein, II.4, [6]1967, p.702; Michels et al., 1958, p.769ff.) calculate now thermal capacities which are everything else but independent of pressure (cf. figure (IV)).

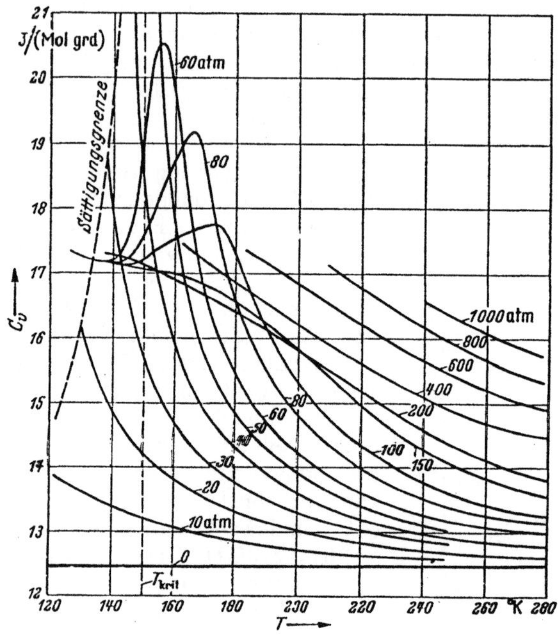

Figure (IV). C_v as calculated for argon.

They base their calculations, in general, on the same material properties relations as Shah and Thodos do also. This example shows, hence, that the material properties relations are in their formulation as partial differential equations in and by themselves problematical, to start with, on account of setting, evidently, the highest requirements for the precision of the approximation of measured P-V-T data.

These circumstances motivated us, in the end, to develop a new method for the complete representation of the material properties while using exclusively P-V-T data which we want to present in the final section of our article.

If, to be sure, the known material properties relations are insufficient in certain circumstances, then it is necessary to look for the reason herefor in an inadequate axiomatization in thermostatics. Since solely two relevant propositions are made on the mathematical level of thermostatics, it turns out that at least one of these must be false. These propositions relate to the existence of total differentials whereby "internal energy" and "entropy" are addressed. Because the material properties relations mentioned are immediately retraceable to the total differential of entropy we must discuss it here.

There is but one reason or argument for the definition of the entropy differential dS as

$$dS = \frac{đQ}{T} \tag{5}$$

This argument is supplied by an assumption of a contradiction, viz. the assumption of material specific integrating denominators for the thermal differential dQ which would result in material specific efficiency factors in one and the same Carnot process. Under such conditions one could convert heat completely into work by means of two oppositely operating ideal Carnot engines. This conclusion is deemed, for reasons whatsever, to be unacceptable. The sole integrating denominator which does not lead to a contradiction with respect to the argument that there is no aggregate which converts heat in a cyclic process entirely into work is given by absolute temperature T. It must be kept in mind that the emphasis in this presentation of arguments is placed on the impossibility of carrying out a thought experiment and not - in any case not directly - on the verifiability of the definition of entropy by way of measurements.

Under these circumstances it will become clear that the contradictions cited will not immediately entice one to rethink the definition of entropy but rather to investigate other sources of error.

We will discuss below the feasibility as to whether a less "trenchant" definition of entropy may not lead to a more adequate representation of material properties. Let us start with a theory immanent criticism of the mathematical version of the second law of thermodynamics.

1. First law: $dU = đQ + \Sigma đW_i = đQ + \Sigma X_i dx_i$ (6)

2. Second law: $dS = \frac{1}{T}dU - \frac{1}{T}\Sigma X_i dx_i$ (7)

Let us first turn to the version of the first law. The differentials of different forms of energy designated by dW_i may be specified thus:

$$đW_1 = -pdV \qquad \text{mechanical volumetric work} \tag{8}$$

$$đW_2 = HdM \qquad \text{magnetic work} \tag{9}$$

For the continued discussion the fact is determinative that - from the phenomenological point of view - the elements of each dyad X_i, x_i is associated with a material specific equation of state:

p,V - "thermal" equation of state p(V,T), e.g. $p = RT/V$
H,M - "magnetic" equation of state M(H,T), e.g. $M = CH/T$

Thus, for any form of energy there is **always** a particular equation of state and without its knowledge the information as to substance would be incomplete. Apparently, however, this does not apply to the heat differential đQ. There is here, to be sure, also factorization into an intensive and an extensive variable, for đQ = TdS, yet there is supposed to be no distinct material specific connection between T and S. The sole connection is established by the defining equation (5) where no material characteristics are found but where only the characteristic of process đQ is encountered.

There are no reasons for explaining this particularity which would only be removed in part by the general definition:

$$dS = đQ/\nu \text{ with } \nu = \nu(T,V) \tag{10}$$

In part, to be sure, because the quantities "entropy" and "integrating denominator" cannot be directly determined as are the other extensive and intensive variables via measurement of mass, length and time. They would be separately incapable of measurement and could be determined quantitatively as product by taking the mechanical heat equivalent as basis. This latter being the essential expression of the first law is also the basis of the method of representation for ν(T,V) to be discussed below.

Precisely for the reason that the quantities ν and S cannot be measured directly, the conclusions resulting from generalisation of the definition of entropy are for this very reason of great importance. This will become apparent when the generalized material properties relation (12) is being compared with the standard relation (11):

$$\frac{\partial U}{\partial V}\Big|_T = T\frac{\partial p}{\partial T}\Big|_V - p \tag{11}$$

$$\frac{\partial \nu}{\partial T}\Big|_V \frac{\partial U}{\partial V}\Big|_T - \frac{\partial \nu}{\partial V}\Big|_T \frac{\partial U}{\partial T}\Big|_V = \nu\frac{\partial p}{\partial T}\Big|_V - p\frac{\partial \nu}{\partial T}\Big|_V \tag{12}$$

With respect to the relation between the functions of state U(T,V) and p(T,V) another function of state plays a part in (12), which turns out to be the integrating denominator of the heat differential.

Why, may one ask, should the information from U and p not suffice to provide for a relation such as (11)? The answer: the function of state U represents merely the prerequisite so that the sum of exchanges of energy of the system in all possible forms between any two states is, independent of the course followed, supposed to be the same at all times. The existence of U signifies, hence, a restriction as to the possible changes of physical quantities and represents, in this sense, no material properties such as e.g. $p(T,V)$. The socalled 'caloric function of state' U is, in the end, merely the warrant for the constancy of the mechanical heat equivalent. The caloric function of state proper is $\nu(T,V)$, the physical meaning and importance of which need be elucidated. Phenomenologically, nothing more can be said than that it is brought into play because the thermal function of state and the existence of a constant mechanical heat equivalent alone do not yet suffice in order to characterize all possible material properties.

In terms of statistical physics we obtain a more profound insight with respect to the function ν. To this purpose we sketch the "bridging" of statistical to phenomenological entropy. In both, entropy S and internal energy U are defined; they are found in a relation of mutually identical structure. Thus, respectively:

$$\left.\frac{\partial S}{\partial U}\right|_V = \frac{1}{\nu} \tag{13}$$

$$\left.\frac{\partial S}{\partial U}\right|_V = k\beta \tag{14}$$

The β from equation (14) constitutes a Lagrange parameter which is brought into play by giving due concern to the constraint of constant total energy U also when there are fluctuations within the system. The customary identification $\beta = 1/kT$ is due to equating (13) with (14) with the supposition of $\nu \equiv T$. In terms of statistical physics there is, presently, no intelligible reason why β is supposed to be essentially identical with absolute temperature. To the contrary, an investigation of the consequences of this one-sided determination allows for the conclusion which we drew already with respect to the integrating denominator of the heat differential: β must be constructed generally to be a material specific function not only of temperature T but also of other physical quantities.

In statistical mechanics we obtain by way of the relation

$$\overline{E}_{kin} = 3N\int\{\frac{m}{2}v^2 \cdot \exp(-\beta\frac{m}{2}v^2\} dv \cdot \{\int \exp(-\beta\frac{m}{2}v^2)dv\}^{-1} \tag{15}$$

the expression

$$\overline{E}_{kin} = \frac{3N}{2\beta} \tag{16}$$

for the mean kinetic energy for N particles. On the other hand, there is the virial therorem:

$$\overline{E}_{kin} = \frac{1}{2}\sum_{i=1}^{N} \underline{K}_i \cdot \underline{r}_i = \frac{3N}{2\beta} \tag{17}$$

wherein $\underline{K}_i$ stands for the total force exerted by the respectively (N-1) other particles of the total. If, now, β is identified as 1/kT, two situations or conditions have been determined at one and the same time.

1. For any empirical temperature T there is precisely **one** form of energy distribution. Hence, it is precluded in principle that the same temperature - being a macroscopic measured quantity or variable - can be provoked by another distribution of energy or velocities at another mean velocity. This is, physically speaking, a nonsensical requirement. It would mean, to be sure, that an ideal and a real gas would have to have the same energy distribution at same temperatures although the kinetic energy for the real gas has a complicated relationship to potential distribution (cf. (17)). An example demonstrating the opposite may be presented thus: when an ideal gas, i.e. an extraordinarily rarefied gas, is subjected to isothermal compression, interaction sets in, of which attraction comes first. On the average an increasingly greater amount of the gas particles will stay in the range of attraction of the potentials of other particles. These are, then, all particles none of which will have zero velocity because they would otherwise not be able to leave the range of attraction. It follows from this that the maximum of the distribution towards higher velocities is shifted with the distribution curve flattening out on the whole (for an example calculated **in extenso** cf. Blöss, 1985, p.43ff.).

2. This problem may be discussed also from the point of view as to which forms or types of interaction potentials are still possible at all under such conditions. We assert that in case of β = 1/kT on the right side of the equation (17) there is, but for the natural constant k and particle number N, only one macroscopic quantity to be found, ie temperature T. In other words, the distribution of loci of the interacting particles must always proceed so that the sum of material specific terms on the left side of equation (17) becomes material unspecific. These terms of the sum comprise explicit material constants because the total forces $\underline{K}_i$ can be generally ascribed to potentials the configuration of which varies from one fluid to another, and this should, in principle, find expres-

sion also on the right side of equation (17). From this it suggests itself, just as expressed under item 1, that a more general function be assumed than - of all things - the Boltzmann's factor 1/kT.

What has been said can also be applied and translated to other forms of energy. Looking at the rotational spectra of molecules which provide indirect information regarding the population densities of the rotational energy levels, we find that such population densities of rotational energy levels E_j are shown to be governed, to the present day, by the Boltzmann's factor. Yet, it is precisely the rotation of macromolecules from which it becomes intelligible that not only temperature affects the population of the levels available but also that it is affected by the density and pressure, respectively, within the fluid. What may rotate unimpeded in the rarefied state does do so no longer, to be sure, at the same empirical temperatures in the compressed state, and it would be premature to hold only the change of the Hamiltonians responsible for changes in population densities (cf. Blöss, 1983, p.32ff.).

Thinking about what governs, properly speaking, the population density of the rotational energy levels or quantum states has led us, when we were starting our investigations, to take a close and extensive look at rotational band spectra. We have found that gas mixtures are particularly interesting in this respect. We based our investigation on the following: if a gas - the rotational spectra of which can be represented when taking as basis the model of the rigid rotator - subjected to isothermal compression give an indication of a shift of the absorption maximum while simultaneously retaining its spacing of absorption peaks, then the population densities need be governed by something more than just temperature. The maximum of absorption is obtained from the formula

$$j_m = \sqrt{\left(\frac{2I}{\beta\hbar^2}\right)} - \frac{1}{2} \qquad (18)$$

wherein j_m is the quantum number of the associated quantum state. It is apparent from equation (18) that it is first the assumption that β were a more general function than 1/kT which will lead one to expect under isothermal compression that there is a shift in the maximum. Inspecting figure (V) will reveal corresponding conditions. The lower series of spectra has been taken with a HCl-Ar mixture. Here, the partial pressure of the buffer gas argon was raised isothermally while holding the partial pressure of HCl constant. It is evident from this that, when increasing the pressure from 9 to 169 amagat, the absorption maximum will shift particularly on the right branch towards the right.

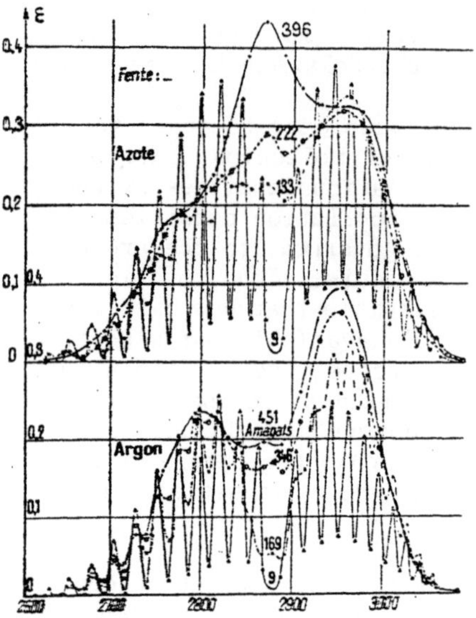

Figure (V). Rotatonial vibration band of HCl (gas) in the mixture with argon (gas), respectively, taken at 300° Kelvin at various partial pressures of both buffer gases (cf. Coulon, 1954, p.642)

This revelation constitutes, of course, but a qualitative indication to the effect that the Lagrange parameter must be "more" than just 1/kT. Yet, we are confirmed in our suspicion that instead of the known material relations in certain ranges of state and, especially, in the very high pressure ranges generalized material relations must be used.

These findings have led us to develop a method of measurement and evaluation which we present here and which can present direct evidence as to whether entropy must in general be defined as

$$dS = \frac{đQ}{\nu(T,V)} \tag{19}$$

or not.

In our reflections we set out from the requirement of doing without genuine caloric measurements so as not to render the measurements needlessly expensive in cost and labor from the start. Thus, a method of evaluation and analysis had to be found so as to glean from thermal measurements sufficient information data in order to be able to answer the question as to the character of the integrating denominator and in order to calculate cyclic processes with regard to efficiency. On account of the fact that in this manner no complete data are, principally, available we had to find a way of bridging this data deficit. Contrary to the standard method wherein the differential material properties relations are used resulting from the existence of the total differential with the assumption $\nu \equiv T$, our method uses the integral version of the first law in the following form:

$$\oint dU = \oint dQ + \oint dW \equiv 0 \tag{20}$$

Moreover, we start from the mere understanding of the **existence** of an integrating denominator for $đQ$; yet, there are no presuppositions taken as to its nature or character. This method will then furnish the function ν by way of an iteration algorithm.

The method is based on measuring the isothermal and isentropic lines*. Its presentation in the p-V diagram furnishes a continous region unfolding or spreading from adjacent Carnot processes with associated areas. In the case of $\nu \equiv T$ a corresponding rectangular process in the T-S diagram may be associated with each of the Carnot processes in the p-V diagram. This becomes readily evident, for the isentropic and isothermal lines in the T-S diagram are, respectively, genuine horizontal and vertical lines. Evidence going counter to the assumption of $\nu \equiv T$ can be examined with relative ease. With the temperature scale in the T-S diagram being fixed, transferring **one** Carnot process from the p-V to the T-S diagram will automatically generate the correct spacing of the corresponding isentrops when presupposing area identity in keeping with equation (20). All other Carnot processes located between both of these isentropic lines must produce, upon areal transfer, always the same spacing for both isentropic lines. If this is the case it is justified to postulate $\nu \equiv T$. If, however, the Carnot processes between two respective isentropic lines produce when transferred divergent isentropic line spacing, then this supposition is in error (cf. figure (VI)).

In this latter case one can only proceed with the assumption that the isotherms are not genuinely horizontals but rather curved trajectories the course of which can be obtained iteratively from the requirement of equation (20). These iterative methods provide a matrix $\underline{\underline{H}}$

$$\underline{\underline{H}} := \begin{pmatrix} h_{11} & h_{12} \\ 0 & h_{22} \end{pmatrix} \tag{21}$$

which distorts the isotherms T' and effects a parallel shifting of the isentrops S' in the figure

$$\begin{pmatrix} T' \\ S' \end{pmatrix} = \begin{pmatrix} h_{11} & h_{12} \\ 0 & h_{22} \end{pmatrix} \begin{pmatrix} \nu \\ \sigma \end{pmatrix} \tag{22}$$

so that the prescription regarding representation

---

* Reasons of patent protection prevent us from entering into a deeper discussion of the method of measurement.

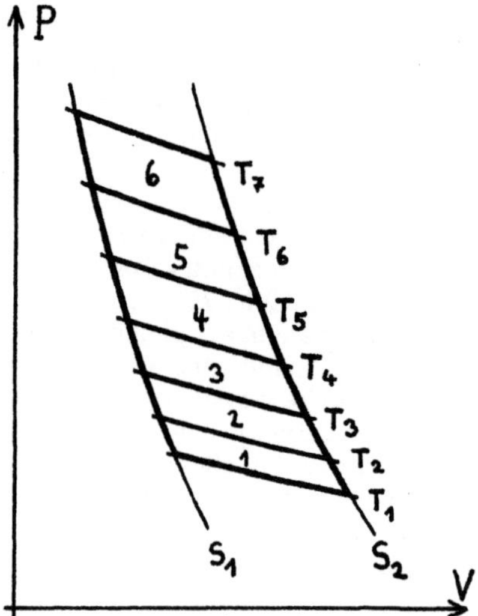

Figure (VI.1)

Figure (VI). The figure (VI.1) shows six Carnot processes unfolded or spreading out by two isentrops and seven isotherms. Transferring it into the T-S diagram (VI.2) would lead to divergent spacings of the isentrops S_1 and S_2. In the figure (VI.3) the isotherms are distorted and the isentrops are shifted in such a manner that sameness of area is retaimed between VI.1 and VI.3.

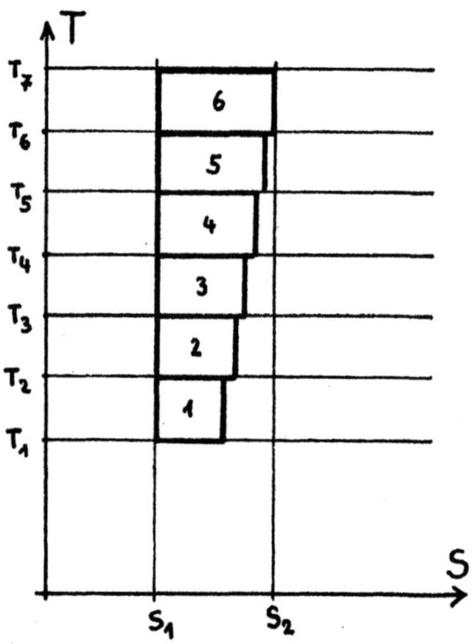

Figure (VI.2)

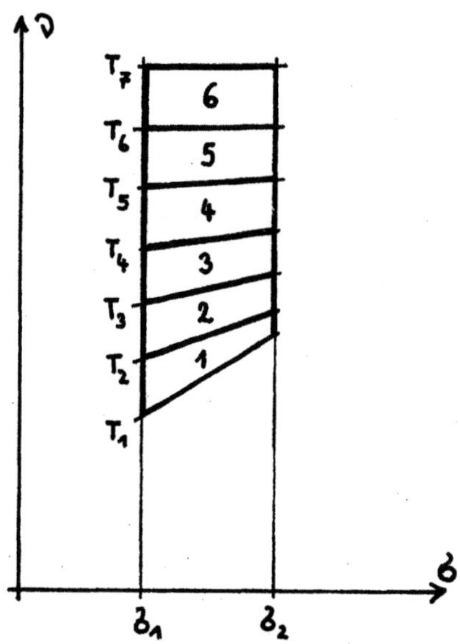

Figure (VI.3)

$$|\oint p dV| = |\oint \nu d\sigma| \tag{23}$$

is, once again, fulfilled for all cyclic processes. The course of isobars and isochores in the figure may be represented in a similar manner. The "true" course is obtained by comparing isobar/isochor-isentropic cyclic processes, the loci of isentrops being already fixed by way of the prior iterative process. Having arrived here, the thermal capacities are obtained by:

$$C_v = \nu \frac{\partial \sigma}{\partial T}\Big|_v \tag{24}$$

$$C_p = \nu \frac{\partial \sigma}{\partial T}\Big|_p \tag{25}$$

We like to emphasize that the representation of the figure requires at no time simple or multiple partial derivations of compensation curves, but merely comparing the areas enclosed thereby. The iterative method itself is, in principle, no source of error since it is applied for such an extent in time until still existing differences of contents of mutually associated areas have been brought down to zero. In addition to these advantages there is another one to bear fruit. Whereas the use of differentially formulated material properties relations will only uncover local incommensurabilities, this method aims, from the start, at global congruency. It is thus possible to discover discrepancies remaining undetected within the capacity of resolution in the local field of investigations; such small local errors may add up, globally speaking, to not unimportant errors.

GERENA Co. of Berlin will finance, over the next two years, a research program based on this method and aimed at solving the problems for the design and construction of a thermal engine for exploiting low temperature heat. It would be to advantage if interested institutions were to cooperate in this venture.

REFERENCES

(1) Abdulagatov, I.M. et al. (1982), *Russ.J.Phys.Chem.*, **56** (1)
(2) Abdulagatov, I.M. et al. (1980), *Russ.J.Phys.Chem.*, **54** (6)
(3) Benzler, H. (1954), 'Zur Zustandsgleichung reiner fluider Stoffe, insbesondere bei überkritischen Dichten', *Abh. Mat.-Natw. Klasse d. Akad. d. Wiss. u. d. Lit.*, Wiesbaden
(4) Blöss, C. ([3]1983), *Empirische Temperatur und Entropie*, Berlin
(5) Blöss, C. (1985), *Der Entropiebegriff*, Bonn
(6) Coulon, R. et al. (1954), *Le Journal de Physique et le Radium*, **15**
(7) Landolt-Börnstein ([6]1967), *Zahlenwerte und Funktionen*, **II** (1); **II** (4), Berlin-Heidelberg-New York
(8) Michels et al. (1958), *Physica*, **24**

ON THE SPONTANEOUS EMISSION FROM TWO-LEVEL ATOMS IN A DAMPED CAVITY

J. Seke
Institute for Theoretical Physics
Technical University Vienna
Karlsplatz 13
A-1040 Vienna, Austria

ABSTRACT. By using a new technique, developed recently by the author, exact analytic and numerical solutions are obtained for the spontaneous emission from N identical two-level atoms confined to a single site (Dicke model) and being initially in the lowest symmetrical Dicke state with only one atom excited. The atoms are placed into a damped cavity with frequencies lying in a narrow range about some given frequency.

1. INTRODUCTION

The recent experiments on the spontaneous emission from one and many two-level Rydberg atoms in damped cavities[1-4] make the testing of simple quantum-electrodynamic models of radiation-matter interaction possible. Here we treat an exact soluble case of spontaneous emission. Our model consists of an assembly of N identical two-level atoms confined to a single site (Dicke model[5]) and placed into a damped cavity with frequencies lying in a very narrow range about some mean frequency. The atoms are excited initially to a symmetrical Dicke state[5] with only one atom excited.

The present model is a twofold generalization of the single-atom spontaneous emission in a single-mode damped cavity. Firstly, it treats an assembly of N atoms and secondly, the radiation field in the cavity is three-dimensional and has many modes.

In Section 2, by using a new method, developed recently,[6-9] we obtain an exact differential equation of the fourth order for the atomic population inversion by successive differentiations of the formal solution of the Liouville equation. In Section 3 we present analytic and numerical solutions. Finally, in Section 4 we give a conclusion.

2. EXACT DIFFERENTIAL EQUATION FOR ATOMIC POPULATION INVERSION

The Liouvillian for N two-level atoms confined to a single site (Dicke model[5]) and interacting with a three-dimensional multimode radiation field in a cavity with frequencies ω_k, lying in a very narrow range

C. W. Kilmister (ed.), Disequilibrium and Self-Organisation, 155–163.

about a given frequency ω_{k0}[7,8], can be written as

$$L = L_0 + L_1 + L_I + i\Lambda_R , \tag{1}$$

$$L_0 = [H_0,\ldots], \quad L_1 = [H_1,\ldots] , \quad L_I = [H_I,\ldots] , \tag{2}$$

with corresponding Hamiltonians ($\hbar = 1$)

$$H_0 = H_A + H_R = \omega(S^z + \sum_{\underline{ks}} a^+_{\underline{ks}} a_{\underline{ks}}) , \tag{3}$$

$$H_1 = \Delta\omega \sum_{\underline{ks}} a^+_{\underline{ks}} a_{\underline{ks}} , \quad \Delta\omega = \omega_{k0} - \omega , \tag{4}$$

$$H_I = (-i)\sum_{\underline{ks}} g_{\underline{ks}}(a_{\underline{ks}} S^+ - a^+_{\underline{ks}} S^-), \quad S^{\pm} = \sum_{i=1}^{N} S_i^{\pm}, \quad S^z = \sum_{i=1}^{N} S_i^z , \tag{5}$$

$$g_{\underline{ks}} = (2\pi\omega_k/V)^{1/2}(\underline{D}_{12}\cdot\underline{e}_{\underline{ks}}) , \tag{6}$$

and the field damping Liouvillian describing the escape of photons from the cavity [10,8]

$$\Lambda_R(\ldots) = \varkappa\{\sum_{\underline{ks}} [a_{\underline{ks}}(\ldots),a^+_{\underline{ks}}] + [a_{\underline{ks}},(\ldots)a^+_{\underline{ks}}]\} , \tag{7}$$

where S_i^z, $S_i^{\pm}$ are the population inversion and dipole moment operators of the ith atom, ω is the frequency of the atomic transition, ω_{k0} is the frequency of the radiation modes $\underline{ks}$, $g_{\underline{ks}}$ is the coupling coefficient, $\underline{D}_{12}$ is the dipole matrix element, $\underline{e}_{\underline{ks}}$ is the polarization vector (s is the polarization index), V is the volume of the field in cavity, and $\varkappa$ is the cavity-damping factor.

For the operators S^z, $S^{\pm}$ and $a_{\underline{ks}}$, $a^+_{\underline{ks}}$ the following realtions hold

$$[S^+,S^-] = 2S^z , \quad [S^z,S^{\pm}] = \pm S^{\pm} , \tag{8}$$

$$S^z|rm\rangle = m|rm\rangle , \quad S^2|rm\rangle = r(r+1)|rm\rangle ,$$
$$S^2 = S^+S^- + (S^z)^2 - S^z , \tag{9}$$

$$[a_{\underline{ks}},a^+_{\underline{k's'}}] = \delta_{\underline{kk'}}\delta_{ss'} \tag{10}$$

where $m = -r, -r+1,\ldots, r-1, r$ and $r = 0,1,2,\ldots,N/2$ for even N, and $r = 1/2, 3/2, \ldots, N/2$ for odd N.

At the initial time $t = 0$ the system A of N atoms is in the lowest symmetrical Dicke state[5] with only one atom excited:

$$\rho_A(0) = |N/2, -N/2 + 1\rangle\langle N/2, -N/2 + 1| . \tag{11}$$

In this paper we will treat the spontaneous emission, and therefore the initial state of the radiation field R is the vacuum state

$$\rho_R(0) = |\{0\}><\{0\}| \ . \tag{12}$$

Moreover, we assume that the systems A and R are initially decoupled

$$\rho(0) = \rho_A(0)\rho_R(0) \ . \tag{13}$$

The statistical density operator for the total system A + R satisfies the Liouville equation, whose formal solution reads:

$$\begin{aligned}\rho(t) &= \exp(-itL_0)\exp[-it(L_1 + L_I + i\Lambda_R)]\rho(0) \\ &= \exp(-itL_0)\rho_I(t) \ ,\end{aligned} \tag{14}$$

where we used

$$[H_0,H_I] = 0, \ [H_0,H_1] = 0, \ L_0\Lambda_R(\ldots) = \Lambda_R L_0(\ldots) \ . \tag{15}$$

As a consequence of the special initial condition, Eqs.(11)-(13), and special form of the Liouvillian, Eqs.(1)-(7), the time evolution of the statistical density operator $\rho(t)$ is restricted to the subspace spanned by the state vectors

$$\begin{aligned}|1> &= |N/2,-N/2 + 1>|\{0\}>, \ |2> = |N/2,-N/2>|\{0\}>, \\ |3(\underline{k}s)> &= |N/2,-N/2>|\{1_{\underline{k}s}\}> \ ,\end{aligned} \tag{16}$$

where in the state $|\{1_{\underline{k}s}\}>$ one photon is in the $\underline{k}s$ mode. The unit operator in this subspace reads

$$I = \sum_{i=1}^{2} |i><i| + \sum_{\underline{k},s} |3(\underline{k}s)><3(\underline{k}s)|, \tag{17}$$

and

$$\mathrm{Tr}\,\rho(t) = \rho_{11}(t) + \rho_{22}(t) + \sum_{\underline{k}s} \rho_{33}(t) = 1 \ . \tag{18}$$

The expectation value of the atomic population inversion operator is then

$$\begin{aligned}s^z(t) = <S^z>_t = \mathrm{Tr}[S^z\rho(t)] &= (-N/2 + 1)\rho_{11} - \\ - N/2(\rho_{22} + \sum_{\underline{k}s}\rho_{33}) &= \rho_{11} - N/2 \ ,\end{aligned} \tag{19}$$

and the squared population inversion

$$\langle S^zS^z\rangle_t = (-N+1)s^z(t) + N/2(-N/2+1), \tag{20}$$

By applying the method of Refs. 6 - 9 we can obtain an exact differential equation of the fourth order for the atomic population inversion $s^z(t)$. That is, by successive differentiations of equation

$$s^z(t) = \mathrm{Tr}\{ S^z \exp[-it(L_1 + L_I + i\Lambda_R)]\rho(0)\} , \tag{21}$$

we obtain

$$\frac{ds^z(t)}{dt} = - \mathrm{Tr}\{\sum_{\underline{k}s} g_{\underline{k}s}(a_{\underline{k}s}S^+ + a^+_{\underline{k}s}S^-)\rho_I(t)\} , \tag{22}$$

$$\begin{aligned}\frac{d^2s^z(t)}{dt^2} = & - N \sum_{\underline{k}s} g^2_{\underline{k}s}[2s^z(t) + N] - \varkappa \frac{ds^z(t)}{dt} - \\ & - 4 \sum_{\substack{\underline{k}s \\ \underline{k}'s'}} g_{\underline{k}s}g_{\underline{k}'s'} \mathrm{Tr}\,[S^z a^+_{\underline{k}s}a_{\underline{k}'s'}\, \rho_I(t)] + \\ & + i\Delta\omega \sum_{\underline{k}s} g_{\underline{k}s} \mathrm{Tr}\,[(a_{\underline{k}s}S^+ - a^+_{\underline{k}s}S^-)\rho_I(t)] ,\end{aligned} \tag{23}$$

where we used the relation

$$\mathrm{Tr}\,[(a_{\underline{k}s})^{\ell}(a^+_{\underline{k}s})^m \Lambda_R(\ldots)] = -(\ell + m)\varkappa\, \mathrm{Tr}\,[(a_{\underline{k}s})^{\ell}(a^+_{\underline{k}s})^m(\ldots)]$$

$$\ell,m = 0,1,2,\ldots \tag{24}$$

A further differentiation gives

$$\begin{aligned}\frac{d^3s^z(t)}{dt^3} = & - 3\varkappa \frac{d^2s^z(t)}{dt^2} - [2\varkappa^2 + 4N\sum_{\underline{k}s} g^2_{\underline{k}s} + (\Delta\omega)^2]\frac{ds^z(t)}{dt} - \\ & - 2\varkappa N\sum_{\underline{k}s} g^2_{\underline{k}s} [2s^z(t) + N] + \\ & + i\varkappa\Delta\omega\sum_{\underline{k}s} g_{\underline{k}s} \mathrm{Tr}\,[(a_{\underline{k}s}S^+ - a^+_{\underline{k}s}S^-)\rho_I(t)] ,\end{aligned} \tag{25}$$

where we have taken into account that

$$\begin{aligned} \mathrm{Tr}\,[a^{+}_{\underline{k}s} a_{\underline{k}'s'} a_{\underline{k}''s''}\, S^{+} \rho_I(t)] &= 0 , \\ \mathrm{Tr}\,[a^{+}_{\underline{k}s} a^{+}_{\underline{k}'s'} a_{\underline{k}''s''}\, S^{-} \rho_I(t)] &= 0 , \end{aligned} \tag{26}$$

which can easily be shown by using the fact that the time evolution of the density operator $\rho_I(t)$ is restricted to the subspace given by Eqs.(16) and (17).

Finally, by differentiation of Eq.(25) we obtain a closed differential equation of the fourth order:

$$\begin{aligned} \frac{d^4 s^z(t)}{dt^4} = &- 4\varkappa \frac{d^3 s^z(t)}{dt^3} - [5\varkappa^2 + (\Delta\omega)^2 + 4N\sum_{\underline{k}s} g^2_{\underline{k}s}] \frac{d^2 s^z(t)}{dt^2} - \\ &- \varkappa\,[2\varkappa^2 + 2(\Delta\omega)^2 + 8N\sum_{\underline{k}s} g^2_{\underline{k}s}] \frac{ds^z(t)}{dt} - \\ &- 4\varkappa^2 N \sum_{\underline{k}s} g^2_{\underline{k}s}\, [s^z(t) + N/2] . \end{aligned} \tag{27}$$

By using the abbreviations

$$g = \Big(\sum_{\underline{k}s} g^2_{\underline{k}s}\Big)^{1/2} , \quad a = \varkappa/gN^{1/2} , \quad b = \Delta\omega/gN^{1/2} , \tag{28}$$

and the new function

$$Z(\bar{t}) = s^z(t) + N/2 , \quad \bar{t} = gN^{1/2} t , \tag{29}$$

the exact Eq.(27) takes the form

$$\begin{aligned} &\frac{d^4 Z(\bar{t})}{d\bar{t}^4} + 4a \frac{d^3 Z(\bar{t})}{d\bar{t}^3} + (5a^2 + b^2 + 4) \frac{d^2 Z(\bar{t})}{d\bar{t}^2} + \\ &+ a(2a^2 + 2b^2 + 8) \frac{dZ(\bar{t})}{d\bar{t}} + 4a^2 Z(\bar{t}) = 0 \end{aligned} \tag{30}$$

with the initial condition (11)

$$Z(0) = 1, \quad \left.\frac{dZ}{d\bar{t}}\right|_{\bar{t}=0} = 0 , \quad \left.\frac{d^2 Z}{d\bar{t}^2}\right|_{\bar{t}=0} = -2 , \quad \left.\frac{d^3 Z}{d\bar{t}^3}\right|_{\bar{t}=0} = 2a. \tag{31}$$

3. ANALYTIC AND NUMERICAL SOLUTIONS

The analytic solution of Eq.(30) can be obtained by finding the roots of the polynomial

$$P_4(r) = r^4 + 4ar^3 + (5a^2 + b^2 + 4)r^2 + + a(2a^2 + 2b^2 + 8)r + 4a^2 = 0 . \tag{32}$$

By simple transformation: $r = x - a$, we obtain the biquadratic equation

$$x^4 + (b^2 - a^2 + 4)x^2 - a^2b^2 = 0 \tag{33}$$

with the roots

$$x_{1,2,3,4} = \pm(q_{1,2})^{1/2}, \quad q_{1,2} = \frac{1}{2}[-c \pm (c^2 + 4a^2b^2)^{1/2}] ,$$

$$c = b^2 - a^2 + 4 . \tag{34}$$

For $a \gg b$ the solution of Eq.(30) reads then:

(i) $a > 2$

$$Z(\bar{t}) = \exp(-a\bar{t})\{A_1 \exp[(a^2 - 4)^{1/2}\bar{t}] + + A_2 \exp[-(a^2 - 4)^{1/2}\bar{t}] + A_3\} , \tag{35a}$$

(ii) $a = 2$

$$Z(\bar{t}) = 2\exp(-2\bar{t})(1/2 + \bar{t} + \bar{t}^2) \tag{35b}$$

(iii) $a < 2$

$$Z(\bar{t}) = \exp(-a\bar{t})\{B_1 \cos[(4 - a^2)^{1/2}\bar{t}] + + B_2 \sin[(4 - a^2)^{1/2}\bar{t}] + B_3\} , \tag{35c}$$

where the constant coefficients are determined by the initial condition (31).

For $a \ll b$:

$$Z(\bar{t}) = \frac{1}{b^2 + 4} \exp(-a\bar{t})\{2 \cos[(b^2 + 4)^{1/2}\bar{t}] + b^2 + 2\} . \tag{36}$$

In Figs. 1-3 we plot the numerical solutions for the atomic population inversion $Z(\bar{t})$ and the radiation rate $I(\bar{t}) = -dZ(\bar{t})/d\bar{t}$ for different values of the cavity damping $a = \varkappa/gN^{1/2}$ and detuning $b = \Delta\omega/gN^{1/2}$. Moreover, in Fig.4 we plot the numerical solutions for the total photon number $N(\bar{t})$ in the cavity modes, which follow from the energy-balance equation

$$\frac{dN(\bar{t})}{d\bar{t}} = -2aN(\bar{t}) - \frac{dZ(\bar{t})}{d\bar{t}} \;, \quad N(\bar{t}) = \langle \sum_{\underline{ks}} a^{+}_{\underline{ks}} a_{\underline{ks}} \rangle_t \;. \tag{37}$$

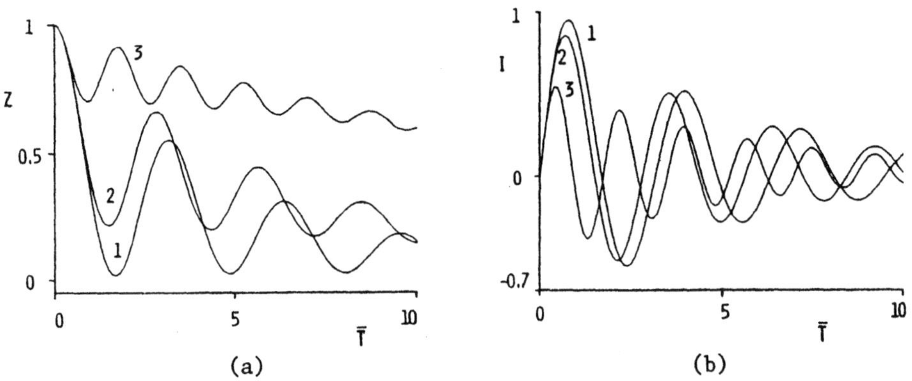

Fig. 1. (a) The atomic population inversion $Z(\bar{t})$ as a function of the scaled time $\bar{t} = gN^{1/2}t$ for the damping $a = \varkappa/gN^{1/2} = 0.2$ and the detuning $b = \Delta\omega/gN^{1/2} = 0,1,3$ (curves denoted by 1 – 3). (b) The time evolution of the radiation rate $I(\bar{t}) = -dZ(\bar{t})/d\bar{t}$ for the same parameters.

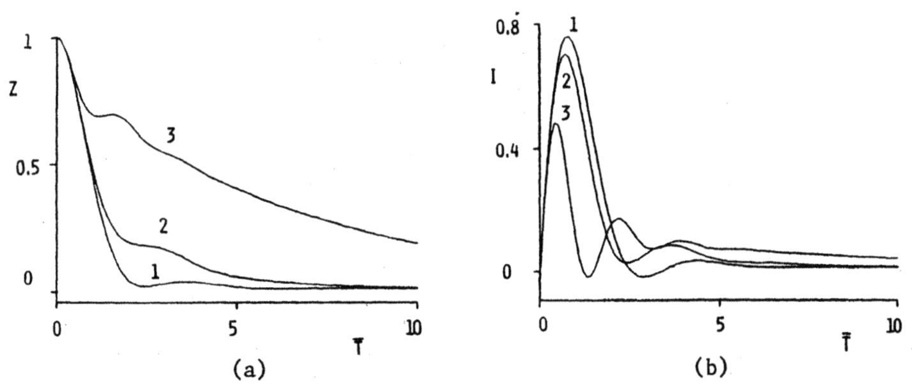

Fig. 2. The same as Fig.1 for a = 1.

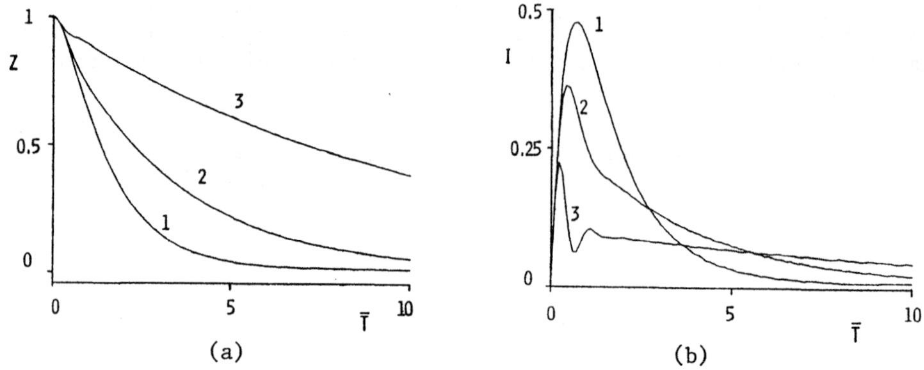

Fig. 3. The same as Fig.1 for a = 3 and b = 0,3,7 (curves 1-3).

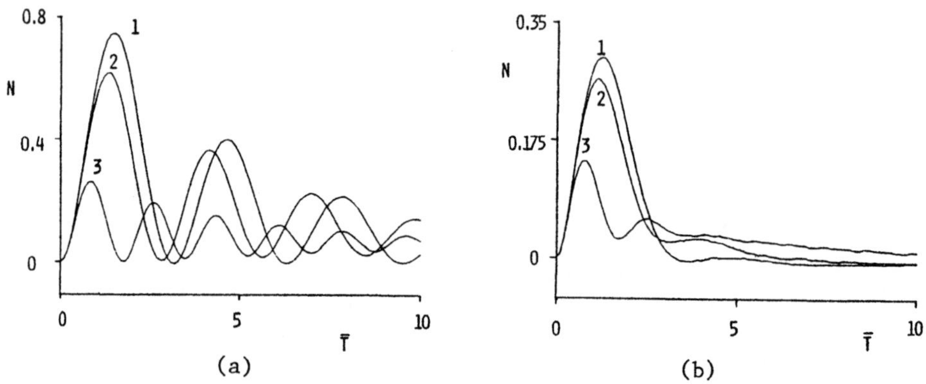

Fig. 4. The time evolution of the total photon number $N(\bar{t})$ for b = 0, 1,3 (curves 1 - 3) , (a) a = 0.2 and (b) a = 1.

4. CONCLUSION

By using our new technique, [6-9] we were able to solve the spontaneous emission from N identical two-level atoms being initially in a symmetrical Dicke state with only one atom excited and interacting with a three-dimensional multimode radiation field in a damped cavity with frequencies lying in a very narrow range about some mean frequency ω_{k0}. This model is a generalization of the single-atom spontaneous emission, since all our results reduce to the single-atom case for N = 1.

As we see from exact analytic and numerical results the spontaneous emission depends significantly on the detuning b. The larger the detuning is (compared to the cavity damping a), the smaller the radiation rate becomes (cf. curves 2 and 3 in Figs. 1 - 4).

For $a \gg b$ the detuning can be neglected and we obtain Eqs. (35), which in the case of low-Q cavities with $a \gg 1$ reduce to

$$Z(\bar{t}) = \exp(-2\bar{t}/a) = \exp(-2g^2 Nt/\varkappa). \tag{38}$$

In this case the emitted photon is not stored long enough in the cavity and cannot be reabsorbed by the atoms. The atomic population inversion decays exponentially increased by a factor of N in the exponent as compared to the single-atom case. This means that the spontaneous emission from an assembly of N atoms shows a collective behaviour.

The detuning b plays a significant role if it is larger than the cavity damping (cf. Figs. 1 - 4). For $b \gg a$ the time-evolution of the atomic population inversion undergoes weakly damped cosine oscillaltions (cf. Eq. (36) and Fig. 1).

From the obtained closed differential equation (30) it can be concluded that the solution for a three-dimensional multimode radiation field with a very narrow frequency range in a damped cavity is analogous to the single-mode case with the difference that the square of the coupling coefficient corresponding to the single mode is replaced by a sum over the square of the coupling coefficients $g_{\underline{k}s}$ corresponding to different modes.

ACKNOWLEDGMENT

This work was supported by the Austrian Fonds zur Förderung der wissenschaftlichen Forschung, under Grant No. P5658.

REFERENCES

1. J. M. Raimond, P. Goy, M. Gross, C. Fabre, and S. Haroche, Phys.Rev. Lett. 49, 1924 (1982).
2. P. Goy, J. M. Raimond, M. Gross, and S. Haroche, Phys. Rev. Lett. 50, 1903 (1983).
3. Y. Kaluzny, P. Goy, M. Gross, J. M. Raimond, and S. Haroche, Phys. Rev. Lett. 51 , 1175 (1983).
4. D. Meschede, H. Walther, and G. Müller, Phys. Rev. Lett. 54, 551 (1985).
5. R. H. Dicke, Phys. Rev. 93 , 99 (1954).
6. J. Seke, Lett. Nuovo Cimento 43 , 103 (1985).
7. J. Seke, J. Opt. Soc. Am. B 2 , 968 (1985).
8. J. Seke, J. Opt. Soc. Am. B 2 , 1687 (1985).
9. J. Seke, Phys. Rev. A 33, No 1 (in print, 1986).
10. H. Haken, 'Laser Theory', in Handbuch der Physik, S. Flügge, ed. (Springer-Verlag, Berlin 1970), Vol. XXV/2c, pp. 51-60.

POLYNUCLEOTIDES - A CLASS OF SELF-ORGANIZING BIOMOLECULES

Peter Schuster
Institut für Theoretische Chemie und Strahlenchemie
der Universität Wien
Währingerstrasse 17
A-1090 Wien / Austria

The notion of self-organizing systems was created in the forties mainly through the theoretical studies by Turing (1936) and von Neumann (1951). Since that time, interest in problems of self-organization has never ceased. What has changed over the years, however, are the preferred objects of interest: the self-replicative automata, more or less abstract concepts in computational physics have been replaced by experimental systems from various fields. The exciting development of molecular biology and biophysics initiated by the ingenious postulate of the double helical structure of DNA through Watson and Crick (1953) revealed more and more data on the structure and properties of biomolecules. Eigen (1971) applied chemical kinetics to molecular self-organization and formulated a general theory of precellular evolution. Thanks to the pioneering work of Spiegelman (1971) self-replication of RNA, biomolecules closely related to DNA which appear as the genetic material of some bacteriophages, can be studied by the experimental techniques used commonly in physics and chemistry. Darwinian evolution may be investigated now at the molecular level.

Polynucleotides, RNA or DNA, are the only known molecules which have an intrinsic capability of self-replication. This property is a direct consequence of their molecular structure which is characterized by three basic and indispensible features:

(1) polynucleotides consist of a repetitive backbone with digits or "bases" occupying structurally equivalent positions;
(2) these bases are complementary in the Watson-Crick base pairs, A=U(T) and G≡C; and
(3) polynucleotides form double-helices with a uniform structure. Every base in this structure is uniquely defined by that in the opposite or vis-a-vis position. Thus, there is a uniquely defined negative for every given polynucleotide.

The molecular geometry of the famous double-helix provides the basis for complementary and for direct replication. Complementary replication starts from a given polynucleotide, the plus strand. A negative copy is synthesized sequentially, i.e. step by step or base by base from the 5' to the 3' end. The negative strand acts as a template

C. W. Kilmister (ed.), Disequilibrium and Self-Organisation, 165–168.

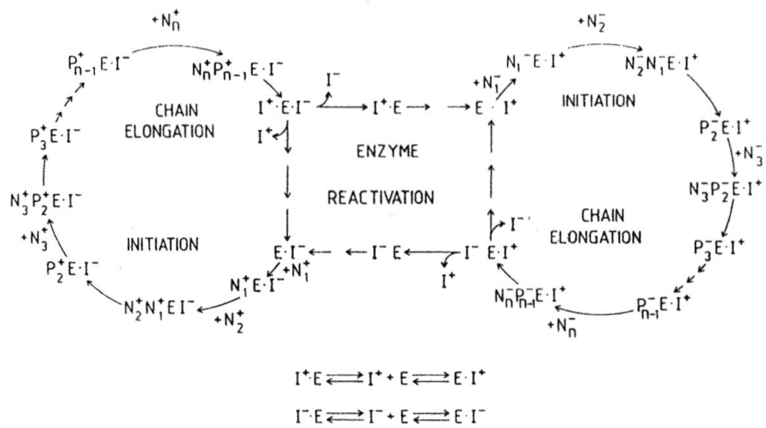

Fig. 1. A cyclic mechanism for bacteriophage RNA replication in vitro as used by Biebricher et al. (1981) to interpret the kinetic results obtained with the Qβ system. E, I^+ and I^- represent free Qβ polymerase, and the plus and minus strand of the RNA to be replicated. N_1^+, N_2^+, ..., N_n^+ and N_1^-, N_2^-, ..., N_n^- are the nucleoside triphosphates in a sequence as they appear in the plus and minus strand respectively. The chain of the newly synthesized molecule grows always from the 5'- to the 3'-end. P_2^+, P_3^+, ..., P_{n-1}^+ and P_2^-, P_3^-, ..., P_{n-1}^- are used as symbols for the growing chains. A nucleation length of two bases is assumed. Enzyme reactivation is necessary because the two polynucleotides are not in suitable positions to restart polymerization: the newly synthesized strand is bound with the correct end (3') but in the wrong site (synthetic site) whereas the old strand is sitting in the correct (reading) site but bound with the wrong end (5'). The many step mechanism thus consists of three distinguishable processes: initiation of RNA synthesis, chain propagation and reactivation of the enzyme.

for the synthesis of a new plus strand molecule. Complementary replication is an example for an autocatalytic process with a unique intermediate, the minus strand. Examples of complementary replication are viral RNA synthesis by specific enzymes (Fig. 1), the mechanism of which has been studied extensively (Biebricher et al., 1981 and 1982) as well as template induced, enzyme free oligo-ribonucleotide synthesis (Lohrmann et al., 1980). Direct replication occurs with DNA. The molecular mechanism is complicated and involves about ten different enzymes. The local topology of the process is summarized in the so-called replication fork (Kornberg, 1980). Despite the highly complex molecular mechanism RNA and DNA replication obey the law of simple autocatalysis under suitable conditions (Biebricher et al., 1981). A characteristic example is shown in Fig. 2.

Chemical kinetics of ordinary, i.e. non-autocatalytic, chemical reactions has been investigated thoroughly by various techniques. In particular we mention the formal mathematical approach by Feinberg (1977) and Horn. Although, their basic formalism is very general, the important theorems are only valid for non-autocatalytic reaction steps. Despite various efforts to develop a fully general theory of chemical kinetics - as one of the most promising approaches we mention the work by Clarke (1980)- no really satisfactory and generally enough treatment of autocatalysis has been presented yet. The reason for this lack is easy to understand: autocatalytic processes lead to highly nonlinear differential equations and these are hard to analyse by the conventional techniques. Despite this lack of a general formal theory, many special systems have been analyzed thoroughly. Examples are given in Eigen and Schuster (1979), Schuster and Sigmund (1983) and Küppers (1983). The approach towards self-organization by chemical kinetics can be extended smoothly to problems of biochemical and biological evolution as well as to the construction of models for the origin of life (Eigen, 1971; Schuster, 1981; Eigen et al., 1981; Eigen and Schuster, 1982). The last paper is added as an appendix to this contribution.

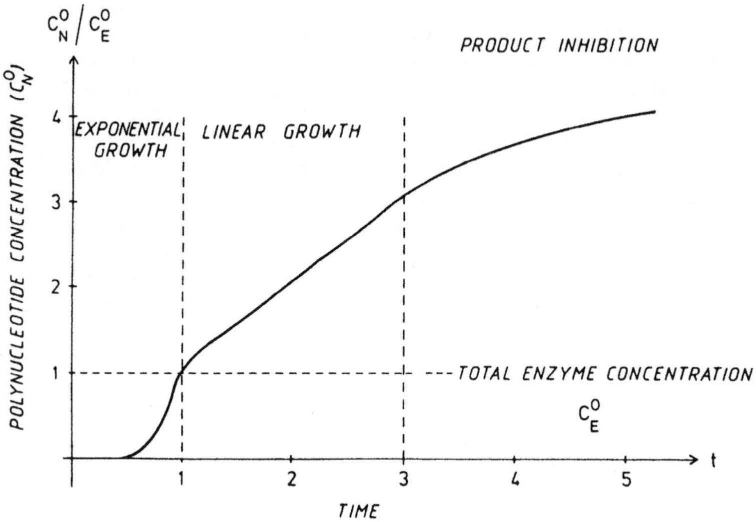

Fig. 2. RNA synthesis in the test tube. A characteristic experimentally recorded curve shows three phases of growth: exponential growth at low template concentration, linear growth above saturation of the enzyme by polynucleotides and further levelling off at still higher concentrations of template when enzyme reactivation becomes the rate determining step.

References:

Biebricher, C.K., Eigen, M., Luce, R.: J. Mol. Biol. 148, 369 and 391 (1981)

Biebricher, C.K., Diekmann, S., Luce, R.: J. Mol. Biol. 154, 629

Clarke, B.L.: Adv. Chem. Phys. 43,1 (1980)

Eigen, M.: Naturwissenschaften 58, 465 (1971)

Eigen, M., Schuster, P.: The Hypercycle. Springer-Verlag, Berlin 1979

Eigen, M., Gardiner, W.C., Schuster, P., Winkler-Oswatitsch, R.: Sci. Am. 244 (4), 88 (1981)

Eigen, M., Schuster, P.: J. Mol. Evol.: 19, 47 (1982)

Feinberg, M.: Mathematical Aspects of Mass Action Kinetics. In: Chemical Reactor Theory, L. Lapidus, N.R. Amundson, eds. pp. 1-78, Prentice Hall, Englewood Cliffs, N.J. 1977

Kornberg, A.: DNA Replication. W.H. Freeman. San Francisco 1980

Küppers, B.O.: Molecular Theory of Evolution. Springer-Verlag, Berlin 1983

Lohrmann, R., Bridson, P.K., Orgel, L.E.: Science 208, 1564 (1980)

Schuster, P.: Prebiotic Evolution. In: Biochemical Evolution, H.Gutfreund ed. pp. 15-87, Cambridge Univ. Press, Cambridge U.K. 1981

Schuster, P., Sigmund, K.: From Biological Macromolecules to Protocells-The Principle of Early Evolution. In: Biophysics. W. Hoppe, W. Lohmann, H.Markl, H. Ziegler eds. pp. 874-912. Springer-Verlag, Berlin 1983

Stages of Emerging Life – Five Principles of Early Organization*

M. Eigen[1] and P. Schuster[2]

1 Max Planck Institut für Biophysikalische Chemie D–34 Göttingen, Federal Republic of Germany
2 Institut für Theoretische Chemie und Strahlenchemie der Universität Wien, A 1090 Wien, Austria

Summary. Five principles underly the evolution of the genetic language: formation of stereoregular heteropolymers, selection through self-replication, evolution of quasispecies towards optimal structures, regulated co-operation between competitors through catalytic hypercycles and evaluation of translation products through compartmentalization. These principles are formulated and illustrated by means of experimental results

Key words: Prebiotic evolution – Polynucleotide replication – Quasispecies – Hypercycles – Compartmentalization

Introduction

New fundamental concepts emerge in the transition from non-living to living matter. Biological information and the machinery to process it originate in this early period of evolution. Ensembles of macromolecules organize themselves and develop the genetic language. In the recent past experiments have been made which examine the forces that have driven this early development and that are still in operation at present under modified and more subtle conditions. The results of these experiments put contraints upon theories of the origin of life and bring them down from the realm of pure speculation to the more solid grounds of experimental tests.

The enormous variety of studies in prebiotic chemistry is reviewed in other contributions to this journal, particularly in the special issues dedicated to the memory of H.C. Urey (Oró and Miller 1981, 1982). Through these systematic investigations the identity of the most abundant molecules on the very early Earth became clear. Most of the natural amino acids were among them. The building blocks of polynucleotides – the four bases, ribose and phosphate – were available too under prebiotic conditions. An enormous amount of low molecular weight organic material and energy-rich compounds was synthesized in those early days. Material was provided from steadily refilling pools for the formation of polymers, among them polypeptides and polynucleotides. The appearance of these molecules brought into operation new principles of organization.

1. Formation of Heteropolymers Introduces an Otherwise Unknown Richness into the Structures and Properties of Molecules

The two most important classes of biopolymers, the polypeptides and the polynucleotides, share a common building principle: a regular, repetitive backbone with variable side chains. The structural concept of such heteropolymers allow a "hyper-astronomically" large number of different molecules to be made even for moderate polymer lengths. (This number is 20^{ν} for natural polypeptides and 4^{ν} for polynucleotides, when ν is the number of segments. For a polynucleotide with $\nu = 300$, a fragment which is just large enough to encode for one of the smallest proteins, 4.10^{180} different sequences are possible, while there are e.g. "only" 10^{11} stars in our galaxy and the total observed matter in the universe corresponds to something like 10^{80} hydrogen masses. In ordinary chemistry we deal with much smaller numbers of different molecules: "Beilstein" lists about 10^{7} organic molecules and Avogadro's number is only 6.10^{23}.

In order to visualize this special aspect of biopolymers thermodynamically we assume a set of polymers with the same chain length ν and the same average com-

* Dedicated to the memory of Harold C. Urey (1893–1981)

Offprint requests to: P. Schuster

Published from, J. Mol. Evol. (1982), 19: 47–61

C. W. Kilmister (ed.), Disequilibrium and Self-Organisation, 169–183.

position. The stabilities of the sequences will differ little, and to this degree only because they have different secondary and tertiary structures. We are thus dealing with a huge number of almost equally stable hypothetical molecules. Every sequence corresponds to a local potential energy minimum, or to a set of minima for different conformations. Minima belonging to different sequences are separated by rather high energy barriers since at least two chemical bonds have to be broken in order to convert one particular sequence into another. At room temperature no interconversions between sequences occur. The material available in a single experiment (or on the entire Earth or even in the universe) would not allow one to synthesize even a single molecule of every possible sequence. Thus, only a tiny fraction of what is theoretically possible is really accessible.

Experimental evidence for polymerization under prebiotic conditions has been provided by several authors. Sidney Fox and coworkers were able to produce polymers by cooking aqueous solutions of amino acids and heating to dryness. These so-called "protenoids" have a limited repertoire of primitive cytalytic properties (Fox 1973). Leslie Orgel and his group have presented a variety of experimental results that show how spontaneous formation of polynucleotides might have occurred under prebiotic conditions. These experiments showed in particular that some divalent metal ions, in particular Zn^{2+}, have strong catalytic effects. Polycondensation of activated ribonucleosides in presence of Zn^{2+} or Pb^{2+} ions gives a much higher yield of regular linkages (3′–5′) than can be obtained in the absence of these ions. In a recent series of experiments they were able to study the influence of a polymer template on polymer synthesis (Lohrmann et al. 1980). They polymerized activated guanylic acid in the presence of a poly (C) template. Again Zn^{2+} ions play an important role. In their presence oligo (G) and poly (G) with chain lengths up to $\nu = 50$ and high predominance of 3′–5′ linkages (85%) wee obtained. These studies on template-induced polymerization of activated nucleotides are a strong indication that polynucleotide replication occurred even in enzymefree systems.

2. Autocatalysis or Self-Replication Under Conditions Far from Equilibrium Introduces Selection into Molecular Ensembles

Polynucleotides are the only known biomolecules which can act as templates for the replication. They can be regarded as "constitutional" autocatalysts since this unique property is a result of their molecular structure. The complementarity of base pairs allows the synthesis of a unique negative copy of each polynucleotide. Replication of this negative then yields a second copy of the original strand. A plus-minus ensemble thus behaves very much like a simple autocatalytic system. The kinetics

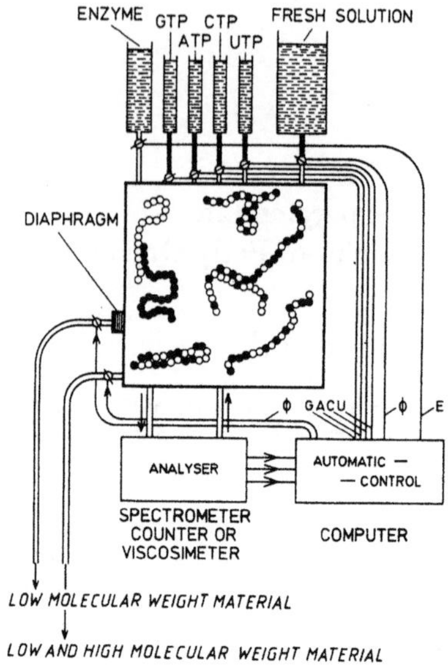

Fig. 1. The evolution reactor. This kind of flow reactor consists of a reaction vessel which allows for temperature and pressure control. Its walls are impermeable to polynucleotides. Energy rich material is poured from the environment into the reactor. The degradation products are removed steadily. Material transport is adjusted in such a way that the concentration of monomers is constant in the reactor. A dilution flux ϕ is installed in order to remove the excess of polynucleotides produced by replication. Thus the sum of the concentrations

$$[I_1] + [I_2] + ... + [I_n] = \sum_i c_i = c$$

may be controlled by the flux ϕ. Under "constant organization" ϕ is adjusted such that $c = c_0$ is constant. The regulation of ϕ requires internal control, which may by achieved by analysis of the solution and data processing by a computer as indicated above

of plus-minus replication has been analysed in detail (Eigen 1971; Eigen and Schuster 1979, p. 13). A straightforward result was obtained: after an initial phase of internal equilibration, the plus-minus ensemble grows as a single autocatalytic entity.

Competition between different self-replicating units, i.e. different sequences of RNA or DNA can be studied most easily under idealized conditions that could be realized in an "evolution reactor" (Fig. 1). An evolution reactor is an open system with adjustable flows of low molecular weight material. The composition of the solution in the reactor can be controlled by means of a general dilution flux ϕ. Suppose n different replicating molecules, $I_1, I_2, ..., I_n$ are present in the reactor at concentrations $[I_1] = c_1, [I_2] = c_2, ..., [I_n] = c_n$. The total concentration is then $c = \Sigma c_i$. The dynamics of processes in the reactor are described by the differential equations

$$\frac{dc_1}{dt} = \dot{c}_1 = (f_1 - d_1 - \frac{1}{c}\phi)c_1 = (E_1 - \frac{1}{c}\phi)c_1$$

$$\frac{dc_2}{dt} = \dot{c}_2 = (f_2 - d_2 - \frac{1}{c}\phi)c_2 = (E_2 - \frac{1}{c}\phi)c_2 \qquad (1)$$

$$\cdot$$
$$\cdot$$
$$\cdot$$

$$\frac{dc_n}{dt} = \dot{c}_n = (f_n - d_n - \frac{1}{c}\phi)c_n = (E_n - \frac{1}{c}\phi)c_n$$

Here f_i denotes the number of molecules I_i synthesized per time unit at unit concentration ($c_i = 1$), d_i similarly denotes the rate of decomposition, usually by hydrolysis. The quantity $E_i = f_i - d_i$ is called the excess production. The rates f_i, d_i and E_i will be constant when the reactor is run under constant internal conditions, i.e. constant temperature, constant pressure and constant concentrations of low molecular weight materials.

Differential equation (1) is non-autonomous since the total concentration c(t) is a function of time that depends on the flux $\phi(t)$ (Eigen and Schuster 1979, p. 31, 32 and 35):

$$c(t) = c(0) + \int_0^t \{\sum_i E_i c_i - \phi(\tau)\}\, d\tau \qquad (2)$$

With a simple transformation we can eliminate the time dependent total concentration c(t): we introduce relative concentrations $x_i = c_i/c$, $(\Sigma x_i = 1)$ and find

$$\dot{x}_i = (E_i - \bar{E})x_i;\ i = 1, 2, ..., n \qquad (3)$$

$\bar{E}$ is the mean excess production $\Sigma E_i x_i$. The solution of equation (3) can be visualized easily (Fig. 2). The relative concentrations of those polynucleotides which fulfil $E_i > \bar{E}$ at a certain instant $t = t_L$ will increase whereas those with $E_i < \bar{E}$ will decrease. $\bar{E}$ increases as a consequence of these concentration changes. At $t > t_1 + \Delta t$ more molecules fall below the critical limit $\bar{E}$. Since the mean excess production $\bar{E}(t)$ is a monotonously increasing function, more and more molecules are taken out of the competition. Ultimately only the polynucleotide with the largest excess production E_m remains. Then $\bar{E}$ also reaches its maximum value $\lim_{t\to\infty} \bar{E} = E_m$, and "selection of the fittest" takes place. Selection brings order into the enormous diversity of the initial formation of heteropolymers.

We did not apply a physically sensible constraint to the time program of the dilution flux. Equation (3) and the results derived from it hold for any $\phi(t)$. Selection between selfreplicating polynucleotides takes place under stationary conditions as well as in growing or shrinking populations. The only assumption which has to be made does not concern $\phi(t)$; it is simple $\bar{E} > 0$, which means that the formation rate has to be larger than the mean decay rate (Eigen and Schuster 1979, pp. 35, 36).

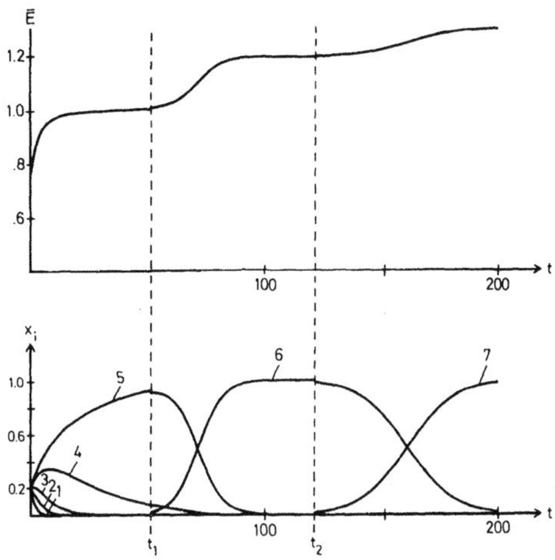

Fig. 2. Selection and optimization in an ensemble of polynucleotides. We present solution curves of the differential equation (3) $\dot{x}_i = x_i (E_i - \bar{E})$, i = 1, 2, ..., 7 with $E_1 = 0.3$, $E_2 = 0.65$, $E_3 = 0.82$, $E_4 = 0.96$, $E_5 = 1.01$, $E_6 = 1.2$ and $E_7 = 1.3$ in arbitrary time units and $\bar{E} = \Sigma E_i x_i / \Sigma x_i$. The initial conditions are: $x_1(0) = x_2(0) = ... = x_5(0) = 0.2$ and $x_6(0) = x_7(0) = 0$. The mean excess productivity $\bar{E}$ starts from an initial value of $\bar{E} = 0.748$ and increases steadily as the population becomes homogeneous: $\bar{E} \to 1.01$ and $x_1, x_2, x_3, x_4 \to 0$ as $x_5 \to 1$. One easily recognizes the cause for the increase of $\bar{E}$: The less efficiently growing polynucleotides are eliminated. They disappear in the same sequence as their excess productivity E_i increases, namely 1, 2, ... 5. At $t = t_1 = 50$ we observe a fluctuation $x_6 = 0 \to x_6 = \delta$. The appearance of I_6 as a favourable mutant leads to a further increase in $\bar{E}$, which approaches the value $\bar{E} = 1.2$ as I_5 disappears and the population becomes homogeneous again: $x_1 = ... = x_5 = 0$, $x_6 = 1$. The same story happens when a more efficient mutant appears at $t = t_2 = 120$. I_6 is replaced by I_7 and $\bar{E}$ approaches the temporary optimum $\bar{E} = 1.3$. This example illustrates the nature of the selection process and the role of the mean excess productivity $\bar{E}$ as the quantity which is optimized

Experimental studies on selection between replicating polynucleotides were performed first by Sol Spiegelman more than ten years ago (Spiegelman 1971). He replicated RNA of the bacteriophage Qβ in serial transfer experiments (Fig. 3). The stock solution contained the specific enzyme Qβ replicase as well as the four nucleoside triphosphates, GTP, ATP, CTP and UTP. Starting out from infectious RNA strands, sequences were obtained through serial transfer which replicated faster and had lower molecular weights than the original molecules. The overall rate of synthesis, comparable to $\bar{E}$, increased steadily during the sequence of transfers. The more rapidly replicating polynucleotides were unable to infect bacteria, since the constraints applied in serial transfer experiments select exclusively for higher replication rates. The

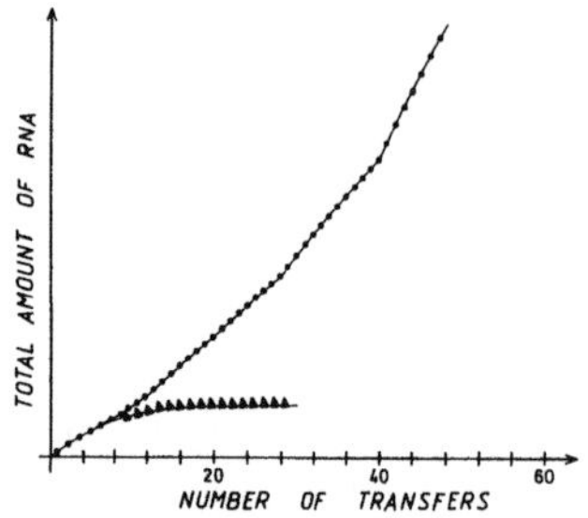

Fig. 3. Serial tranfer experiments on RNA replication by Qβ- replicase according to Spiegelman (1971) RNA replication in the test tube is initiated by addition of Qβ RNA. Two measurements are recorded: the number of infectious particles (▲) and the amount of nucleoside triphosphate incorporated into RNA, a measure of total polynucleotide synthesis (●). The rate of RNA synthesis increases spontaneously in jumps. The percentage of infectious polynucleotides, however, becomes smaller and smaller. New RNA molecules were formed which grow faster but are no longer infectious. Since serial transfer selects for excess productivity only, infectiousness is readily lost. Several spontaneous, stepwise increases in the rate of RNA synthesis are observed until an optimal value is reached.

mechanism through which the more rapidly replicating variants were formed leads us to the next principle.

3. Replication Errors Lead to Mutant Distributions Called Quasispecies. The Accuracy of the Replication Process Sets a Limit to the Amount of Genetic Information that Can Be Transmitted

No physical process can proceed with ultimate accuracy. Hence, we have to account for replication errors in polynucleotide reproduction. The incorporation of mutation into the deterministic kinetic equations of selfreplication is straightforward (Eigen 1971; Eigen and Schuster 1979, pp. 8–15). We account for error copies by means of a quality factor Q, which is used here in a manner slightly different from our earlier papers. It can be expressed as a matrix

$$Q = \{Q_{ij}\};\ i,j = 1,...,n \qquad (4)$$

which characterizes the synthesis of a distribution of polynucleotides from each template. Q_{ii} is the fraction of copies of template I_i which are free of errors and Q_{ji} the fraction of molecules I_j obtained as mutants of I_i. A conservation law of probabilities holds: $\sum_j Q_{ji} = 1$, as every copy has to be either correct or erroneous. The rate constants for error-free replication and mutant formation are obtained by multiplication with the total synthetic efficiency of the template (for template I_i we use f_i as before). Additionally, we account for degradation (d_i) in the diagonal terms

$$\omega_{ii} = f_i Q_{ii} - d_i \text{ and } \omega_{ji} = f_i Q_{ji} \text{ for } i \neq j$$

The kinetic equations for polynucleotide replication under the condition of an evolution reactor are then of the form

$$\dot{x}_i = (\omega_{ii} - \bar{E})x_i + \sum_{j \neq i} \omega_{ij} x_j;\ i = 1, ..., n \qquad (5)$$

The mean excess production $\bar{E}$ is as before $\sum_i (f_i - d_i)x_i$. Note that equation (5) converges to the simple selection equation (3) if Q approaches the unit matrix or, in other words, in case the replication process becomes absolutely correct[1].

Equation (5) can be converted into a linear differential equation by means of suitable integral transformaton (Thompson and McBride 1974; Jones et al. 1976). In case we are interested in relative concentrations as we are here, common factors cancel and the system is completely determined by the differential equation

$$\dot{z}_i = \sum_{j=1}^{n} \omega_{ij} z_j;\ i = 1, ..., n \qquad (5a)$$

with $x_i(t) = z_i(t) \exp\{-\int_0^t \bar{E}(\tau)\, d\tau\}$. This equation can be studied by standard techniques of linear algebra. Ultimately we are dealing with an n-dimensional eigenvalue problem. Apart from kineticly degenerate cases there always exists a largest eigenvalue λ_{max} which is related to the largest selective value ω_{mm} (ω_{ii}; i = 1, ..., n). As in the error-free case an optimization process $\bar{E}(t) \to \lambda_{max}$ takes place. The eigenvector corresponding to λ_{max} represents the "long-lived" mode of the differential equation (5). After long enough time all the other modes have died out and the mixture of polynucleotides or mutant distribution, described by this eigenvector is all that remains. We have called it a "quasispecies" in order to indicate the analogy to the notion of a species in biology.

In order to learn more about the nature of the "quasispecies" distribution we distinguish two cases: (1) high and (2) low accuracy of the replication process. At sufficiently high accuracy the quasispecies is the stable long-term distribution of polynucleotide sequences. This distribution is dominated by the most efficiently replicating molecule, that is the master sequence I_m, with $\omega_{mm} = \max(\omega_{ii})$, which is accompanied by its most frequent and most efficient mutants.

Approximate formulas for λ_{max} and the relative frequencies of master sequence and mutants in the steady state have been obtained by perturbation theory (Eigen and Schuster 1979, p. 13; Thompson and McBride 1974;

---

1 Part of this section is more technical than the previous ones. The reader not interested in these details may continue with the paragraphs following equation (9)

Jones et al. 1976). These approximations describe ensembles of polynucleotides above a certain chain length ($\nu > 20$) quite well (Swetina and Schuster 1982). The most important result for our present concern is the existence of a sharply defined threshold for the accuracy of replication

$$Q_{mm} > Q_{min} = \sigma_m^{-1} \tag{6a}$$

The superiority σ_m of the master sequence contains the relevant kinetic parameters

$$\sigma_m = \frac{f_m}{d_m + \bar{E}_{-m}} \tag{6b}$$

wherein

$$\bar{E}_{-m} = \sum_{\substack{i=1 \\ i \neq m}}^{n} E_i x_i \Big/ \sum_{\substack{i=1 \\ i \neq m}}^{n} x_i$$

is the mean excess production in absence of the master sequence.

Now let us study the quality factor of the replication process more closely. Template induced replication proceeds digit per digit[2]. Therefore we can assign a single digit accuracy to every propagation step of the growing chain. For a polymer with ν segments the quality factor Q can be written as a product of ν individual factors. For the correct replication of a given polynucleotide say I_k we obtain:

$$Q_{kk} = q_1^{(k)} \cdot q_2^{(k)} \dots q_\nu^{(k)} = \bar{q}_k^{\nu} \tag{7}$$

Herein q_1 is the accuracy for the incorporation of the first base, q_2 that for the incorporation of the second base etc. In general the single digit accuracy depends on the nature of the base to be incorporated, on the neighboring base pairs, on the mechanism of replication and on environmental factors. Nevertheless, we can define a mean single digit accuracy $\bar{q}_k$ which accounts for all these influences implicitly and which is characteristic for a given sequence I_k. For long enough natural polynucleotides with similar base compositions these mean single digit accuracies will mainly depend on the mechanism of replication since specific neighbour effects cancel out in long sequences. For a given mechanism of replication the limiting case of a minimum accuracy is equivalent to a maximum chain length ν_{max}:

$$Q > Q_{min} = \bar{q}^{\nu_{max}} = \sigma^{-1}$$

2 For the sake of simplicity we dispense with a detailed discussion of other sources of replication errors besides point mutations. In principle deletions or insertions could be incorporated into the model as well

or

$$\nu_{max} = -\frac{\ln \sigma}{\ln \bar{q}} \cong \frac{\ln \sigma}{1 - \bar{q}} \tag{8}$$

Equation (8) has been used to explain the lengths of genomes in primitive organisms like RNA viruses and bacteria (Eigen and Schuster 1979, pp. 17–24). In the case of RNA viruses, moreover, the existence of nucleotide sequence heterogeneities as predicted by the concept of a quasispecies has been proved through cloning experiments (Domingo et al. 1978, 1980; Ortin et al. 1980; Fields and Winter 1981).

In order to apply equation (8) to the development of early replicating systems under prebiotic conditions we may use the results obtained in Leslie Orgel's laboratory (Lohrmann et al. 1980). He and his coworkers determined the accuracy of enzyme-free oligo (G) synthesis on a poly (C) template. They used a 50 : 50 mixture of activated A and G and found an incorporation of A:G in the ratio 1 : 200, which is equivalent to a single digit quality factor for G of $q_\nu \cong 0.995$ under these conditions. What they have determined, of course, is not the mean single digit accuracy ($\bar{q}$) of enzyme-free replication which we would like to know, but an upper limit for it: the interaction of the GC base pair is stronger than that of the AU pair and hence we have to expect a lower quality factor in the latter case. The maximum chain length for replication in enzyme-free systems thus can be estimated to lie around $\nu_{max} \cong 100$ bases or somewhat lower. The molecules which can be replicated with sufficient fidelity are of the size of the tRNA's.

Equation (7) is also useful for estimating the frequencies of mutants. For this we have to assign single digit mutation frequencies to the various base exchange processes. A systematic way to do this is shown in Table 1. Again we have to assume that neighbouring base pairs influence the muation frequencies, which then are specific for the replacement of the nucleotide at position "i". Following Table 1 we denote the three possibilities by α_i, β_i, and γ_i. In absence of insertions and deletions we have the conservation relation

$$q_i + \alpha_i + \beta_i + \gamma_i = 1$$

which expresses the fact that one of the four bases must occupy position "i" in the replica. For the correct copy and the three possible mutations at position "i", say $I_k \to I_l$, $I_k \to I_m$, $I_k \to I_p$, we thus have the following distributions of quality factors and mutation frequencies

$$I_k \to I_k : Q_{kk} = q_1 q_2 \dots q_i \dots q_\nu$$
$$I_k \to I_l : Q_{lk} = q_1 q_2 \dots \alpha_i \dots q_\nu$$
$$I_k \to I_m : Q_{mk} = q_1 q_2 \dots \beta_i \dots q_\nu \quad \text{and}$$
$$I_k \to I_p : Q_{pk} = q_1 q_2 \dots \gamma_i \dots q_\nu$$

Table 1. Systematics of point mutations

	Base in the mutant					
Original base	Purine-Purine Pyrimidine-Pyrimidine (Transition)[a]	Frequency	Intrapair exchange (Transversion)[a]	Frequency	Intrapair exchange (Transversion)[a]	Frequency
G	A	α (G)	C	β (G)	U	γ (G)
A	G	α (A)	U	β (A)	C	γ (A)
C	U	α (C)	G	β (C)	A	γ (C)
U	C	α (U)	A	β (U)	G	γ (U)

a Notation of Volkenstein (1979)

Now let us consider a realistic situation as observed with RNA replication in bacteriophages. In this particular example purine-purine or pyrimidine-pyrimidine exchanges are the most probable mutations by far. We may account for this situation by the approximative assumption

$$1 - q \cong \alpha >> \beta, \gamma$$

In this case it is possible to calculate the frequency factor Q_{jk} for a certain mutation of this kind ($I_k \rightarrow I_j$) from the Hamming distance D_{jk} of these two sequences. (The Hamming distance is the smallest number of base exchanges with which it is possible to convert a given sequence into another.) Again we use mean single digit accuracies ($\bar{q}_k$) and mutation frequencies ($\bar{\alpha}_k \cong 1 - \bar{q}_k$)

$$I_k \rightarrow I_j : D_{jk} = d \rightarrow Q_{jk}\ \bar{q}_k^{\nu-d}\ \ \bar{\alpha}_k^d \cong \bar{q}_k^{\nu-d}\ (1 - \bar{q}_k)^d$$

Remembering that $\bar{q}$ is close to one and hence $1 - \bar{q}$ is very small we realize a coarse graduation of mutation frequencies: one error mutations are formed with higher frequencies than two error mutations, three error mutations have still lower frequencies, etc.

The concentration of any mutant in the quasispecies is determined by a superposition of the corresponding mutation frequency and the relevant rate constants. Perturbation theory leads to an approximate formula for the ratios of concentrations within the quasispecies

$$\bar{x}_i = \bar{x}_m \cdot \frac{Q_{im} \cdot f_m}{\omega_{mm} - \omega_{ii}}\ ; \ i = 1, .., n;\ i \neq m \qquad (9)$$

where the stationary concentrations of the master sequence and some mutant are denoted by $\bar{x}_m$ and $\bar{x}_i$ respectively. By "m" we denote tha master sequence as above. Before we turn to the low accuracy case we consider the fraction of a master sequence within this quasispecies as a function of the accuracy of replication (Fig. 4). We realize a steady decrease of this fraction ($\bar{x}_m$) until it becomes exceedingly small at the critical value Q_{min}.

At low replication accuracy the relative frequencies of the individual mutants are no longer determined by the kinetic constants. The selective values ω_{ii} then have no influence on the long term polynucleotide distributions. These are determined instead exclusively by the

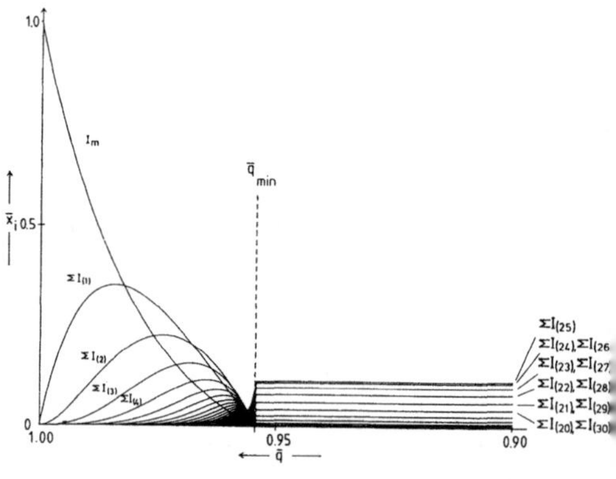

Fig. 4. Stationary mutant distribution for the prevailing conditions (quasispecies) in an replication ensemble of polynucleotides. The curves represent the relative concentration of the master sequence (I_m), the sum of the relative concentrations of all one error mutants ($\Sigma I_{(1)}$), of all two error mutants ($\Sigma I_{(2)}$) etc. as functions of the mean single digit replication accuracy ($\bar{q}$). In this example the polynucleotides length is $\nu = 50$. The formation rate constant for the master sequence is chosen to be $f_m = 10$, all other sequences $f_1 = f_2 = ... = 1$ in arbitrary time and concentration units. All decomposition rate constants including that of the master sequence are set equal and hence do not enter into the differential equation (5). From these parameter values we calculate $\sigma_m = 10$ and a critical single digit accuracy of $\bar{q}_{min} = (Q_{min})^{1/\nu} = 0.945$ according to equation (8). With decreasing replication accuracy a pronounced decrease in the relative concentration of the master sequence is observed. One-error mutants, then two-error mutants dominate the polynucleotide distribution. The critical value $\bar{q}_{min}$ calculated by perturbation theory (8) is an excellent approximation to the exact solution. Below the critical accuracy the concentrations are exclusively determined by the statistical weight of the corresponding sequences. Hence, the sum of the concentrations of 25-error mutant ($\Sigma I_{(25)}$) is largest, followed by 24- and 26 error mutants ($\Sigma I_{(24)}$, $\Sigma I_{(26)}$) etc. (For details see Swetina and Schuster 1982)

statistics of mutations. In case there are no statistical degeneracies all sequences are present at equal concentrations (see e.g. Fig. 4). In such a system replicating at a low level of accuracy there is no inheritance. Following a consecutive series of replica from generation to generation there is no conservation of the polynucleotide sequence. Such a series of polynucleotides thus approaches a random walk in an abstract space of all possible sequences.

So far we have studied our problem by means of ordinary deterministic kinetic equations. This approach assumes that all molecules under consideration are present in large numbers of copies. What "large" means depends on the particular reaction mechanisms studied. In most cases a few thousand molecules are sufficient to guarantee for all practical purposes, coincidence of the deterministic solution and the corresponding stochastic process. Facing the reality of polynucleotide replication we have to account for the "hyperastronomically" high numbers of different sequences. These numbers, as we pointed out earlier, exceed by far the numbers of molecules available in any realistic population. It is impossible to interpret the kinetic equations naively to include all possible sequences, because the dimension n is so large that most variables take exceedingly small values. (A meaningful concentration could really not be less than one molecule per earth's volume of water.) The differential equations thus have to be restricted at least to those sequences which are actually realized in the system.

Let us visualize the evolution of a replicating system by means of an abstract but nevertheless illustrative model. We construct a point space by assigning a point to each sequence. The points are arranged such that sequences which differ by single base exchange, i.e. pairs of sequences with Hamming distance D = 1, are first (closest) neighbours. The Hamming distance, thus, defines a kind of metric on this point space. Due to the hyerastronomically large number of possible sequences this set of points is giant. A quasispecies occupies a small part of it only which can be understood as a connected subset or subspace[3]. We distinguish three different scenarios:

(1) The replication process is very accurate, so that only a few mutants are present. The quasispecies then corresponds to a stationary or practically stationary subset in a low dimensional subspace of the giant space of possible sequences. The appearance of mutants outside this occupied subspace is a rare event and consequently the system is described well by deterministic differential equations.

(2) The replication process is of low accuracy. Then the evolution of the system corresponds to a random walk through the giant space. No stationary distribution of polynucleotides can be reached since frequently occurring replication errors create a high percentage of new sequences all the time.

(3) The replication process is just sufficiently accurate to sustain a master sequence at low stationary concentration. Replication in this case operates close to the minimum accuracy (Q_{min}) and we encounter a great variety of mutants. Eventually a sequence turns up that replicates more efficiently than the master and thus will replace it. The ensemble of polynucleotides as a whole follows a trajectory through the giant point space along which the efficiency of replication is optimized. This process resembles the schematic diagram shown in Fig. 2.

Case (3) is most relevant for evolution since it allows the polynucleotide structures to optimize with respect to properties relevant for replication speed and accuracy. At first glance it is not at all obvious that such a "hill climbing process" really leads to any respectable peak, and we may well ask why optimization does not soon stop at some relatively "low hill". If by a similar procedure we wanted to reach any of the major peaks on the Earth we soon would have to carry out practically forbidden large jumps between lesser peaks.

The optimization procedure based on RNA- or DNA-replication, however, is not well represented by a hill climbing process in an ordinary mountainous landscape. Consider a sequence having ν positions. Any of them can undergo three possible changes at any instance, e.g. from A to G, U or C (see Table 1). Our metapher, "hill climbing in some mountain-range" considers the optimization of height within a landscape erected upon a two-dimensional grid. We have two degrees of freedom to chose a step of given length. In polynucleotide replication we have 3^ν degrees of freedom for single base exchange in a given polynucleotide of length ν (in case the three exchanges are not equivalent the number of degrees of freedom may reduce to ν with some peculiar constraints). The topologies of two- and high-dimensional space – ν is the range of a few thousand for viral RNA already – are quite different. One of the most dramatic consequences of this is the reduction of (average) distances. With ν positions there are alltogether 4^ν possible sequences. If these 4^ν sequences are mapped onto a linear space, i.e. arranged in a row, the average distance of two sequences is very large. In an ν-dimensional space with unit edge length, the largest possible distance – e.g. between some arbitrary initial and the optimal final sequence – never exceeds ν. Hence the landscape in ν-dimensional space is much more bizarre most points being saddle points that allow change to continue uphill in at least one direction. Among the accessible points one might thus find – for virtually all initial positions – always some respectable nearly peak. Moreover, if the

3 "Connected" means in this context that the subset does not contain isolated members, i.e. points which have no other member of the subset in a neighbourhood of Hamming distance D = 1

dimension ν is not too large and it is possible to populate the mutant spectrum with k-error mutants, k being larger than 1, one might always jump from any local peak to a quite different environment. How far one can get in this way, whether one reaches the absolute maximum, depends largely on the distribution of peaks. Theoretical studies of this question with various topologies of protein structures are underway. Experiments carried out with "de novo" synthesized RNA chains using virus replicases show that RNA populations really adapt quickly to optimal performance of replication in a changing environment (e.g. increasing salt concentrations or inhibitors of replication such as ethidium bromide, Sumper and Luce 1975; Biebricher et al. 1981).

4. Cooperation Between Otherwise Competitive Self-Replicating Elements is Introduced by Higher Order Catalytic Action in the Form of Positive Feedback Loops Called Hyercycles

Self-replication inevitably leads to competition and selection of the most efficiently replicating species. The combination of selection and mutation allows structures and properties to optimize in the Darwinian sense. The accuracy of the replication process, however, sets a limit to the size of the optimizable structures. The lengths of the genetic message that can be transmitted from generation to generation is limited by the properties of the replication machinery. In evolution such a replicating system thus runs into a dead end when it approaches the critical polynucleotide length λ_{max}. In order to build a more accurately replicating machinery more genetic information is needed, but more information can be processed only when a more accurately operating machinery is available. How can this machinery – i.e. better replicase function – be made? There is a straightforward answer to this problem: several information carriers must be present at the same time and in controlled relative amounts. In order to be sufficiently general, control has to be exerted internally by a suitable reaction mechanism. Let us consider the mathematical problem first. If we apply mass action kinetics the analysis is simplest: first order template-induced replication leads to selection and, hence, we have to consider second-order terms in order to search for cooperation between self-replicating elements. The differential equation to be integrated is thus of the form

$$\dot{c}_i = (k_i + \sum_j k_{ij} c_j - \frac{1}{c}\,\phi)\,c_i;\quad i = 1,2,\ldots,n \qquad (10)$$

For sake of brevity we dispense with terms which arise from mutations. The rate constant k_i, in general, will be equal to the excess production of template I_i, i.e. the difference in rate constants between first order formation and degradation: $k_i = E_i = f_i - d_i$. The total concentration c is defined as in equation (1) to be $c = \sum_i c_i$. Note that we use concentrations instead or relative concentrations (x_i) here, because the dynamics of the system depends on the total concentration c. In order to visualize this dependence we adjust the flux ϕ in such a way that it leads to the condition of constant organization, $c = c_o$ = constant, and study equation (10) as a function of c_o. At low total concentration ($\lim c_o \to 0$) the first order terms will dominate and we are back to equation (1), which leads to selection. In the high concentration limit the first order terms can be neglected. The dynamics of the system then may be studied appropriately by means of the simpler equation

$$\dot{c}_i = c_i(\sum_j k_{ij} c_j - \frac{1}{c}\,\phi);\quad i = 1,\ldots,n \qquad (11a)$$

As we have shown (Eigen and Schuster 1979; pp. 31, 32) the topology of equation (11a) is independent of c, since the kinetics of the system is determined by a homogeneous function of the concentrations ($\sum_{ij} k_{ij} c_i c_j$ is homogeneous of degree 2) and therefore we may formulate the problem in relative concentrations without loosing generality

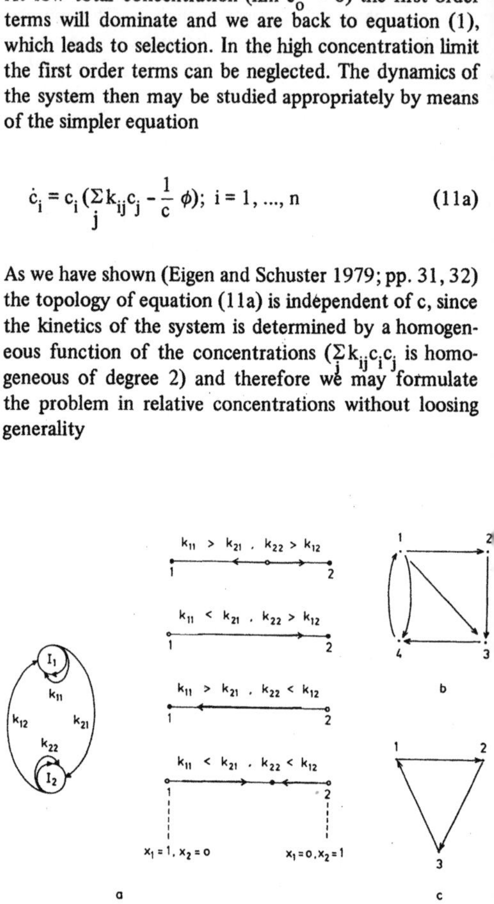

Fig. 5 a–c. Catalytic action and cooperation. a Different qualitative behaviour of the differential equation (11b) with n = 2. We distinguish four cases: (1) $k_{11} > k_{21}, k_{22} > k_{12}$ leads to selection of either I_1 or I_2 depending on the initial concentrations; (2) $k_{11} < k_{21}, k_{22} > k_{12}$ leads to selection of I_2; (3) $k_{11} > k_{21}, k_{22} < k_{12}$ to selection of I_1; and (4) $k_{11} < k_{21}, k_{22} < k_{12}$, finally, is the only case where both I_1 and I_2 are present at the stable stationary state. The relations of rate constants for cooperation ($k_{11} < k_{21}, k_{22} < k_{12}$) can be subsumed under the general formulation that mutual enhancement has to exceed self-enhancement; b A graph with a Hamiltonian arc: 1 → 2 → 3 → 4 → 1; c The graph corresponding to a hypercycle (n = 3). Note that it consists only of a single Hamiltonian arc

$$x_i = x_i(\sum_j k_{ij}x_j - \phi);\ i = 1, ..., n \quad (11b)$$

Equation (11b) has been studied in great detail (Hofbauer et al. 1980). For present purposes we require two general results. Cooperation between self-replicating elements is observed in systems in which the elements are connected by a positive feedback loop of catalytic actions (Fig. 5). In graph theory such a closed loop is called a Hamiltonian arc. In low dimensional systems (n = 2, 3 and 4) the existence of such an Hamiltonian arc is a necessary condition for cooperation. Independently of the dimension n another general theorem of cooperation could be proved (Schuster et al. 1979; Hofbauer et al. 1981): the elements of a system show cooperative behavior if the network of catalytic action consists exclusively of such a positive feedback loop. In context of the rate constants in equation (11b) this means that only particular off-diagonal elements of the matrix $K = \{k_{ij}\}$ may by different from zero.

$$K = \begin{bmatrix} 0 & 0 \ \dots\dots\dots & 0 & k_{1n} \\ k_{21} & 0 \ \dots\dots\dots & 0 & 0 \\ 0 & k_{32} \ \dots\dots & 0 & 0 \\ & & & \\ 0 & 0 \ \dots\dots\dots & k_{n,n-1} & 0 \end{bmatrix}$$

Equation (11b) than simplifies further ($k_{i,i-1} = \gamma_i$ all indices are understood modulo n: n + 1, 0 = n):

$$x_i = x_i(\gamma_i x_{i-1} - \phi);\ i = 1, ..., n \quad (11c)$$

A system which is described by an equation of type (11c) is called an elementary hypercycle (Eigen and Schuster 1979; pp 41 ff.).

For a physical interpretation of the theorem of cooperation we consider the simple two-dimensional case n = 2 (Fig. 5). Co-operation requires that mutual catalytic enhancement exceeds self-enhancement. In terms of rate constants this requirement implies $k_{12} > k_{22}$ and $k_{21} > k_{11}$. Higher dimensional systems (n > 2) have to meet additional requirements: mutual enhancement has to be such that it sustains a closed catalytic feedback loop.

Do we find hypercyclic organization in nature? In order to start from simple systems let us consider the repli-

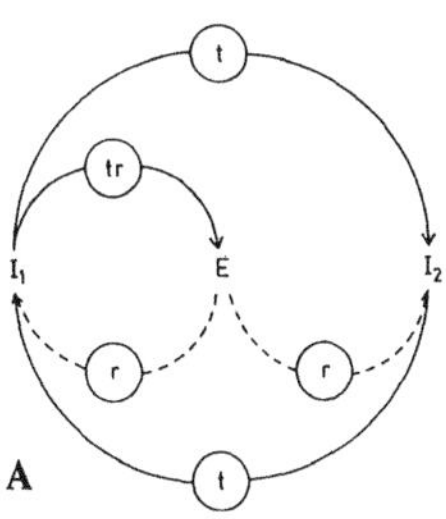

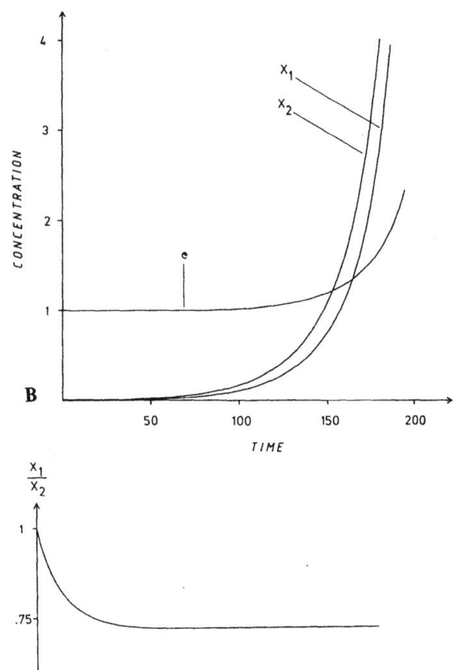

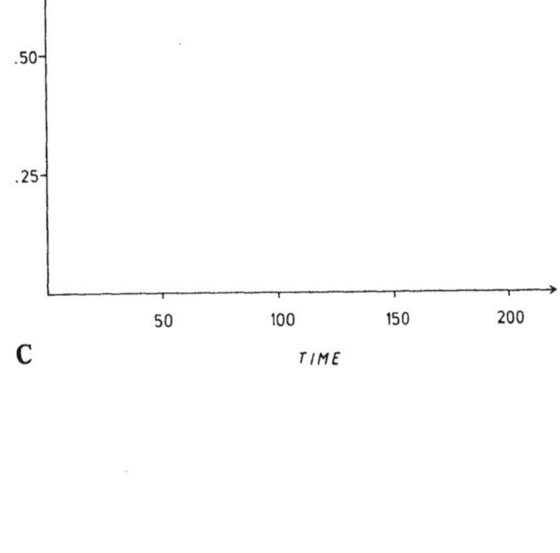

Fig. 6. RNA-phage infection of a bacterial host, as a simple hypercyclic process. Using the translation machinery (tr) of the host cell the infectious plus strand (I_1) first instructs the synthesis of a protein subunit (E) which associates with three host proteins to form a phage-specific RNA-replicase. This replicase complex (r) exclusively recognizes both phage-RNA strands, plus and minus, and replicates them. The result is a burst of phage-RNA production which follows a hyperbolic growth law. We show solution curves obtained by numerical integration of the differential equations which describe the most simple reaction mechanisms of this kind:

$$I_1 \xrightarrow{k} E + I_1 \qquad [I_1] = x_1$$

$$I_1 + E \xrightarrow{f_2} I_2 + I_1 + E \qquad [I_2] = x_2$$

$$I_2 + E \xrightarrow{f_1} I_1 + I_2 + E \qquad [E] = e$$

The rate constants and initial concentrations were: k = 0.01, f_1 = 0.02, f_2 = 0.04 and $x_1(0)$ = 0.01, $x_2(0)$ = 0.01, e(0) = 1 in arbitrary time and rate constant units. Despite the fast growth of x_1 and x_2 their ratio soon approaches the stationary value $(f_1/f_2)^{1/2}$

cation of a simple RNA bacteriophage like Qβ in a host cell (Fig. 6). Although this system represents a rather trivial example of a hypercycle – the plus-minus ensemble is one replicating entity and thus, the cycle contains only a singe element – it is illustrative with respect to the nature of catalytic action. The plus and minus strands both act as templates in replication. Catalytic action upon the replication process is the origin of the second factor (x_j) in equation (11b). RNA molecules lack the functional groups needed to serve as catalysts, so it is not hard to understand that almost all catalysis in biochemistry is done by proteins. The role of RNA in this particular catalysis (aside from the template function) is an indirect one, to be seen in the action of the plus strand as the messenger for translation into the specific replicase. More precisely, the plus strand codes for one of the four subunits of Qβ-replicase. Positive catalytic feedback of RNA on RNA replication thus occurs via translation. Polynucleotide translation is a very complicated, many step process. Under suitable conditions, however, it can be described well by a simple "over-all" kinetics which is closely related to the dynamics of differential equation (11b) (see also Eigen and Schuster 1979, pp. 50 ff.).

Considering the growing plus-minus ensemble in Fig. 6 we observe a faster than exponential increase in the concentration of RNA, which we denote as hyperbolic growth (Eigen and Schuster 1979, pp. 29, 30). Moreover, we recognize internal regulation of the ratio of concentrations of plus and minus strand: after a short initial period this ratio approaches the square root of the corresponding ratio of rate constants (Eigen 1971).

Some higher RNA viruses, like influenza, have a genome split into independent RNA molecules (Palese 1977). RNA synthesis in such viruses has to be organized in such a way, that the individual molecules are present in equal molar amounts for packing into the new virus particles. Although the mechanism of this regulation is not known, it seems to be a good candidate for some kind of hypercycle.

Other examples of hypercyclic organization are found with higher organisms when different species cooperate in symbiosis. In these systems again mutual enhancement has to exceed self-enhancement in order to stabilize the symbiontic organization.

Among other examples from the biology of higher organisms we mention one from sociobiology that has been analyzed in detail by different mathematical models. Often we observe a kind of self-organization of behaviour in animal societies which can be analyzed formally by a differential equation of type (11b). The mathematical basis for co-existence of two or more strategies of behaviour is precisely the same as in hypercycles: mutual support has to exceed self-enhancement (Schuster et al. 1981; Hofbauer et al. 1982).

Let us return to the question of how hyercycles may have been realized under prebiotic conditions. Replicating polynucleotides organize themselves into quasispecies under conditions such that the error limit for replication is rather low. According to the experiments by Leslie Orgel and coworkers (Lohrmann et al. 1980) it lies around $\nu_{max} \cong 100$ or somewhat lower in A,U-containing molecules. The RNA molecules available for further evolution were thus of the size of present day tRNAs. Polynucleotides are not efficient catalysts, so in order to improve catalytic efficiency instructed protein synthesis is required. Randomly polymerized polypeptides with the desired catalytic properties help very little. They can exert their catalytic function only temporarily, because they are hydrolysed after some time and as a consequence of the "hyperastronomically" large number of different sequences the same molecule will never be formed again. How could translation or a primitive version of it then arise? A detailed mechanism of this process is not known yet. Several models have been suggested (see e.g. Crick et al 1976; Eigen and Schuster 1979, pp. 62–67) and these have to be exposed to experimental tests now. We do not intend to go into too much detail here and to repeat the arguments that have been discussed extensively elsewhere (Eigen and Schuster 1979; pp. 62–67; Schuster 1981; Eigen and Winkler-Oswatitsch 1981) but we would like to stress some features of a "minimum requirement system" for primitive translation. Such a minimal system definitely requires more than a single RNA sequence. With reduced demands on structural and functional perfection we can assert multiple functions for primordial polynucleotides: a straightforward suggestion in this direction is that the primordial adaptors, the tRNA precursors, were also the first genes, and that eventually the plus strands had one, the minus strands the other function (Eigen and Winkler-Oswatitsch 1981). Nevertheless, one would need at least as many RNA molecules as there are amino acids to be incorporated into the primitive code. Four different amino acids might well have been sufficient to build up the first instructed polypeptides. Accordingly, we would require at least four polynucleotides to be present at the same time. Polynucleotides can be available for longer periods in case they are members of the same quasispecies. Four polynucleotide sequences might have originated from a single quasispecies distribution in which the relative frequences of the individual molecules were determined exclusively by their capability to replicate and by the frequencies of mutation. After an initial phase, during which the simultaneous presence within the same quasispecies was sufficient, the primitive translation system needed more reliable control on the relative concentrations of its constituents. This can be seen from a straightforward consideration: Let us assume that the system succeeded in building a primitive replicase. Then, catalyzed replication soon will outweight the uncatalyzed process and the rate constants of replication are no longer the same as for enzyme-free replication. Another quasispecies forms. The new and the old mutant distri-

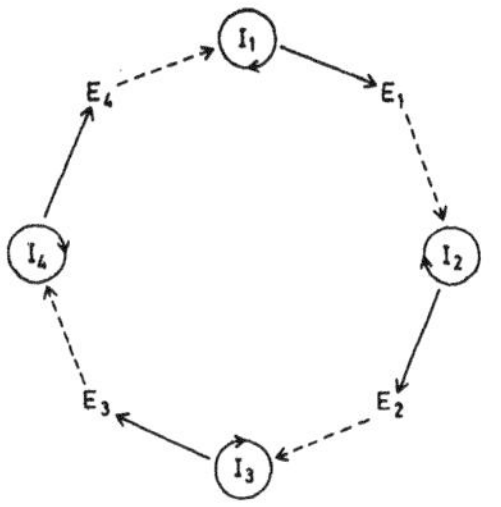

Fig. 7. Hypercycle with translation. The polynucleotides I_i are carriers of two kinds of instruction, one for their own replication, due to their capability to act as template, and the other for their translation into the proteins E_i, which are the effective catalysts. The proteins act as specific replicases or specific subunits in a multi-subunit enzyme

butions may differ substantially. Eventually, the former master sequence is replaced by another one which is better adapted to the enzyme-catalyzed reaction. Very likely the system loses control of the relative concentrations of the four essential polynucleotides. This deficiency in regulation is a substantial handicap for further evolution of the primitive translation system. Hypercyclic organization avoids this difficulty: once a catalytic feedback loop has been formed the mechanism of catalyzed replication controls internally the relative concentrations of all the constituents. One of the major problems concerns the origin of hypercyclic coupling. The conceptually simplest suggestion, but perhaps not that which is easiest to realize physically, introduces specificity into RNA replication: the translation products of the primordial genes act as specific catalysts in the replication of other RNA molecules (Fig. 7). Deriving all of the RNA molecules from a single quasispecies can be of advantage in this context: the individual RNA-molecules and their translation products differ only slightly in the beginning. Thus, they may have similar properties and closely related structures.

How can hypercycles evolve after they are formed and have passed through an internal stabilization period? A system with hypercyclic organization approaches either an internal equilibrium or a state with oscillating concentrations. Thereafter we can consider the whole system as an integrated functional entity. In competition with the external world such a hypercycle behaves like a single individuum growing with a non-linear rate. In the simplest case we deal with quadratic growth rates. Competition between hypercycles then is determined by the differential equation

$$C_i = k_i C_i^2 - C_i \cdot \frac{\phi}{C} \qquad (12)$$

Here C_i is the total concentration of hypercycle H_i, i.e. the sum of the concentrations of all polynucleotides belonging to H_i, and $C = \Sigma C_i$. Equation (12) is of some interest since it describes an example of restricted optimization. In an initial phase during which several hypercycles are present in comparable concentrations, selection of the system with the largest value of k_i takes place. After on hypercycle has won this initial competition the chance of replacing it is practically zero for any new although more efficient hypercycle. Even large fluctuations in concentrations are not enough to defeat the established system. There is no optimization of properties through adaption by successive replacement of less efficient competitors between hypercycles like that we observed with competitive polynucleotide sequences.

One way how hypercycles can nevertheless evolve is shown schematically in Fig. 8. Suitable mutations can be incorporated into the system when they fit properly into the network of catalytic reactions. Stepwise enlargement of the system through the incorporation of mu-

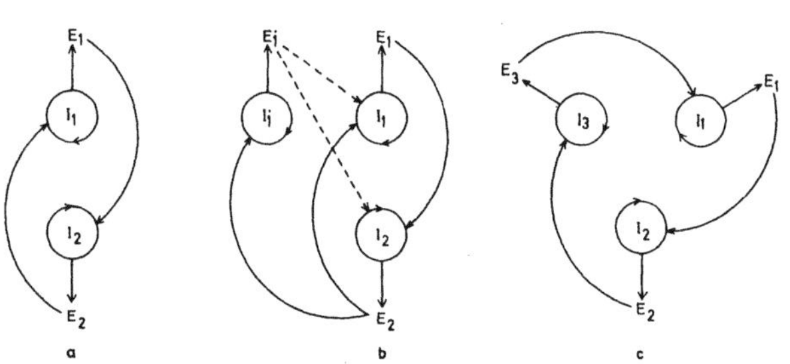

Fig. 8 a–c. Evolution of hypercycles. A mutation – incorporation mechanism allows the stepwise extension of hypercycles. a A two membered hypercycle with translation; b A mutant of I_1 here denoted by I_1' appears. It has a certain catalytic effect on the replication of I_1 and I_2 by means of its translation product. c Extension of the hypercycles by one member occurs in the case the mutant fulfils the following two conditions: (1) I_1', the later I_3, is recognized better by E_2 than I_1 is and (2) the translation product of I_1' (= I_3), the protein E_1' (= E_3) is a better catalyst for the replication of I_1 than E_2 is. In other words, catalytic coupling has to fulfil the criterion of a three membered hypercycle: mutual enhancement along the cycle 1 → 2 → 3 → 1 has to prevail over self-enhancement

tants is thus a highly conservative process: what has been established once is kept in the future. We may speak of "once for ever" decisions.

How can one test models of the origin of tranlation? Primarily, further experimental research on polynucleotide-polynucleotide and polynucleotide-polypeptide interactions will help to bring order into the various suggestions by making some more, others less plausible. The study of specific aggregates has to be carried on. How can we hope to speculate constructively on the origin of translation when our knowledge of the ribosome, its structure and function as well as the details of subunit interactions within it, is still far from being complete? In paleontology, research on the course of biological evolution is based on the fossil record. Prebiotic chemistry is guided by the results of simulation experiments in which reactions are studied under physico-chemical conditions which mimic the most likely environment on the primordial Earth. Research on the origin of translation is in a much worse situation: there are no fossils available for this very early period of emerging life and the molecules involved in this development are too complicated to allow straight forward simulation. So one has to look for the basic features of possible reaction mechanisms first. How can one check these predictions? Did this important stage of the origin of life pass by without leaving any remnants or traces for us? Attempts have been made to search for "intellectual fossils", remnants of early evolution which are hidden in present day structures and functions of biological macromolecules. Particularly wellsuited candidates for such investigations are the sequences of polynucleotides in regions where the restrictive demands on their functional properties are low or where redundancies in the genetic code allow some freedom in the choice of bases. Some of the recent sequence studies (Eigen and Winkler-Oswatitsch 1981; Cedergren et al. 1981; Sheperd 1981) are conclusive with respect to regularities in the present polynucleotides. Studies were performed on tRNA's and on DNA from various sources. These regularities suggest that primordial polynucleotides preferred a regular pattern of repeated triplets of the type -RNY- where R stands for a purine base, Y for a pyrimidine and N for any of the four bases. This pattern agrees well with the postulates of repetitive patterns made in models of primitive translation (Crick et al. 1976; Eigen and Schuster 1979, pp. 62–67). Stability considerations suggest a high GC content of the early information carriers. Combining the various arguments we end up with a suggestion for the first four codons in the primordial translation system: these four triplets and the corresponding amino acids are GGC = glycine, GCC = alanine, GAC = aspartic acid and GUC = valine. It is worth noting that these four amino acids are the most abundant found under simulated prebiotic conditions.

Further search for intellectual fossils and conclusive experimental studies on the dynamics of polynucleotide-polypeptide systems will bring more light into this exceedingly important and fascinating period of early evolution.

5. Formation of Compartments and Individualization Allows Efficient Evaluation of the Relevant Functional Properties of Translation Products

The fifth and last principle of emerging life we discuss here is ubiquitous and easy to observe in living matter: life is everywhere cellular. The most obvious question that comes to our mind then is: why and when did the cellular structure arise in the course of early evolution? Let us concentrate on the "why" first and visualize a system which consists of polynucleotides and polypeptides and which is capable of replication and translation. Is such a system also capable of precellular evolution in the Darwinian sense? Selection occurs at the level of polynucleotides, as has been verified in great detail experimentally. What about their translation products? In order to visualize this problem, consider two idealized classes of mutations which we denote as "phenotypic" and "genotypic" (Fig. 9). The major effect of a phenotypic mutation is a change in the tertiary structure of the molecule. Accordingly, the polynucleotide becomes a better or a worse target for replication (as well as it gains or loses in stability towards hydrolysis). Selection immediately sets in and the phenotypic properties are readily improved step by step until an optimal rate of replication is achieved. The properties of the translation product may be changed very little in case of a typical phenotypic mutation. In the language of molecular genetics an extracistronic mutation would be an example of a phenotypic mutation. Genotypic mutations behave oppositely in this respect: they lead to a substantial change in the translation product without having much of an effect on the tertiary structure of the polynucleotide. Let us assume that the modified translation product is more efficient than its precursor. In a homogeneous medium this advantageous polypeptide does not specifically favour the production of the mutant gene. It favours the synthesis of different polynucleotides or leads, eventually, to a general enhancement of polynucleotide and polypeptide synthesis. The result of a favourable genotypic mutation thus will be a local increase in the rate of synthesis and also a local increase of concentrations provided diffusion is sufficiently slow. At the steady state we shall find a concentration gradient and all molecules from the more productive volume element will spread slowly together with the favourable mutant gene. This scenario suggests a possible solution to the problem: in spatial isolation the favoured volume element would have the most pronounced advantage.

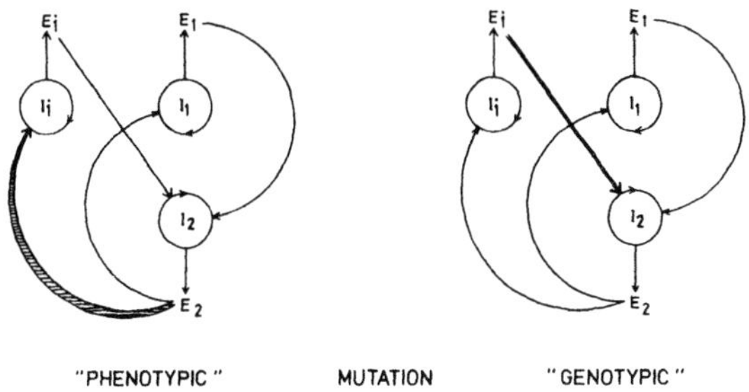

Fig. 9. Two idealized classes of mutations in a primitive replication-translation system. The "phenotypic" mutations lead to mutants (I'_1) which are better targets for the specific replicase, whereas the properties of their translation products (E'_1) are about the same as in the wild type (E_1). The "genotypic" mutant, in contrast, is characterized by a better translation product but more or less unchanged recognition by the replicase

Phenotypic and genotypic mutations are abstractions which we have used here for the purpose of illustration. In practice, every change of the sequence of a primordial gene alters both the tertiary structure of the polynucleotide and the properties of its translation product. In most cases one effect will dominate and a classification is meaningful. In principle, the whole argumentation can be given also with four classes:

(1) phenotypically favourable – genotypically favourable,
(2) phenotypically favourable – genotypically deleterious,
(3) phenotypically deleterious – genotypically favourable and
(4) phenotypically deleterious – genotypically deleterious.

The picture we obtain in this case is much more detailed but the general conclusions are the same.

Slow diffusion limits spatial spreading and has a weak isolating effect. Nevertheless, it cannot really support selection of favourable genetic mutation effectively. The real solution to the problem, spatial isolation and individualization, is the fifth principle of self-organization. In primitive systems this is achieved by the formation of compartments. A compartment can be defined as a volume which is separated from the surrounding medium by structurally stable, spatial barriers. These barriers consist e.g. of ampholytic organic material with lipid like properties and thus resemble present day cell membranes. Barriers built from other material are also thinkable. The compartment includes a set of macromolecules. Some of these molecules are polynucleotides undergoing replication. In the compartment the fate of the mutant gene is essentially coupled to the efficiency of its translation product, because the entity on which selection acts is the whole compartment. Replication of RNA in compartments is shown schematically in Fig. 10. The evolving system is back to Darwinian evolution. Full optimization of all properties which are relevant for compartment replication is possible now.

Isolation of genes in compartments solves some problems that cannot be solved in homogeneous solution in a simple way. However, it also creates problems that do not exist in homogeneous solution. Apart from the capability to select for advantageous genotypic mutations, the major positive effect of compartments is to maintain high local concentrations of macromolecules. The membranes surrounding the compartment act as diffusion barriers for all high molecular weight materials. Other advantages that may result from the presence of membranes are catalytic action of surfaces, activation of monomers through processes coupled to membrane transport, etc. The problems created by compartment formation can be summarized in four groups:

(1) Material transport has to be organized in some way. The growing system requires an input of the activated building monomers. Hydrolysis products have to be re-

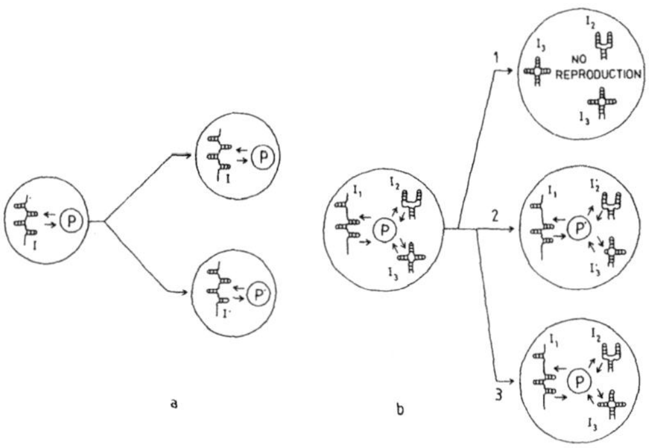

Fig. 10 a and b. Replication in compartments. Compartmentation puts selection pressure on genotypes. Suppose information carrier I encodes machinery (P) for translation and replication a After mutation of I to I′ and translation of I′ to P′, the system may be recompartmentalized into I/P and I′/P′ daughter compartments. Such compartmentation allows selection of the gene encoding the more efficient machinery. For selection to occur, the superiority of the machinery must compensate for errors made in each round of replication. Compartmentation by itself, however, does not suffice for selection of superior information divided among several information carriers. b Daughter compartments with incomplete sets of components are not viable (1). Compartments with sets that are complete but include mutations (2) allow evolutionary change. Even if the original set is the optimal one (3), the rate of proliferation required to maintain stability is much higher than it would be in a homogeneous solution without compartmentation. The net effect, in other words, is a severe sharpening of the error-threshold condition.

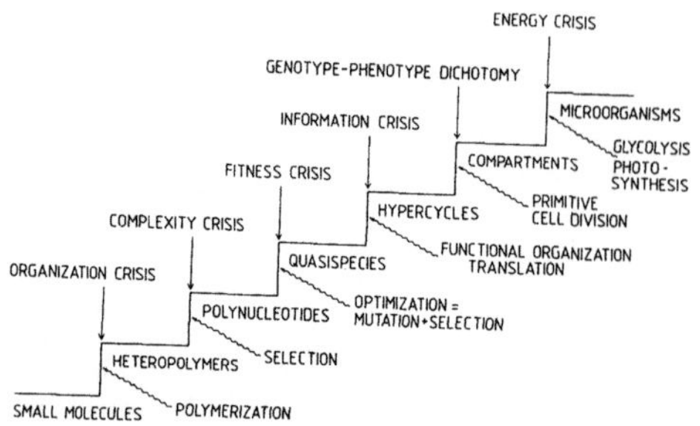

Fig. 11. Six critical steps during early evolution. On its way from small molecules to microorganisms the evolving system had to pass at least the six "crises" indicated by arrows. A "technical" innovation or new principle helped to overcome the obstacle the system had run into. The new concept is indicated by wavy arrows

moved or reactivated to enter into polymers synthesis anew. This problem requires e.g. membranes which sustain the diffusion of low molecular weight compounds by means of gates or pores which are small enough to hinder diffusion of macromolecules.

(2) Formation and division of compartments lead to packing problems. A complete set of molecules is necessary to make a compartment viable.

(3) The synthesis of macromolecules and the formation or division of compartments have to be synchronized in order to guarantee stationary concentrations of polymers inside the compartment.

(4) The replicating set of macromolecules in the compartment is subject to an error propagation problem similar to that we have discussed before.

Compartments which are able to master problems (1) to (4) would appear to require a rather elaborate molecular organization. At present it is difficult to suggest plausible models for this phase in prebiotic evolution. Based on the arguments presented above we suggest that compartments were incorporated into the evolving system as soon as necessary, but as late as possible (The interested reader is referred to a recent summary: Eigen et al. 1981).

The critical stages of early evolution are summarized in Fig. 11. We use the expression "crisis" to indicate that a new principle of organization had to come into operation in order to overcome an obstacle into which the previous development had run. The various crises represent a logical sequence of steps in prebiotic evolution from a mixture of small molecules to the first prototypes of procaryotic cells. Each of these steps requires the achievements of the previous ones, although two or even more might have been taken at nearly the same time during the historical course of prebiotic evolution. It is necessary to stress one point in order to avoid misunderstanding. Some of the critical steps incorporate new materials into the evolving system. For example, polypeptides came under the control of the replicating system after the information crisis had been overcome, membranes after the problem of genotype-phenotype dichotomy had been solved, and the utilization of sunlight after the primordial energy crisis had been surmounted through the development of the photosynthetic machinery. Emphasis is laid on the formulation "came under control of the replicating system". We indicate thereby that polypeptides, lipids and membranes, energy rich compounds must have existed first as necessary environmental factors and were used as resources before the replicating polynucleotides "learned" to get control of these essential constituents of present day life. This control consisted of the instruction of polypeptide synthesis through translation, of membrane synthesis and cell division, and of the devlopment of diverse primitive metabolic pathways in order to utilize environmental resources.

Acknowledgements. The authors are indebted to Professor W.C. Gardiner for many suggestions and careful reading of the manuscript. The work in Wien has been supported financially by the Fonds zur Förderung der Wissenschaftlichen Forschung (Project no 3502 and 4506).

References

Biebricher CK, Eigen M, Luce R (1981) J Mol Biol 148:369–390 and 391–410

Cedergren RJ, Sankoff D, LaRue B, Grosjean H (1981) CRC-Crit Rev Biochem 11:35–104

Crick FHC, Brenner S, Klug A, Pieczenik G (1976) Orig Life 7: 389–397

Domingo E, Davilla M, Ortin J (1980) Gene 11:333–346

Domingo E, Sabo D, Taniguchi T, Weissmann C (1978) Cell 13: 735–744

Eigen M (1971) Naturwissenschaften 58:465–526

Eigen M, Schuster P (1979) The hypercycle. Springer Verlag, Berlin Heidelberg New York

Eigen M, Winkler-Oswatitsch R (1981) Naturwissenschaften 68: 217–228 and 282–292

Eigen M, Gardiner W, Schuster P, Winkler-Oswatitsch R (1981) Sci Am 244(4):88–118

Fields S, Winter G (1981) Gene 15:207–214

Fox SW (1973) Naturwissenschaften 60:359–368

Hofbauer J, Schuster P, Sigmund K (1981) J Math Biol 11:115–168

Hofbauer J, Schuster P, Sigmund K (1982) Biol Cybern 34: 51–57

Hofbauer J, Schuster P, Sigmund K, Wolff R (1980) SIAM Appl Math C 38:285–304

Jones BL, Enns RH, Ragnekar SS (1976) Bull Math Biol 38: 15–28

Lohrmann R, Bridson PK, Orgel LE (1980) Science 208:1464–1465

Oró J, Miller SL (1981) J Mol Evol 17:263–306

Oró J, Miller SL (1982) J Mol Evol 18:149–220

Ortin J, Najera R, Lopez C, Davilla M, Domingo E (1980) Gene 11:319–331

Palese P (1977) Cell 10:1–10

Schuster P (1981) Prebiotic Evolution. In: Gutfreund H (ed) Biochemical Evolution, Cambridge Univ Press, Cambridge, pp 15–87
Schuster P, Sigmund K, Wolff R (1979) J Diff Equations 32: 357–368
Schuster P, Sigmund K, Hofbauer J, Wolff R (1981) Biol Cybern 40:1–25
Sheperd JCW (1981) Proc Natl Acad Sci USA 78:1596–1600
Sheperd JCW (1981) J Mol Evol 17:94–102
Spiegelman S (1971) Quart Rev Biophys 4:213–253
Sumper M, Luce R (1975) Proc Natl Acad Sci USA 72:162–166
Swetina J, Schuster P (1982) Biophys Chem, in press
Thompson CJ, McBride JL (1974) Math Bioscience 21:127–142
Volkenstein MV (1979) J Theor Biol 80:155–169

Received May 19, 1982/ Revised July 29, 1982

THE PHYSICAL MATRIX FOR SELF-ORGANISATION

R.K. MISHRA
DEPARTMENT OF BIOPHYSICS
ALL-INDIA INSTITUTE OF MEDICAL SCIENCES
NEW DELHI -110029. India.

ABSTRACT. Living organisms constitute the most outstanding example of self organisation. They are truely exceptional. Several views of self organisation which have been emerging in the last few decades have provided some conceptual models (1), and have even led to experiments in order to test their applicability to biological systems. However, the mass, the solidity and viscosity of organs and their components would seem to suggest that these schools of thinking, many of which require Navier-Stokes fluid would offer all but a partial explanation. Some models for oscillatory phenomena, information transfer, immune response, enzyme activity, muscle contractions have been offered. But for a more general description which will be in consonence with most of the observables in living systems like influence of radiation, density fluctuations, specificity of interaction, morphogenesis, an alternate approach is to be conceived. With this end in view the concept of "living matter" is proposed. A semi-qualitative account of this state, the living State, based on interacting bosons of diverse kinds is submitted.

I. Background

Some fifty years back the dominant view amongst biophysicists perplexed by the extra-ordinary phenomena exhibited by living systems was that one needed " new physics"or "new mathematics",meaning new laws. Here was a system hierarchical in structure, a system of subsystems, which adapts to internal and external forcings, undergoes development and differentiation and then decays. It has unique specificity of molecular interactions and works by otherwise trifling amounts of energy like a few millical per action potential or a few Kcal per mole for a high energy bond. On the other hand membranes separate such

C. W. Kilmister (ed.), Disequilibrium and Self-Organisation, 185–195.

high potentials as approximately 1 mV/Å. The system reacts to a single photon and survives within a very narrow zone of temperature. In 1950's one expressed his pessimism about the possibility of a proper understanding of such a system by expressions like "there is nonlinearity" or that "the system is in nonequilibrium". But new applications of mathematics and physics were made and the nature of inadequacies of classical theory became apparent. Irrever--sibility and thermodynamics in far from equilibrium realm were all investigated with better physical insight. In the last two decades many other concepts like order parameter, slaving principle, catastrophe theory, imperfection sensitivity of perfect structures, fractal geometry, solitonic propagation of energy, autocatalysis were applied to structure formation and to the question of dynamic stability, oscillation and other features and these were utilised for understanding biological structure and function.

But when examined all seem to beg one question - will they be applicable to a solid which is an organised assembly of bodies,masses or domains which might be viscous, solid, crystalline, liquid crystalline, semiconductors or piezoelectric? Yet this would be an average description even locally. A more fundamental level of structure has to be sought to enable comprehension of dynamic transformations. The dynamism is manifested by contraction, relaxation, polarisation and depolarisation of surfaces,information transfer, transduction of stimuli and a generalised pattern of function which controls all this autopoiesis and autocriny. Something more is needed: a description of the physics of living matter,indeed in other words, a theory of Living State. This must needs be an elaborate task. But we can try a formal and a semiqualitative approach to the essential peculiarity : the interacting boson fields. We emphasize that this is only formally so, some of the 'bosons' so considered could equally be dealt with as fermions with some constraints. We describe the interaction between photons, phonons and " orbital excitons" whereby one can describe chemical potential of these excitations and derive Mclachlan Longuet-Higgins relation(2). This relation leads to a proper description of atom-atom, ion-ion, ion-dipole, dipole-dipole and higher multipole interaction. It may also serve as a basis to explain coherence, self induced transparency, Bose-Einstein condensation, amplifi--cation of signals and transfer of information. Experimental data which can be interpreted to be in accord with this view are presented by Fröhlich(3), Popp(4) and Mishra(5),(6), Smith(7) and Webb(8).

The basic notion is that 'bosons' so considered provoke a state as summarised in this paper. At body temperature fluctuations ensue automatically. These are modified by molecular force fields. Fluctuations amplify and dissipate. Bifurcative and morphogenetic consequences result. In short, functional and adaptive dynamism for global needs becomes apparent. Boson pool has large number density. They are gaslike in some respects but they cohere to drive structures to produce macroscopic results. The role of atoms and bonds chosen by Nature is to provide specificity. These latter aspects are subject matter of other communica--tions(9-15).

II. Qualitative Features of the Living State:

1. 'Loose Structure'. This is central to the concept. By this we mean an assembly of molecules such that energies of intermolecular association are approximately similar to energies for intramolecular motion like bending, stretch--ing and vibration. Thus small energies which change conform ation could govern intermolecular interaction and cause changes in the liquidity or solidity of assembly. Vice versa, changes in compaction will bring about different conformations and states of excitation. This permits one to consider the organism as an approximate many-component viscous mass and not an immutable solid. Thus a viscous liquid like behaviour from a solid becomes feasible. This follows even from a classical analysis. Various data from bioenergetics, enzymology, immunology, NMR, EPR, sound absorption, dilatometry, Kerr effect, Raman spectroscopy, ORD, MORD can be cited in support. While we have been emphasizing loose gathering of flexible molecules, the non-rigid molecules have been subject matter of much more attention recently(16).

2. In a "loose structure", using classical interaction and nonquantum mechanical formalism it can be shown that in an N-body aggregate, slight variation in the magnitude of interaction as by pulsation and mass or momentum transfer will lead to anisotropy and varying metastability, and symmetry breaking in space and time (17),(18).

3. Another principal of the assemblage is that molecules are arranged in order of their interaction energies so that least abrupt change of local force fields is ensured.

4. Not all types of atoms and bonds are chosen and these invoke specific features which have been discussed, but are not of any great bearing in the context of this communica--tion, which deals with elementary excitations, without considering specific atoms.

5. Although much can be achieved without invoking quantum theory there are certain aspects which demand such an application. These are recognition, desired specificity in molecular structures, "homing" of drugs, hormones and other molecular interactions. These may be involved in vision, photosynthesis, electron transport and enzyme action

Cooperative action of elementary excitations is therefore examined and formulated.

III. Elementary Excitations and the Living State.

Types of Bosons considered and their reservoirs:
Photons. A living organism is subject to numerous photons from the environment with a flux of over 10^{23} photons/cm^2/sec. Some compounds like the anticancer hematopophyrin derivatives fluoresce when exposed to light. NAD may be the molecule to fluoresce when electron transfer reactions are occuring. Luciferin-Luciferase system is accompanied by emission of light. Thermally induced luminiscence is also possible. Biophotons (10^3/cm^2/sec)have been demonstrated by Popp[4], in far from equilibrium concentration, reflecting metabolic activity. Since transformations(3),(4), in infrared, microwave and radio-frequency range are constantly occuring at body temperature the crossing of energy barriers is accompanied by to and fro transfer of suboptical photons causing energy exchange. The photons therefore contribute major packages of energy in the living systems. These are bosonic in nature.
PHONONS: Individual long molecular chains in lipids, nucleic acids, proteins and polysaccharide are close approximation to one dimensional lattices seating phonons.
EXCITONS: The interaction of apolar molecules by dispersion forces was proposed by London on the basis of dipoles induced by virtual excitations. The latter occur because of Heinsenberg's uncertainty principle applied to energy. The Lamb shift in hydrogen atoms in microwave region and its theory and experimental verification (19),(20) show that interactions of this type by radiation, say of $\approx$ 1000 mc, can occur. In a body filled with RF field their stability would be further enhanced. On that basis the assembly can be recognised in the framework of a collection of "orbital" excitons. London(21),(22) relations and McLachlan-Languet-Higgins(2) relation can then be derived, which in themselves bear experimental proof.

Green and Triffet(23) derived the excited states of excitation of axons and its oscillating propagation by considering creatinn and annihilation of ions. These are not considered in the communication.

Again, we consider only these excitons and with such densities that Bose commutation relation is obeyed and the Keldysh-Kozlov(24) restriction is respected.

The heterogeneous boson field with the three components can be written as

$$\Psi = \prod_{i=1} \psi_i^{(1)} \prod_{j=1} \psi_j^{(2)} \prod_{t=1} \psi_t^{(3)} \qquad [1]$$

or rather

$$\Psi = \Psi(1)\,\Psi(2)\,\Psi(3)$$

If the bosons interact we get the interaction representation (25)

$$H = \frac{1}{2m}\int \nabla\Psi^{\dagger}\cdot\nabla\Psi\, dr + \frac{1}{2}\langle \Psi^{\dagger}(x)\,\Psi^{\dagger}(y)\,|\,V(x-y)\,|\,\Psi(x)\,\Psi(y) \qquad [2]$$

From this the general Hamiltonian after interaction can be stated in terms of field operators of elementary excitations taken as interacting members. The interaction potential between excitons uses Haken's displaced oscillar wave function(26) which is given by

$$\Psi'(x) = C_0\left[\Psi(x) + \gamma a_m + \Psi(x)\right] \qquad [3]$$

and γ is defined by $\gamma^2 = -\frac{8\pi|d|^2}{E_0 V}$

The exciton-photon interaction is represented using transformation

$$\vec{P} = \left(\vec{p} - \frac{e}{c}\vec{A}_0\right);\quad V_{ex-photon} = -\frac{e}{mc}\vec{A}_0\cdot\vec{P}$$

$$\vec{A}_0 = \sqrt{\frac{2\pi\hbar c^2}{V\Omega_m}}\;\vec{e}_k$$

while exciton-photon and exciton-phonon interaction in first approximation is represented respectively,

$$V_{ex-photon} = \sum_{m,l,k} (\lambda'\alpha C_0)\, a_m^{\dagger}\, a_l\, (\rho_k + \rho_k^{\dagger})$$

$$V_{ex-phonon} = \sum_{m,t,q} (g\alpha C_0)\, a_m^{\dagger}\, a_t\, (b_q + b_q^{\dagger})$$

λ' and g are coupling strengths of respective systems and

αC_0's occur in the transformation function:

$$f^{\dagger}(b_q) = \alpha C_0\, e^{\alpha b_q}, \quad f(b_q^{\dagger}) = \alpha C_0\, e^{-\alpha b_q^{\dagger}}$$

$$f^{\dagger}(\rho_k) = \alpha C_0\, e^{\alpha \rho_k}, \quad f(\rho_k^{\dagger}) = \alpha C_0\, e^{-\alpha \rho_k^{\dagger}}$$

Constant α is derived as per Fröhlich, namely

$$\nabla \rho = \operatorname{div} A = -c' \frac{\rho^2}{V^2} ; \qquad C' \text{ is a constant.}$$

By using Gauge transformation $\vec{A}' = A - \nabla A$

and using velocity of sound $v = 10^5 \text{cm/sec}$

and net charge density $= \rho + \nabla \rho$ and $\alpha \approx \rho' + \nabla \rho$

for $\rho \approx 1$, $\alpha \approx 0.8$ and λ' and g as obtained are comparable with well shown relations of exciton-photon and exciton-phonon coupling

$$\alpha C_0 \lambda' = \langle \Psi_m | \lambda | \Psi_\lambda \rangle$$

$$\alpha C_0 g = \sqrt{\frac{\hbar^2}{2 m N_3 \omega_q}} \left\langle \Psi_m \left| \left(\frac{\partial V}{\partial x} \right)_{x = R_0} \right| \Psi_t \right\rangle$$

Now the chemical potential of interacting Bosons is:

$$\mu_0 n_0 = \left\langle T \left\{ \left[H_{int}, \tilde{\xi}_0(t) \right] \tilde{\xi}_0^{\dagger}(t') \right\} \right\rangle$$

n_0 is particle density while $\tilde{\xi}(t), \tilde{\xi}(t')$ are operators for Bose-Einstein condensed system in Heisenberg's interaction picture. Above formula is rewritten with the irreducible self-energy parts

$\sum_{11}(0)$ and $\sum_{02}(0)$ as

$$\mu_0 = \sum_{11}(0) - \sum_{02}(0), \quad \vec{p} = 0$$

This relates to chemical potential. In diagramatic language in Σ the number of ingoing zigzag lines for $\tilde{\xi}^{\dagger}$ is same as number of outgoing lines for $\tilde{\xi}_0$ while in $\sum_{02}$ the number of ingoing lines exceeds by two the number of outing zigzag lines

Thus

Reflecting on this number, one can see that such situation leads to morphogenesis due to interacting Bosons in terms of B_K and B_q, which occurs in $\sum_{02}$

$$B_q = \alpha^2 \cos \alpha \, (1 + 2 n_q)^{\frac{1}{2}} = \alpha^2 \cos\left\{ \left(\coth \frac{\hbar \omega_q}{2KT} \right)^{\frac{1}{2}} \alpha \right\}$$

$$B_k = \alpha^2 \cos \alpha \, (1 + 2 n_k)^{\frac{1}{2}} = \alpha^2 \cos\left\{ \left(\coth \frac{\hbar \Omega_k}{2KT} \right)^{\frac{1}{2}} \alpha \right\}$$

n_q and n_k are thermal average for phonon and photon respectively.

If the assembly is put in an electromagnetic field, dielectric constant and polarisation become manifest. The response to external electromagnetic field follows by considering average potential of external field A $(\vec{r}, \Omega)$, frequency using Ω , and using Maxwell's equation for electromagnetic waves. Dyson's equation for Geen function for polarization operator enables us to calculate dielectic constant, being its real and imaginary part, respectively.

Thus

$$\epsilon' = 1 + \frac{4\pi}{\hbar^5 \Omega_n^2 c^2} \sum_k (V\Omega_k) \Big[\sum_m \lambda^2 \int_p \frac{1}{\Omega_n - \Omega_m - \frac{1}{\hbar}\sum_l V_{ml} \langle n_l \rangle} + \sum_{m,l,q} \lambda'^2 \beta_q \Big\{ \int_p \frac{\langle n_l \rangle}{\Omega_n - \Omega_m + \Omega_l} + \int_p \frac{\langle n_m \rangle}{\Omega_n - \Omega_m - \Omega_l} \Big\} \Big]$$

$$\epsilon'' = 1 + \frac{8\pi^2}{\hbar^5 \Omega_n^2 c^2} \sum_k (V\Omega_k) \Big[\sum_m \lambda^2 \delta \big(\Omega_n - \Omega_m - \sum_l V_{ml} \langle n_l \rangle \big) + \sum_{m,l,q} \lambda'^2 \beta_q \Big(\langle n_l \rangle \delta (\Omega_n - \Omega_l + \Omega_m) + \langle n_m \rangle \delta (\Omega_n - \Omega_m + \Omega_l) \Big) \Big]$$

We thus see that the following properties which enter at different levels into the works on self organisation by Prigogine(27), Haken(28), and Fröhlich (29) can be derived by the above approach :

a. Chemical potential and internal pumps due to interacting bosons.
b. Frequency dependent dielectric constants and adaptation to electromagnetic fields.
c. Relation to sound pressure and adaptation to gravity fluctuations since the chemical potential will be affected by gravity and oscillation of sound pressure.
d. Fluctuation.
e. Dissipation.

This approach may thus serve to define a state of matter which can support self organisation.

IV. Speculation on Some Biophysical Consequences.

(A) Role of coherence of elementary excitations may be stressed with the following consequences:

a. Coherence lowers the energy barriers between different states and permits reactions and associations, prohibited otherwise. Since the energy is lowered during the reactions or or association of molecules the partners cannot move apart. Thus, these are responsible for molecular recognition, specificity and 'homing'. Since the aspects like induced Bose-Einstein condensation are a function of rate of radiation a fine gradation of specificity occurs.

(B) For self-organisation a pump is needed <u>within</u> the system. Because of exothermic reactions, a thermal bath and narrowly set zone of temperature, radiation of thermal photons occurs. When Bose-coherent or even Bose condensed in the limit, these may be sufficient and efficient to drive changes in conformations and associations of molecules, leading to altered phenomena.

The pump, a coupled system, and a bath, are the required aspects that drive the systems to order and coperative phase transitions. These can all be conceived on the present basis.

Correspondence with experimental work is widespread and it may be stated that amongst the various types of assemblies of matter, as understood currently, lyotropic liquid crystallinity(30) with internal pumps and fluctuations due to density would be closest to the desired model of the living state. The Table I expresses the physical end-mechanisms underlying morphogenesis and function.

Table I

Localised properties exhibited by living organism mastered by time course of any specific phenomenon

Observables	Physical processes	End mechanisms
1. Movement	Transformation to altered physical state by chemical, mechanical or electrical stimuli. Phase transformation like contraction relaxation; charging discharging of opposite aspects of surfaces etc.	Phase transition
2. Growth involving accretion, ingestion, digestion, metabolism, influenced by enzymes, hormones, allergens, oncogens, pathogens, drugs.	Formation of covalent bond, catalysed by enzymes, with the help of coenzymes and cofactors.	Directed diffusion, recognition, specific covalent bonding using elements of water and radical containing phosphorus and sulfur; cooperative effects leading to variations in conformation and association.

4. Responsivity	Neuromuscular perturbation leading to contractile, secretory, exoretory, activity, changes in absorptive, adsorptive properties; phase transformation and propogation of altered states following stimulation by light, smell (molecules), teste(molecules), vibration(sound), gravity, stretch, pain, temperature, touch.	Transformed physical state of molecular assemblages locally following stimulation. A local phase transition.
5. Emotion and higher mental function	Special case of above transduction, involving brain and its extension.	Interalising and modelling of environmental data into specific physical patterns, followed by systematisation of data and responses in a hierarchical patterns governed by the principle of the more general mastering over the less general. These features are catalysed in a positive or negative sense by emotive energy. The principle of enslaving by order parameter is (31) followed.

Cooperative effects in shaping of molecular assemblies and effects like phase change appear to be ubiquitous. The above analysis offers the necessary ingredient and quality to the living matter for this purpose.

Note

Details of mathematical aspects were presented in a paper by R.K. Mishra and S.K. Dube in the International Seminar on the Living State II in Bhopal, November 13 to 19,1983.

The present paper is dedicated to the memory of late Dr. S.K.Dube.

References

1. R.K. Mishra, Ed. The Living State, Wiley Eastern Ltd., New Delhi, 1984, and The Living State II, World Scientific Publishing Co. Pvt. Ltd., Singapore,1985.
2. H.C. Longuet-Higgins. Intermolecular Forces. Faraday Soc. Symp. P.7,1965.
3. H. Fröhlich and F. Kremer. Coherent Excitations in Biological Systems, Springer, Heidelberg,1983.
4. F-A Popp. Biologie des Lichtes, Paul Parey, Berlin,1984.
5. R.K. Mishra. Molecular Crystals and Liquid Crystals. 10,85,1970.
6. R.K.Mishra. The IVth International Liquid Crystals conference, Kent, Ohio, 1972 and Mol.Crystals and liquid Crystals. P.25, 201, 1975.
7. C.W. Smith. IEEE Pub. No. CH 2058, 6/84/000-0176,176 1984.
8. S.Webb, M.E. Stoneham and M. Fröhlich. Physics Letters, 63A, 407, 1977.
9. R.K. Mishra, K. Bhaumik, S.C. Mathur and S. Mitra. I.J.Q.Chem. 16,69,1979
10. R.K. Mishra, K. Bhaumik, A. Srivastava, S.S. Chaudhary. Int. J. Quant. Chem.20,337,1981.
11. R.K. Mishra, The Living State, Wiley Eastern Ltd., New Delhi, P.35, 1984.
12. R.K. Mishra, G.C. Shukla and K. Bhaumik. Physics in Medicine and Biology. Proceedings of International Conference on Physics in Medicine and Biology. International Centre for Theoretical Physics,Trieste, World Publishing Co. Pvt. Ltd., P.319,1981.
13. R.K. Mishra and K. Bhaumik. Int.J.Quant. Chemistry, 23(4),1579,1983.
14. R.K. Mishra, S.K. Dubey and R.S. Tyagi. Int.J.Quant. Chemistry, 27,303,1984.
15. R.K. Mishra. Living State II. World Scientific Publications Pvt. Ltd., Singapure, P.492, 1985.
16. J. Maruani and J. Serre. Symmetries and Properties of Non-Rigid Molecules, Elsevier, Amsterdam, 1983.
17. G.C.Shukla, K. Bhaumik and R.K. Mishra. The Living State I (R.K. Mishra,Ed), Wiley Eastern, New Delhi P.70, 1985.

18. R.K. Mishra and G.C. Shukla. The Living State (R.K. Mishra, Ed.), Wiley Eastern, New Delhi P.64,1984.
19. H.A. Bethe. Physical Rev. 72,339, 1947.
20. W.E. Lamb, Jr. and R.C. Retherford. Physical Rev. 73, 412,1947.
21. R. Eisenschitz and F. London Z.f.Phy. 63, 1929.
22. F. London, Naturwiss. 17, 516,1929.
23. H.S. Green and T. Triffet. Int. J. Q. Chem. Quantum Biol. Symp. 289, 1975.
24. L.V. Keldysh and A.N. Kozlov. Soviet Phys. JETP 27,521, 1968.
25. P.R. Gross. Unified Theory of Interacting Bosons in Many Body Problems. J.E. Percus, Ed. Interscience N.Y. P.275, 1963.
26. H. Haken. Quantum Theory of Solids. North Holland Publishers, Amsterdam, P.40, 1976.
27. G. Nicolis and I. Prigogine. Self Organisation in Nonequilibrium Systems. John Wiley and Sons, N.Y. 1977.
28. H. Haken. Rev. Med. Phys. 47,68,1975.
29. H. Fröhlich. Riv. Nuovo Cim. 7,393,1977.
30. R.K. Mishra. III International Cong. on Liquid Crystals, Berlin. Abstr. S6.2,1970.
31. R.K. Mishra. Int. Jour. Quant. Chem. Suppl. 11,45,1984.

SPONTANEOUSLY BROKEN SYMMETRIES AND DISSIPATIVE STRUCTURES

E. Del Giudice, S. Doglia, M. Milani
Physics Department of the University of Milano
Istituto Nazionale di Fisica Nucleare - Sezione di Milano
Via Celoria, 16 - 20133 MILANO
Italy

and

G. Vitiello
Physics Department of the University of Salerno
Istituto Nazionale di Fisica Nucleare - Sezione di Napoli
84100 SALERNO
Italy

ABSTRACT. Living structures appear as time-dependent ordered structures. These structures are also generally unsymmetric. On the other hand basic interactions, such as electromagnetic interactions, which could reasonably be responsible for living structures are highly symmetric. Quantum field theory predicts that ordered structures are likely to appear when the symmetry of the basic Lagrangian is broken by choosing an unsymmetric vacuum. We suggest that long-range correlations could appear in living matter as a consequence of the breakdown of rotational invariance in the set of electric dipoles, which are the basic components of that matter. Moreover we discuss the behavior of electromagnetic field in the region of correlated dipoles and show that it gets confined inside narrow filaments, which could be at origin of cell architecture.

Spontaneous creation of order in matter - in particular, in living matter - has been the object of many investigations. In this paper we discuss a point of view, which could be called a pure physicist's point of view. We do not consider as a starting point the existing highly ordered structures with their shapes, sizes, orientations, which are highly unsymmetric. On the contrary we start from a very symmetric situation: an homogeneous soup of many objects, governed by a Lagrangian, and consequently by equations of motion as symmetric as possible. We investigate then, on purely physical grounds, how structures can arise and consequently asymmetry can emerge in space and in time from the initial homogeneous soup where possibly all processes are reversible.

C. W. Kilmister (ed.), Disequilibrium and Self-Organisation, 197–205.

This approach must reproduce observations only at the very end. The intermediate steps can be looked at in two different ways. Looking at these steps as to a chain of historical facts, this approach could provide the description of the historical evolution which produced the system. On the other hand, if we consider the system as it stands today independently of its previous history, the intermediate steps provide the inner fabric underlying the system's appearance. In this sense physicists speak of a "basic dynamics".

For instance, the description of an elementary particle in terms of quarks could correspond to the actual formation of particles in the early Universe. On the other hand, in the modern accelerators, particles are seen to emerge from the inelastic scattering of other particles, without any appearance of quarks. Nevertheless, the description in terms of quarks provides the "basic dynamics" which only "explains" the full set of observable data.

The same could be said about the description of a molecule in terms of nuclei and electrons or about the description of a macroscopic system in terms of "elementary" components.

A beautiful description of this approach, as referring to the problems of life, has been presented by P.W. Anderson and D.L. Stein[1].

In the above approach, dynamical and phenomenological levels are not absolute concepts. For instance, a nucleon stands at phenomenological level respect to quarks, but nucleons endowed with interactions form the dynamical level from which nuclei emerge. In biology, cells appear as the phenomenon emerging from the working of the different cell ingredients, but cells are also the dynamical elements that, endowed with suitable interactions, are responsible for tissues, organs, plants and animals. The concept of such a universal hierarchy appears also among the requirements at the origin of "hypercicles" introduced by Eigen and Schuster[2] and of closely related concepts of synergetics (H. Haken)[3] and "dissipative structures" (I. Prigogine)[4].

As a conclusion an object appearing on the phenomenological description is a free object, whereas at dynamical level, its relations with other objects, say with the environment, are the most important thing. In other words, the ingredients of the dynamical description look like elementary objects interacting among themselves, while the ingredients of the phenomenological description appear as collective variables connected with the whole system.

The above line of thinking has found a mathematical formulation in quantum field theory[5]. In this scheme a physical system is described by space-time distributions of quantities, named fields, mathematically represented by operators acting on a suitable Hilbert space.

The elements of the space are the physical states, built up by applying the creation operators deduced from the phenomenological fields to a particular element, the fundamental state, or "vacuum", which is the minimum energy state of the system. Actually a system could admit several energy minima, so that we could deal with different vacua, giving rise to different Hilbert spaces, which would describe different phases of the same system. The transitions among the different vacua allow the phase transitions of the system.

Coming back to the operators, they are the "solutions" of operator valued differential equations deduced from an operator-valued Lagrangian. The two levels, the dynamical and the phenomenological, correspond to two different sets of fields, say ψ_i and ϕ_i respectively, and their equations of motion. According to the definition the equations for ψ_i are non linear equations

$$\Lambda_i(\partial)\psi_i(x) = J_i(\psi_k(x)) \tag{1}$$

where $\Lambda_i(\partial)$ are kinetic operators and J_i are usually complicated non linear functions of the fields, which account for the interaction among them. On the contrary, the phenomenological fields ϕ_i are required to obey to free equations:

$$K_i(\partial)\phi_i(x) = 0 \tag{2}$$

where only kinetic terms are involved.

Of course, the two sets of equations are supposed to describe the same system. Consequently we assume the existence of a relationship between each matrix element of the "dynamical" operators and a corresponding matrix element of the "phenomenological" ones:

$$\langle a|\psi_i(x)|b\rangle = \langle a|\Psi_i[\phi_j(x)]|b\rangle \tag{3}$$

where Ψ_i is the functional defined by the assumed relationship. Eq. (3) is called "dynamical map" of the system and traduces just the way how the inner dynamics produces the observable features of the given system. In a mathematically constructive approach the existence of eq. (3) must be proved by the machinery of the equations of motion, but in a physical approach we give for granted that each phenomenon corresponds to a deeper level of dynamics and, consequently, we exploit eq. (3) to compare the different symmetries holding at the different levels of dynamics and phenomenology.

The strategy is usually chosen to take the "dynamical" equations (1) invariant under the largest possible group G of symmetry transformations. This choice makes these equations more easily solvable. Of course, this choice could not be shared by nature, whose structures are generally unsymmetric. To reproduce Nature, it is necessary to introduce asymmetry at the phenomenological level. The only possibility is to introduce asymmetry through the fundamental state or "vacuum". This is done by letting the vacuum invariant under a smaller symmetry group than G, say H. This fact is referred to as "spontaneous breakdown of symmetry".

Through the "dynamical map" (3) each element g of G corresponds to an element g_0 of the group G_0, which leaves "phenomenological" equations (2) invariant. It has been proven by general arguments[6] that G_0 contains a subgroup H_0 isomorphic to H and a subgroup T of transformations which act as "translations" for a subset $\{\phi_i\}$ of the fields ϕ_i, which are massless bosons. The generators of G_0 are as many as the generators of G. It means that the number of "translations" equals the number of broken transformation of G. Consequently, at the phenomenological level, there are as many massless fields $\bar{\phi}_i$ as the asymmetries introduced in the vacuum. This is the Goldstone theorem[7].

Those massless fields $\bar{\phi}_i$ are responsible for the structure of the asymmetric vacuum measured by the so-called order parameter, which vanishes in the symmetric limit. This number measures the correlation among the components of the system in the fundamental state; the correlation agents are the massless quanta of the fields $\bar{\phi}_i$, which have been shown[8] to form a Bose condensate. The group T of the "translations" induces such condensations. Moreover, the states constructed starting from such a vacuum - which coincide with the physical observable states - are coherent respect to the $\bar{\phi}_i$ quanta. The masslessnes of $\bar{\phi}_i$ implies the long range of the correlation.

It should be stressed that the "translations", since are elements of the group G_0 which leaves equations (2) invariant, connect states which are energy degenerate, even if their respective degree of order is different. Consequently the ordering of the system is "spontaneous" in the sense that no energy is required. This conclusion rests on the existence of an unsymmetric vacuum. But the vacuum, namely the minimum energy state of the system, is unsymmetric because some physical agent is at work. Consider, for instance, a piece of unmagnetized iron; its fundamental state is then symmetric under rotations of the atom spins. To induce order and align the spins in a correlated way, we make the fundamental state unsymmetric by switching on an externally supplied magnetic field, which required some energy to be kept on. But this energy is a sort of bail, which stands there and must be returned when the field is switched off; otherwise we would get the iron heated and consequently disordered. But in this case the decrease of magnetic strength would change the vacuum, spoiling the above reasoning.

Once the external field is on, the transition from unmagnetized to magnetized iron, namely from a symmetric to an unsymmetric vacuum, requires no energy. Moreover the amount of ordering we get is not related to the amount of energy of the external field, once a threshold is overcome.

We can see that a self-ordered system needs to be an open and dissipative one. The system is open because its fundamental state is kept unsymmetric by an external physical agent. Moreover, the involved energy is not transferred to the system components, but simply triggers the transition of the system to an ordered state. This energy therefore is not lost, but is completely recovered. In this way the open system is described by a Lagrangian depending on its internal structures and by a vacuum where the effects of the whole external world are taken into account as nonvanishing expectation values of appropriate fields. This procedure is different from the usual approach, where the open system is a part of a larger closed system, whose Lagrangian contains explicitly the interaction between the system we are interested in and its complement to the larger closed system. In our approach we don't need to know the detailed coupling of the open system with the environment.

The influence of the environment on the system is summarized by the nonvanishing values of the order parameter.

Let us now analyze a living system in the above framework. Living matter can be regarded as a dieletric, whose properties are very peculiar and extraordinary[9]. We assume[10] then that living matter, which appears as a highly unsymmetric system, has emerged from a symmetric set of many species of molecular electric dipoles, described by an a priori quite complicated Lagrangian.

The only constraint we can reasonably prescribe on that Lagrangian is invariance under dipole rotations. On the other hand we know that the final result of dipole dynamics is a biological system unsymmetric under rotations. This reduction of symmetry can occur through a rotationally unsymmetric vacuum, namely we choose a fundamental state, where a privileged direction is introduced by a net electric polarization kept by some agent. According to the above analysis, this symmetry reduction, via the Goldstone theorem, gives us automatically a correlation among the system dipoles. The messengers of this correlation are the massless Goldstone bosons, which correspond to the broken symmetry transformations and whose quantum numbers are the generators of such transformations.

In living matter, the flow of externally supplied metabolic energy produces, through a multi-step dynamics[11], a net electric polarization directed along a given direction. This situation, without any further expense of energy, induces the ordering of the system. The agents of the ordering are quanta of the electric polarization field, which form a Bose condensate. At macroscopic level, the above dynamics appears in the form of coherent polarization modes, which were predicted by H. Fröhlich[12] long ago. These quanta are "quasi-particles" as phonons in crystals or magnons in ferromagnets.

It is worthwhile to compare this result with the conclusions drawn in the framework of the other approach to open systems we mentioned before, namely the open system as a part of a larger closed system. In this case it is recognized[13] that coherent polarization modes appear in the open system for peculiar couplings with the environment. In our approach, we say that the common feature of the "favorable" couplings is to ensure inside the open system a breakdown of the dipole rotation invariance of its sub-Lagrangian.

So far we have considered an infinite volume. But we know that living systems, for instance cells, have a finite size. Then Goldstone bosons cannot be massless, because a finite correlation length implies a non-vanishing mass of the messenger "quasi-particle". The modification of the Goldstone theorem in this situation has been discussed in Ref. 10.

The confinement of those massive Goldstone bosons in a finite domain implies, through the quantum uncertainty relations, a non vanishing temperature.
Actually:

$$(\Delta p)^2 (\Delta q)^2 \geq \hbar^2/4$$

(4)

$$(\Delta p)^2 = \overline{(p-\bar{p})^2}, \quad (\Delta q)^2 = \overline{(q-\bar{q})^2}$$

For a confined particle $\bar{p} = 0$, so that:

$$\frac{\overline{p^2}}{2m} \geq \frac{\hbar^2}{8m(\Delta q)^2} \tag{5}$$

The l.h.s. of inequality (5) is the quantum average of the kinetic energy of each component of mass m present in the system. The temperature of the system, which is the average kinetic energy of the population of the particles in the system, cannot be less than the average of each particle. Consequently the r.h.s. of inequality (5) is also the minimal temperature of the system

$$T_{min} = \frac{2}{3} \frac{\hbar^2}{8Km(\Delta q)^2} \tag{6}$$

where K is the Boltzmann constant. Eq. (6) traduces the third principle of thermodynamics, which actually is a consequence of quantum theory. A physical, i.e. finite ($\Delta q < \infty$), system will never reach T = = 0. For a fixed size of the system to be ordered, if $T < T_{min}$, we cannot produce the Goldstone quanta of mass m without violating Heisenberg uncertainty relations. So we cannot order the system until T exceeds T_{min}. This quantum field conclusion matches the conclusion drawn by Prigogine[(14)] on thermodynamic grounds that, in a dissipative system, order could be induced by heating.

The above considerations, however, are not yet satisfactory because we have introduced the finite size of the system by hand. A more correct approach would get the nonvanishing mass of the "quasi-particle" from dynamics and deduce the size from it.

This analysis will be presented elsewhere[(15)]. Here we sketch the main points.

In the above approach we did not take into account the long-range electric interactions, which, by amplifying thermal **fluctuations,** would spoil the ordering of the dipoles. It is well known that Goldstone theorem does not hold when long-range fields play an important role. In this case a mechanism is at work, which has been introduced in electrodynamics by Anderson[16] and in non-Abelian gauge theories by a number of authors[17]; it is commonly known as "Higgs mechanism". Let us summarize it in our case.

It is known that the electromagnetic four-potential $A_\mu(x)$ appears as a consequence of the requirement that Lagrangian is invariant under a local phase shift of the field $\psi(x)$ describing the material components of the system.

However, if we correlate, as in our case, the phases of the basic fields, the phase transformation invariance (or gauge invariance) is lost and usual electrodynamics breaks down. The system organizes itself by trying to exclude electromagnetic field from its inside (Meissner effect) and in this way the ordering of the components is preserved.

Electromagnetic field, when its strength overcomes a threshold, succeeds in penetrating the ordered region along narrow filaments, whose diameter is controlled by a mass factor M, whose value is proportional to the square root of the order parameter.
Transversally to the field propagation, we have:

$$A_\perp = A_{0\perp}\, e^{-\frac{Mc}{\hbar} z} \tag{7}$$

In our case, if P is the non-vanishing electric polarization of the unsymmetric vacuum:

$$M^2 = \text{const.} \cdot P \tag{8}$$

Eq. (8) shows that in a symmetric theory ($P = 0$) M is vanishing and the electromagnetic field is free to invade all the available space. But, when symmetry is broken, the nonvanishing vacuum polarization, via eqs. (7) and (8), squeezes the electromagnetic field into narrow filaments, where ordering is suppressed, whereas it holds in the space among filaments.

Strong electric field gradients are expected on the side boundaries of those filaments. We have discussed elsewhere[18] the role of these gradient forces in the cell biochemical organization. We stress that filamentous structures, which appear dynamically produced, emerge in physiological conditions inside cells (cytoskeleton), on the membrane (flagelli) and also among cells. Contractil fibrils have been shown[19] to appear between adjacent metabolically active erythrocytes.

References

1. P.W. Anderson and D.L. Stein: Broken symmetry, emergent properties, dissipative structures, life: are they related? in the book: P.W. Anderson "Basic Notions of Condensed Matter Physics" Benjamin, Menlo Park, 1984, p. 263-285.

2. M. Eigen and P. Schuster: The Hypercycle: A principle of Natural Self-Organization - Springer-Verlag, Berlin, 1979.

3. H. Haken: Synergetics: An Introduction Springer-Verlag, Berlin, 1979.

4. P. Glansdorff and I. Prigogine: Thermodynamic Theory of Structure, Stability and Fluctuations Wiley, New York, 1971.

5. H. Matsumoto, M. Tachiki and H. Umezawa: Thermo-field dynamics and condensed states North-Holland, Amsterdam, 1982.

6. H. Umezawa: Nuovo Cimento, 40, 450-475 (1965).
   L. Leplae, R.N. Sen and H. Umezawa: Nuovo Cimento 49B, 1-31 (1967).
   H. Umezawa: in "Renormalization and Invariance in Quantum Field Theory (E.R. Caianiello, ed.) p. 275-328, Plenum Press, New York (1974).

7. J. Goldstone: Nuovo Cimento 19, 154-164 (1961).
   J. Goldstone, A. Salam and S. Weinberg: Phys. Rev. 127, 965-970 (1962).

8. H. Matsumoto, H. Umezawa, G. Vitiello and J.K. Wyley: Phys. Rev. 9D, 2806-2813 (1974).
   M.N. Shah, H. Umezawa and G. Vitiello: Phys. Rev. 10B, 4724-4736 (1974).
   C. De Concini and G. Vitiello: Nucl. Phys. 116B, 141-156 (1976).

9. H. Fröhlich: Proc. Nat. Acad. Sci. Usa, 72, 4211-4215 (1975).

10. E. Del Giudice, S. Doglia, M. Milani and G. Vitiello: Phys. Lett. 95A, 508-510 (1983);
    Nucl. Phys. B 251 (FS 13) 375- 400 (1985).

11. S. Celaschi and S. Mascarenhas: Biophysical Journal 20, 273- 278 (1977).
    J.B. Hasted, H.M. Millany and D. Rosen: J. Chem. Soc. Faraday Trans. 77, 2289-2297 (1981).

12. H. Fröhlich: Rivista del Nuovo Cimento 7, 399-418 (1977).

13. Fr. Kaiser, in "Interactions between Electromagnetic Fields and Cells" eds. A. Chiabrera, C. Nicolini and H. Schwan, (Plenum, New York, 1985) p. 131.

14. G. Nicolis and I. Prigogine: Self-Organization in Non-Equilibrium Systems (Wiley, New York, 1977).

15. E. Del Giudice, S. Doglia, M. Milani and G. Vitiello - Milano Preprint (1986).

16. P.W. Anderson: Phys. Rev. 110, 827 (1958); 130, 439 (1963).

17. P.Higgs: Phys. Rev. 145, 1156 (1966).
T.W.B. Kibble: Phys. Rev. 155, 1554 (1967).

18. E. Del Giudice, S. Doglia and M. Milani: Physica Scripta, 26, 232-238 (1982);
Phys. Lett. 90A, 104-106 (1982)
in "Coherent excitations of biological systems"
(H. Fröhlich and F. Kremer eds.) pp. 124-127, Springer, Berlin (1983).
in "Nonlinear electrodynamics in biological systems"
(W.R. Adey and A.F. Lawrence eds.) pp. 477-488, Plenum Press, New York (1984).

19. S. Rowlands: in "Coherent Excitations of Biological systems"
(H. Fröhlich and F. Kremer eds.) pp. 145-161, Springer, Berlin (1983).

ON THE COHERENCE OF ULTRAWEAK PHOTONEMISSION FROM LIVING TISSUES

Fritz-Albert Popp
Technology Center
Opelstraße 10
6750 Kaiserslautern 25

ABSTRACT. At present it is generally accepted that all living systems exhibit a very weak photonemission of a few up to some hundred photons per second and square centimeter of surface area, ranging at least from ultraviolet to infrared. At a first view it appears likely that this "low-level luminescence" corresponds to a chaotic, spontaneous chemiluminescence. However, its temperature dependence and the manifold correlations to physiological and biological functions, as, for instance, radical reactivity, oxygen consumption, stress, cell proliferation and differentiation, biological rhythms, even DNA conformations, point to a regulatory activity of these "biological" photons ("biophotons").

Moreover, a careful analysis of the decay behaviour of photonemission after exposure of the living tissue to external light-illumination indicates that "low level luminescence" orginates from an electromagnetic field with a surprisingly high degree of coherence, as compared to that of technical fields (laser). The basis of this conclusion, namely the results of photo count statistics as well as the relaxation dynamics within the framework of unstable quantum systems under ergodic conditions, is extensively discussed in this paper.

INTRODUCTION

Today it is well accepted that photons of different wavelengths trigger certain biological functions, as, for instance, photorepair (1), phototaxis (2), photoperiodic clocks (3), cell divisions (4), and multiphoton events (5). For a long time, however, it was on the contrary less accepted that all living tissues <u>themselves</u> emit a quasi-continous photo radiation. This phenomenon of "ultraweak" photon emission from living cells and organisms (see e.g.Fig.1) which is different from bioluminescence, exhibits an intensity of a few up to some hundred photons per second and per square centimeter of surface area. Its spectral distribution ranges at least from infrared (at about 900 nm) to ultraviolet (up to about 200 nm). Recently, the most modern aspects of "biological luminescence" were extensively discussed (6).

C. W. Kilmister (ed.), Disequilibrium and Self-Organisation, 207–230.

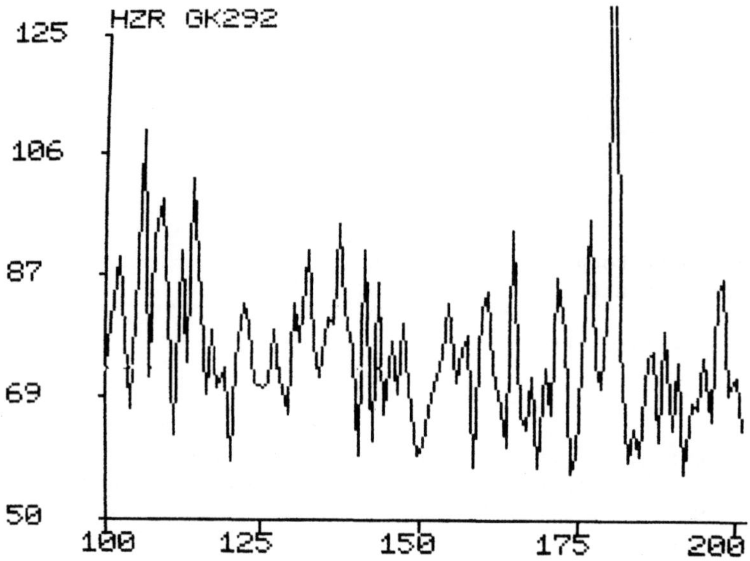

Fig.1a: "low-level luminescence" of cucumber seedlings in counts per second (corresponding to approximately photons per second) during time t from 100 to 200 s.

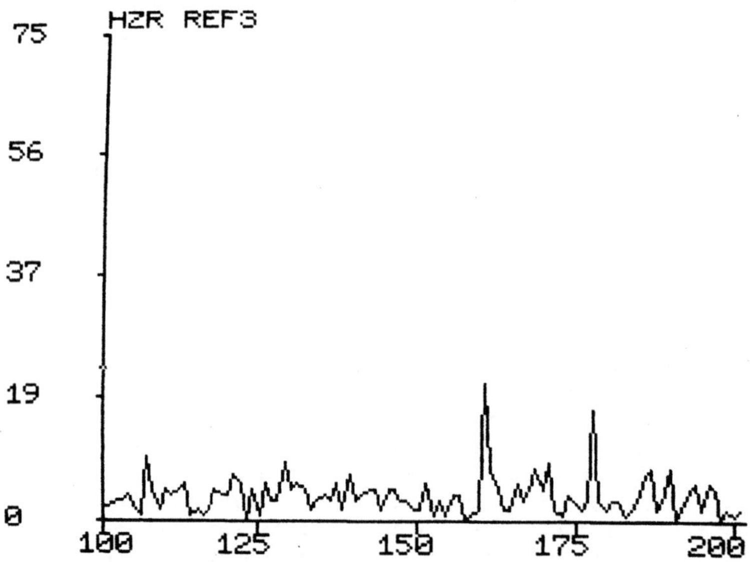

Fig.1b: reference of measurement of Fig.1a, namely the photon emission from the quartz cuvette without cucumber seedlings. The intensity is again presented in counts per second from 100 s (after putting in the sample) to 200 s.

Mainly the weakness of this radiation, however, corresponding to the intensity of a candle at a distance of about 10 kilometers, provoked the opinion that low-level luminescence can only be understood in terms of metabolic "imperfections" (7), orginating from spontaneous chemiluminescence (8), and being associated with the permanent trial of the living state to return after metabolic excitation to thermal equilibrium. Actually, some correlations between the intensity of low-level luminescence and quasi-spontaneous biochemical reactivity, mainly oxidative radical reactions, have been found (9,10).

On the other hand, there are distinct correlations of photon intensity and conformational states of DNA (11), or DNase activity during meiosis (12). At present there exists no further doubt about the physiological character of biophoton emission, since it exhibits just the same temperature dependence as it is characteristic for most of the physiological functions (13,14). Despite of its obvious weakness, the wellknown hysteresis effects of low-level luminescence (15,16) indicate the non-linear character of this emission. The comparison of the "ultraweak" intensity of these biological photons with the expected intensity of black-body radiation at physiological temperatures within the same wavelength range (from infrared to ultraviolet) gives evidence for optical couplings within living tissues "far away" from thermal equilibrium. Actually, the spectral intensities of biophotons amount to magnitudes that are up to about 10^{40} times higher than those of thermal equilibrium at physiological temperatures (14). There is no question that, in view of the Arrhenius factor, this fact alone provides the capacity of a much higher chemical reaction rate than it is possible in thermal equilibrium systems. Thus, biophotons have a much more powerful potency of regulating biochemical reactivity, already by considering the availability of the necessary activation energy, than enzymes alone which enhance biochemical reactivity by means of lowering the activation energy due to complex binding to the substrate (17).

A further characteristic property of this biological radiation turns out to provide a biological basis of distinctive importance, namely the rule

$$f(\nu) \cong 10^{-20} = \text{constant}$$

what is probably generally valid for living systems (18,19). $f(\nu)$ is the probability that photons of energy $h\nu$ occupy the (vacuum) phase space.

This rule does not only mean that biological "matter" is far away from thermal equilibrium which, as is well known, in contrast is governed by the law

$$f(\nu) = \exp(-\frac{h\nu}{kT})$$

where kT represents the mean thermal energy (see Fig.2).

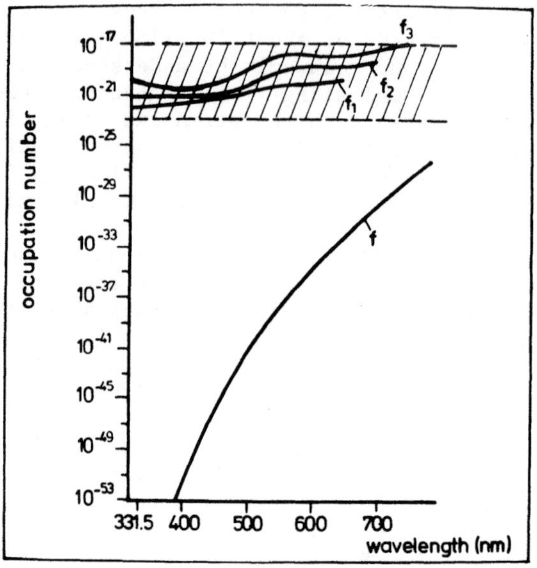

Fig.2: Compared to thermal equilibrium (f) the occupation of the electronic levels of living biological systems is several orders of magnitude higher. At the same time, the "biological" occupation (examples f1, f2 and f3 according to measurements on cucumber seedlings) does not considerably depend on the energy.

f (ν) = constant indicates an "ideal" open system, not subjected to any constraint. It is supplied with always sufficient energy (20), hence representing an absolute maximum of entropy. This does not mean that the entropy of biological systems has to be extremely high. On the contrary, it may become relatively low (theoretically even zero) by the only possible way in those systems, namely the reduction of the degrees of freedom. Such a system has thermodynamically a variety of important properties, for instance noiseless behaviour, which is easy to understand when one optimizes the signal/noise-ratio with respect to the frequency, delivering just f (ν) = constant. Under definite circumstances this system can even amplify the incoming signals (17). Moreover, the rule f (ν) = constant corresponds to a multimode-laser at threshold, since the probability of absorption always equals that of emission between any two excited energy levels, giving thus rise to a theoretically vanishing absorbance or amplification. Variations around this state allow the use of both amplification and absorption of field amplitudes. That this assumption is a realistic one, is demonstrated, for instance, by the surprising results of Mandoli and Briggs (21). They have shown that biological material can guide coherent light without significant loss over distances of some centimeters. In supporting these results, our own transparency experiments (22) indicate that the surprisingly high optical transparency of tissues has not only to be associated with the state of the material, but has to be assigned to a considerably high degree of coherence of the "biophotons" themselves, at least in just our experiments.

In order to show that the coherence of low-intensity laser light is not essential when biological objects are affected, V.V. Lobko et al. (23) assumed the biological matter is in thermal equilibrium and only larger macroscopic entities are relevant. These suppositions are not valid with respect to the real non-equilibrium state, the natural biological distances in significant units of cell-diameters or even smaller units and, in particular for our case, in view of the quasi-stationarity of biophoton-fields.

In fact, recent results of W.B. Chwirot et al. (24) have given evidence of periodic oscillations of low-level luminescence after exposure of synchronized larch microsporocytes to weak quasi-manochromatic light irradiation. The amplitudes and frequencies, which are of the order of a few minutes, depend in these experiments on the wavelength of the irradiated light. Chwirot et al. emphasize that these results are predicted by the electromagnetic model of differentiation (25) that is based on coherent biophoton emission.

A further promising starting point uses the following fact: The temporal intensity distribution i(t) of coherent-state emission displays complete similarity to its Fourier transform $f(\nu)$ (26), where $f(\nu)$ represents the probability of the system to emit its photons with the emission frequency ν. Hence, by comparing the "periodogramm" $f(\nu)$ with the original intensity distribution i(t) one gets some criteria of the coherence. Actually, this transformation of biophoton emission i(t) (see, for instance Fig.1) provides always an "image" that can be brought to some coincidence with the original i(t) by the transformation $\nu \rightarrow t$ and by appropriate linear scaling. Thereby the periodogramm displays a further remarkable feature that is typical for "life" in a quasi-stationary system (though it can be produced also by technical arrangements): Let us consider at first a time interval t from t_1 to $t_1 + \Delta t$ with $\Delta t \gg (\dot{n})^{-1}$, where $\dot{n}$ represents the total count rate. The Fourier transform $f_1(\nu)$ displays within this time interval a pattern of, say, N_1 $(t_1, \Delta t)$ almost equally occupied modes between $0<\nu<(\Delta t)^{-1}$. Then, after taking the time interval from $t_1+\Delta t<t<t_1+2\Delta t$, the Fourier transform $f_2(\nu)$ will now exhibit $N_2(t_2,\Delta t)$ again almost equally occupied modes between $0<\nu<(\Delta t)^{-1}$. As a result, N_2 is generally not equal to N_1 (and all the following values $N_i(t_i,\Delta t)$, i = 3,4,...). Rather, the N_i oscillate with fairly high amplitudes significantly around an average value (that clearly is constant for quasi-stationary conditions). The number N of modes thus reflects some kind of "breathing" which can be assigned to the postulated periodic changes of the number of degrees of freedom in an ideal open system. The Figs.3a and 3b display typical examples of the "mode breathing" and at the same time to similarity of the temporal course of the signals with its "periodogramm".

There is, in addition, a molecular model which explains the phenomena of low-level luminescence. It is based on metabolically "pumped" exciplex formation in the DNA (27-29).

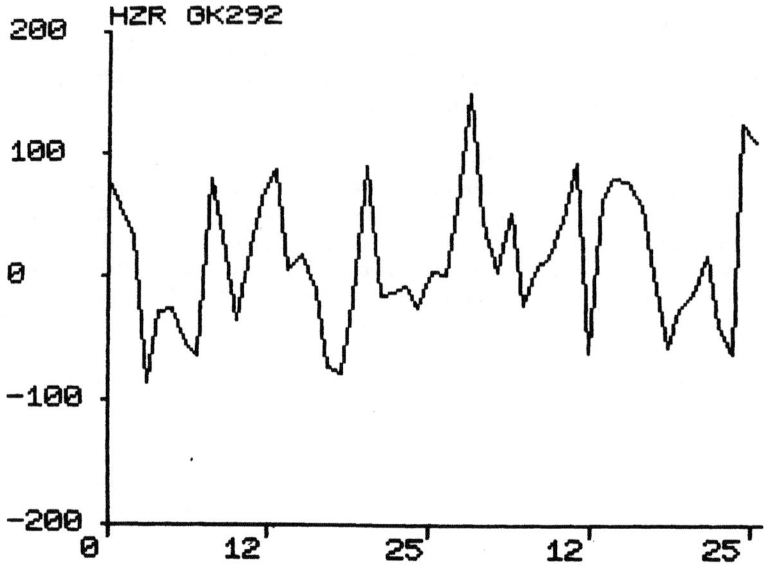

Fig.3a Fourier components (for cosinus from 0 to 25, for sinus from there to 25 again) of the photon intensity of Fig.1a from 100 to 150 s.

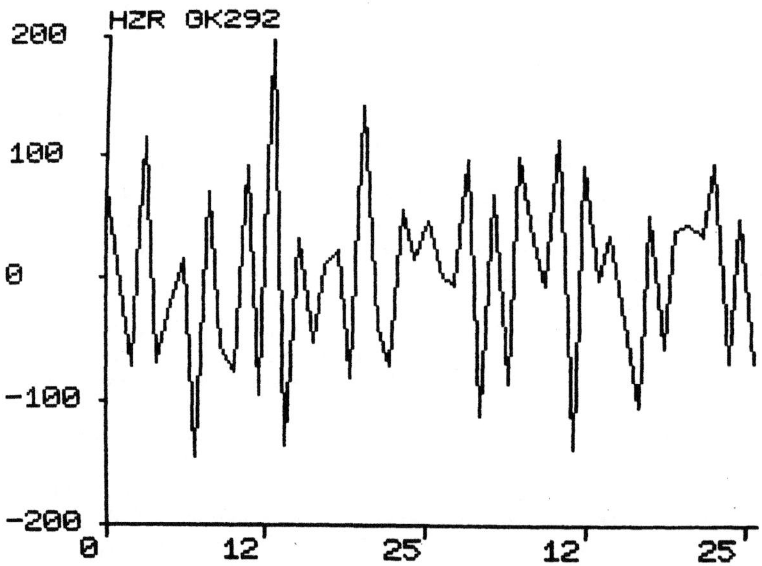

Fig.3b The same as in Fig.3a but now for the photon intensity of Fig. 1a from 151 to 200 s.

All these findings and considerations concentrate more and more to an essential question: Are biological systems paradigms of coherent states in such a sense that evolution has naturally "selected" them just according to a definite coherence-rule (for instance $f(\nu)$=constant),or does coherence, if at all, play only a very limited role?

According to my knowledge, the idea of coherence in biology is commonly neglected or even rejected mainly due to the general concept of biochemistry and linked disciplines, including more local approaches of the living state. On the other hand, Fröhlich's model (30), as well as the concept of dissipative structures (31) and recent papers on the role of Bose condensation in biology (32-35) point to a fundamental importance of coherence in living systems.

In this paper the question of optical coherence in biological systems will be discussed

(1) from a more speculative point of view, considering some principles of the physical background, e.g. the problem of coherence at very low radiation intensities, and
(2) by presenting experimental results on the relaxation dynamics that indicate a surprisingly high degree of coherence within the living state under ergodic conditions.

Some biological and medical consequences have been discussed elsewhere (25, 35-37).

THE QUESTION OF OPTICAL COHERENCE IN BIOLOGICAL SYSTEMS

The degree of coherence has been defined, for instance, by Glauber (38). Coherence to n-th order can be expressed in terms of the expectation value of the operator

$$O(n) = \vec{E}_+(\vec{r}_1, t_1)\ \vec{E}_+(\vec{r}_2, t_2)\ \ldots\ \vec{E}_+(\vec{r}_n, t_n)\vec{E}_-(\vec{r}_n, t_n)\ldots\vec{E}_1(\vec{r}_1, t_1)$$

in the actual state of the electric field.

$\langle O(n)\rangle = Tr(\rho O(n))$, where ρ represents the density operator of this state. $\vec{E} \pm (\vec{r}_i, t_i)$ are the mutually adjoint electric field operators at space-time point $\vec{r}_i$, t_i. If $\langle O(n)\rangle$ is completely factorizable such that

$$\langle O(n)\rangle = \vec{A}^*(r_1, t_1)\ \vec{A}(r_1, t_1)\ \ldots\ \vec{A}^*(r_n, t_n)\ \vec{A}(r_n, t_n)$$

where the $\vec{A}^*\vec{A}$ are the square values of the classical field amplitudes, the field is by definition a fully coherent one to n-th order. For example, the eigenstates $|\alpha\rangle$ of the annihilation operator a according to

$$a|\alpha\rangle = \alpha|\alpha\rangle$$

satisfy the coherence condition. They are called "coherent states".

This fundamental dependence of coherence on factorization of field operators shows that the degree of coherence is neither a question of the intensity of emitted photons nor of the narrowness of their spectral bands. Coherence may occur at all levels of field amplitudes and for all bandwiths including a continuous spectrum (39).

From a classical point of view, however, the term coherence makes only sense, if at least t w o photons are present in the field, since otherwise there could not be interference at all. This objection does not hold from a quantum theoretical point of view, since a coherent state may take the expectation value of particle number o n e or even z e r o. Classical physics requires that the intensity $\dot{n}$ (= number of emitted photons per unit of time) times the coherence time τ shall exceed at least the number 1:

$$\dot{n}\tau > 1 \qquad \text{equ.1)}$$

For decreasing intensities ($\ddot{n} < 0$) the coherence time τ has to increase more and more. This condition seems to contradict experience of technical laser physics, where a considerably long coherence time τ (of the order of 10^{-2}s) is achieved by extraordinary high field amplitudes (pumping power). For low-level luminescence $\dot{n}$ is of the order of a few up to some thousand photons/s (see Fig.1). Consequently, a coherence time up to the order of minutes is required for multiphoton coherence from a classical point of view. This condition seems to be unrealistic or, in other words, it postulates a device that is not available in technical physics, at present. Could this condition, however, be realized in biology, even and possibly in particular by weak fields?

Before presenting sufficient experimental results and their discussion, let us recall that active biological systems actually contain manifold excited states that are really populated. Considering a stationary "white noise" it appears evident that, besides allowed states, also "forbidden" states will be occupied according to the wellknown equ.2):

$$\dot{n}\tau \cong N \qquad \text{equ.2)}$$

N is the actual number of excited states, τ the coherence time of and n the intensity, originating from these states with coherence time τ. As soon as $N > 1$, the conditional equ.1) is always satisfied. Providing a stationary state of a relatively high excitation of matter, from equ.2) we learn that a low intensity $\dot{n}$ may then rather indicate a state of h i g h coherence than that of a vanishing one.

Although this argumentation does certainly n o t p r o v e the coherence for weak effects, it shows that low intensities can never deal as a basis to reject a realistic possibility of even high coherence. In contrast, a quasi-stationary state far away from thermal equilibrium exhibits a natural tendency for coherence.

It is, however, principally difficult to examine coherence at low intensities experimentally, for the following reasons:

(1) The usual methods of interferometry must fail in view of the low amplitudes,
(2) the signal/noise ratio may become rather low,
(3) from a classical point, at low intensities only long coherence times make sense, but it is difficult to keep stationary conditions during the consequently long observation times, in particular for living systems.

Apart from these technical problems, the photo count statistics (PCS) that provide the only possible tool for examining coherence of low-level luminescence, fail even for stationary states, namely, if the number of the degrees of freedom is not known. This becomes evident by the following considerations.

The photo count distribution $p(n, \Delta t)$, which represents the probability of registering n counts in a preset time interval Δt, takes for a chaotic field the form

$$p(n, \Delta t) = \frac{<n>}{(1+<n>)^{n+1}} \qquad \text{equ. 3a)}$$

while a fully coherent field is subject of

$$p(n, \Delta t) = \exp(-<n>) \frac{<n>^n}{n!} \qquad \text{equ. 3b)}$$

where $<n>$ represents the average number of registered photons during Δt. For a derivation of equs.3) see, for instance, ref. (40).

A consequence of (3) is that the variances associated with equs.3a) and 3b) are

$$<(\Delta n)^2> = <n>(1+<n>) \qquad \text{equ. 4a)}$$

and

$$<(\Delta n)^2> = <n> \qquad \text{equ. 4b)}$$

respectively.

To take an example, let us consider the count rate of cucumber seedlings shown in Fig.1 (case 1) and those which are poisoned by acetone, in order to achieve a high count rate (case 2). After the seedlings reached a maximum emission, they remain for some minutes at a quasi-stationary state, where approximately the following count rates (in cps = counts per second) were registered:

TABLE I

Case	$<n>$	$<(\Delta n)^2>$
reference	50 ± 10 cps	60 ± 20 cps
1	150 ± 30 cps	180 ± 70 cps
2	45 400 ± 100 cps	45 700 ± 500 cps

The preset time interval was chosen in case 1 and case 2 as $\Delta t = 0.1$ s. The errors have been estimated by taking into account the dark count rate of about 50 $\pm$ 20 cps including unavoidable deviations from stationarity.

Comparing this result with equs.4a) and 4b), respectively, it appears that only a fully coherent field can account for the experimental values, while a chaotic field could be significantly rejected.

However, equ.4a) is only valid for an one-mode field. In the case that multiple different and independent modes contribute to the total intensity, equ.4a) has to be considerably modified, while equ.4b) remains valid. If M represents the number of modes that may participate (for simplicity, with always the same intensity), we obtain instead of equ.4a)

$$<(\Delta n)^2> = <n>(1 + \frac{<n>}{M}) \qquad \text{equ. 4a*)}$$

For a derivation of equ.4a*) see, for instance, ref. (41).

A comparison of equ.4b) for a fully coherent field and equ.4a*) for a chaotic field shows that the difference in terms of photo count statistics of a stationary field disappears more and more with increasing M. Since M represents the ratio of measuring time interval Δt to the coherence time τ, equs.4) allow only to estimate the possible upper limit of the coherence time for a chaotic field. Evidently, we can exclude $M < 10^5$ for our example (case 2), where we tacitly assumed that the radiation originates from a chaotic field. With respect to the measuring time interval $\Delta t = 0.1$ s, we learn from equ.4a*) that the coherence time should be smaller than 10^{-6} s in case of spontaneous chemiluminescence. This would exclude strongly forbidden transitions as a possible source. Unfortunately, M is up to now not determined with sufficient accuracy. Consequently, without further knowledge it is impossible to decide whether low-level luminescence corresponds to a chaotic field of coherence times $< 10^{-6}$ s, or to a fully coherent field.

There are two ways of solving this crucial problem, namely

(1) an extented spectral analysis of PCS with the aim to limit M to its actual boundaries, and

(2) an examination of non-stationary states, e.g. the relaxation dynamics under external stationary conditions (ergodicity).

Actually, by analysing the decay behaviour of different spectral modes of biophotons after exposure of plant seedlings to light illumination we found an extremely strong mode-coupling indicating that M is of order 1 (14). In this case, low-level luminescence corresponds evidently to a fully coherent field. However, in order to confirm these indications of coherence of "low-level luminescence" it is at the actual state of discussion important to investigate the relaxation dynamics of biophoton emission more detailed. This includes the point of view that "coherence" of biological systems may become only evident under non-stationary conditions.

RELAXATION DYNAMICS OF LOW-LEVEL LUMINESCENCE

Let us imagine an unstable system in which a definite part of kinetic energy of the decay products (e.g. photons) is restored again by rescattering within the source system (e.g. excited molecules). By confining ourselves at first to a classical oscillator model we have consequently

$$<\ddot{x}x> = (1 + \kappa)<\dot{x}\dot{x}>, \qquad \text{equ. 5)}$$

where x represents the amplitude of the oscillator, and $\kappa = 0$ accounts for the usual chaotic rescattering, wh,ere the average values of kinetic and potential are equal, while $\kappa \neq 0$ is due to the re-storage effect of coherent rescattering (42,43). Of course, for $\kappa = 0$ the solution of equ.5) is exponential (equ.6a), whereas in case of coherent rescattering ($\kappa \neq 0$) the solution of equ. 5) takes the form of a hyperbolic function (equ.6b).

$$\langle x \rangle = x_0 \exp(-\beta t) \qquad \text{Equ. 6a)}$$

β is a (generally complex) constant.

$$\langle x \rangle = x_0' \cdot (t + t_0)^{-\beta'} \qquad \text{equ. 6b)}$$

Hence, under the same (ergodic) conditions, an exponential decay turns into a hyperbolic one as soon as a chaotic rescattering is substituted by a coherent rescattering of the decay products to their source.

Unfortunately, the theory of unstable quantum systems, which could confirm this result generally, is not developed to such an extent that coherent rescattering can be described already without some puzzling problems. At present, even the generally accepted experimental evidence of an exponential decay in the case of chaotic rescattering is not yet clearly established theoretically.

However, as Ersak (42) has demonstrated, possible deviations from exponential decay are always due to a coherent rescattering of the decay products to their source. Clearly, by separating the time evolution of the actual state $|\psi\rangle$ of the system according to $H|\psi\rangle = i\hbar|\dot{\psi}\rangle$

$$\exp(-iHT)|\psi(0)\rangle = A(t)|\psi(0)\rangle + |\phi(t)\rangle, \qquad \text{equ. 7)}$$

where H is the Hamilitonian of the system between any two observation points, and $|\phi\rangle$ represents the dynamical state of the decay products, we have consequently

$$\langle\psi(0)|\phi(t)\rangle = 0 \qquad \text{equ. 8)}$$

After multiplication of equ.7) with the bra $\langle\psi(0)/\exp(-iHt')$, the relation

$$A(t+t') = A(t)A(t') + \langle\psi(0)|\exp(-iHt')|\phi(t)\rangle \qquad \text{equ. 9)}$$

is obtained.

Since a semi-group law $A(t + t') = A(t) \cdot A(t')$ is a necessary and sufficient condition of exponential decay, equ.9) indicates that a non-random rescattering of the decay products to their source suffices for deviations from exponential decay. As Fonda et al.(43) and Davies (44) have shown, the exponential decay can be generally derived from

$$i\hbar\dot{\rho} = [H,\rho(t)] - i\hbar\lambda_1(\rho(t) - \sum_j P_j\,\rho(t)P_j), \qquad \text{equ. 10)}$$

where ρ is the density operator associated to $|\psi\rangle\langle\psi|$ of equ.7) λ_1 is a constant that describes the frequency of randomly distributed rescattering processes of the decay products with respect to their source. Hence, the quantum description of rescattering refers to measurement processes where the P_j are the projectors onto the eigenmanifolds of the corresponding observables.

It is easy to show that, if instead of λ_1 = constant a coherent rescattering according to

$$\lambda_1 = a_j^{\kappa}$$

is chosen, where κ is a constant ($\neq 0$) and a_j is the probability amplitude of the excited state $|\psi>$ of the source associated to a projector

$$P_j = |\phi_j><\phi_j| \ ,$$

again a hyperbolic decay is obtained. At the same time, the uncertainty

$$[\dot{\rho}, \ \Sigma P_j \rho P_j]$$

is then minimized, in accordance to the general property of coherent states (45).

A possibly more interesting approach to the problem originates from the fact that rescattering depends on the number of reductions that occur during decay (43). From this an apparent Zeno's paradoxon arises: the more reductions (observations) are taken into account, the more improbable it becomes that the unstable state decays at all. This problem has been investigated by several authors (46-48). However, as Bunge and Kalnay (49) have shown, one cannot hypothesize that measurements which lead to the reduction of the state under investigation can be carried out in infinitely short time intervals.

This interesting result is confirmed also by the following considerations which deliver a further possibility of differentiating chaotic and coherent fields.

Take the identity

$$\int_0^{\tau} \psi^* \dot{\psi} t dt = \tfrac{1}{2}[\psi^2 t]_0^{\tau} - \tfrac{1}{2}\int_0^{\tau} \psi^2 dt \qquad \text{equ. 11)}$$

If $\psi^2 = 1$ for all t, the RHS of equ.11) vanishes. Consequently, the LHS should vanish, too. Since according to the Schrödinger equation

$$i\hbar\dot{\psi} = H\psi$$

we then obtain after substitution into equ.11)

$$\frac{1}{i\hbar} \int_0^{\tau} (\psi^* H\psi) t dt = 0 \qquad \text{equ. 12)}$$

In case of a Hermitian operator, this is obviously correct for the real part. However, for the imaginary part the uncertainty relation comes into play

$$\Delta(\psi^* H\psi) \cdot \Delta t \gtrsim h$$

such that for times

$$\tau < \frac{h}{\Delta(\psi^* H \psi)}$$

a deviation from equ.12) is allowed.

For an unstable system, on the other hand, the real part of the LHS of equ.11) cannot vanish. Consequently, for time intervals

$$0 < t < \tau$$

where τ is the coherence time, the real part of the RHS of equ.11) cannot vanish, too. This means that an uncertainty in evaluating the RHS of equ.11) has to be taken into account. For $0 < t < \tau$ $\psi^2 t$ cannot be identical to

$$\int_0^\tau \psi^2 \, dt$$

This leads to a fundamental difference of evaluating equ.11) for chaotic and coherent fields, respectively. Providing ergodic conditions, for both coherent and chaotic rescattering, the value of the RHS increases proportionately to the observation time. For a chaotic field we then have

$$t \cong \tau_{ch} << \Delta t = \tau$$

such that

$$\int_0^\tau \psi^* \dot{\psi} t dt \cong \Delta t \cdot \tau_{ch} \langle \psi^* \dot{\psi} \rangle \qquad \text{equ. 13a)}$$

This leads obviously to an exponential decay.

In the case of a fully coherent field, on the other hand, we have

$$\tau_{ch} > \tau = \Delta t \geqq t$$

and consequently

$$\int_0^\tau \psi^* \dot{\psi} t dt = \lambda \cdot \tau \psi^2 \neq 0; \; \lambda = \text{const.} \qquad \text{equ. 13b)}$$

Equ.13b) is generally valid if, and only if $\dot{\psi} \propto \frac{\psi}{t}$

Hence we obtain the general result

$$|\dot{\psi} \rangle\langle \psi| = \frac{\lambda}{t} |\psi \rangle\langle \psi| \qquad \text{equ. 13b*)}$$

that describes again the expected hyperbolic decay law.

In order to confirm the argumentation, the PCS theory can be extended to an ergodic unstable quantum system.

By definition we have

$$\langle (\Delta n)^2 \rangle = \sum_i (\dot{n}_i - \langle \dot{n} \rangle)^2 \, \Delta t^2 \qquad \text{equ. 14)}$$

where n_i is the count rate of the i-th measuring interval Δt of finite and constant length. Let us imagine that by suitable choice of the number of ensembles under investigation Δt can always be kept at a value of the order of the coherence time of the field.

In order to determine $<\ddot{n}>$, we may either keep Δt constant and register $<\dot{n}>$ for all t, or we change at all t slightly the length of Δt and register the alteration of $<\dot{n}>$. Since t and Δt are independent quantities for an ergodic field, it doesn't matter what method is prefered.

Noting that the derivation of $(\dot{n}_i - <n>)$ does not contribute to

$$\frac{d<(\Delta n)^2>}{d<\dot{n}>}$$

we obtain from equ.14)

$$\frac{d<(\Delta n)^2>}{d<\dot{n}>} = 2\,\frac{<(\Delta n)^2>}{\Delta t}\cdot\frac{d<\Delta t>}{d<\dot{n}>} \qquad \text{equ. 15)}$$

An ergodic system provides the homogeneity of t and Δt. Hence, the relation between d (Δt) and dt for constant

$$\left|\frac{d<(\Delta n)^2>}{d<\dot{n}>}\right|$$

must be a linear one, if the time average can always be represented by the ensemble average. Consequently, we have

$$d(\Delta t) = \beta dt \qquad \text{equ. 16)}$$

β is a (generally complex) constant for a preset time interval Δt, for definite coupling parameters and a fixed number of ensembles, including $\beta = 0$ for a stationary system that represents a special case of an ergodic field.

It is well known that a Gaussian field obbeys the relation

$$<(\Delta n)^2> = <\dot{n}>^2\,\Delta t^2 + <\dot{n}>\,\Delta t \qquad \text{equ. 17a)}$$

while a coherent field is subject of

$$<(\Delta n)^2> = <\dot{n}>\Delta t \qquad \text{equ. 17b)}$$

These relations are valid at any instant for any Δt. The degree of freedoms does noch play a decisive role, as we will see later. Hence, let us confine at first to a single-mode field. By calculating

$$\frac{d<(\Delta n)^2>}{d<\dot{n}>}$$

of equ.17a), substituting the general relation equ.15) into these derivations and taking into account equ.16), we then arrive after straightforward calculations at

$$\frac{d<\dot{n}>}{dt} = \frac{\beta}{\Delta t}\left(\frac{<\dot{n}>}{1 + 2<\dot{n}>\Delta t}\right) \qquad \text{equ. 18a)}$$

for chaotic fields and

$$\frac{d\langle\dot{n}\rangle}{dt} = \frac{\beta}{\Delta t}\langle\dot{n}\rangle \qquad \text{equ. 18b)}$$

for coherent fields, respectively.

Again the differences of equ.18a) and equ.18b) would disappear for increasing number M, since the term $2\langle\dot{n}\rangle\Delta t$ in the denominator of equ.18a) vanishes for $M \to \infty$ in case of a multimode field or $\Delta t >> \tau$, respectively (41). However, this does not bother the remarkable differences of the relaxation dynamics of a chaotic and a coherent field. In fact, in case of a chaotic field we have

$$\Delta t >> \tau_{ch}$$

This means that after taking into account

$$\langle\dot{n}\rangle\, \tau_{ch} << 1$$

from equ.18a) a relation

$$\frac{d\langle\dot{n}\rangle}{dt} \cong \frac{\beta}{\Delta t}\langle\dot{n}\rangle \qquad \text{equ. 19a)}$$

is obtained that delivers in view of $\beta/\Delta t$ = constant an exponential decay law. However, in case of a coherent field it is allowed to extend $\Delta t \to t$ as long as t is smaller than the coherence time τ. Consequently, we than have

$$\frac{d\langle\dot{n}\rangle}{dt} = \frac{\beta}{t}\langle\dot{n}\rangle \qquad \text{equ. 19b)}$$

that yields again the hyperbolic decay law.

EXPERIMENTAL BACKGROUND AND AN APPLICATION IN CANCER RESEARCH

In a previous paper (14) it has been shown that living tissues exhibit significant deviations from exponential decay after exposure to light illumination, while the agreement to a hyperbolic decay law is excellent, even, and in particular, for the decay of single modes that can be observed by using interference filters.

Fig.4 displays a further example, where the total emission from cucumber seedlings after exposure to a 10-second illumination of a Halogen lamp (150 W) at a distance of 20 cm was observed, subjected to the same technique as referred in (14).

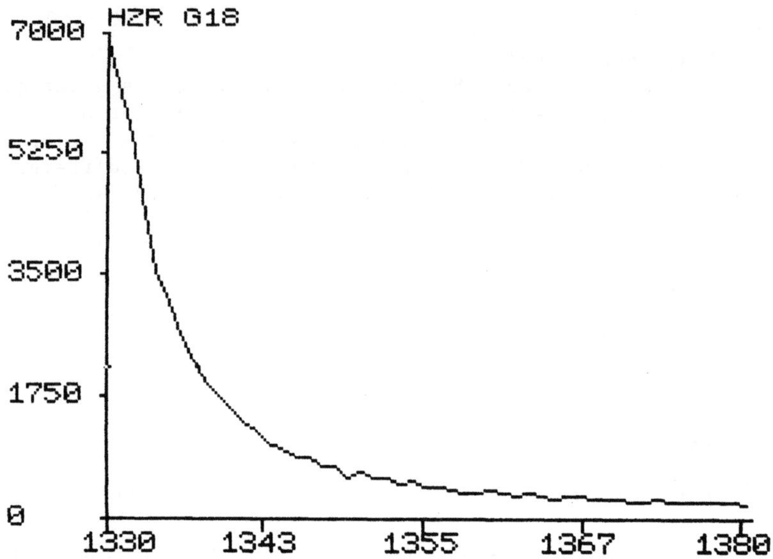

Fig.4a: Photonemission from cucumber seedlings after exposure to weak white-light illumination (in counts per 0.5 s) for 300 measuring intervals (150 s).

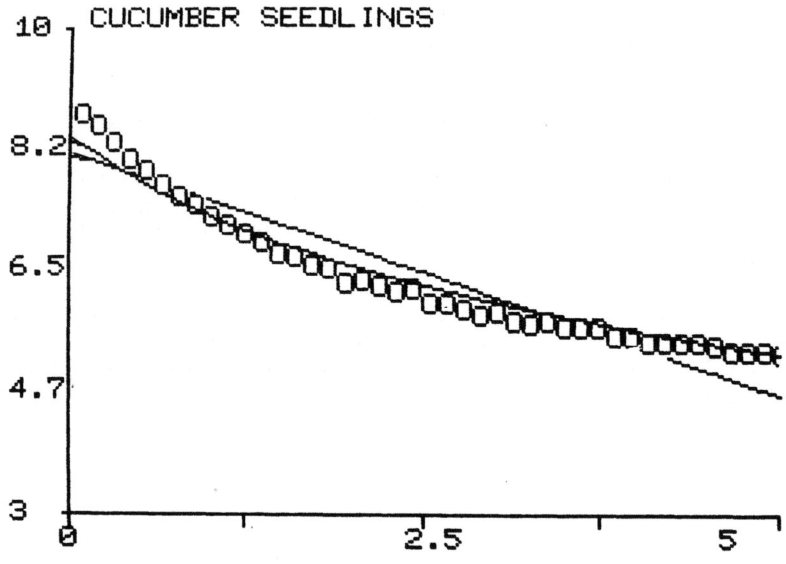

Fig.4b: Logarithmic scale for the ordinate of the measurements of Fig.3a, where the measured values (000...) were approximated by a hyperbolic law and an exponential one, comparably. The abscissa displays the time in arbitrary units.

Chwirot et al.(50) have demonstrated that synchronized cell cultures at meiosis exhibit a more or less hyperbolic decay after exposure to weak white-light illumination. The agreement to the hyperbolic law is there correlated to the cooperativity within the different stages of the cell cycle, appraised from the biological point of view.

In ref. (14) it was already shown that the relaxation dynamics of normal and corresponding malignant tissues display significant differences, which can be associated to diminished cooperativity in tumours.

Recently, Schamhart et.al. (51) have shown that the total number of counts which are emitted by cell cultures after exposure to white-light illumination

(1) increases with increasing cell densities for malignant liver cell cultures, and

(2) decreases with increasing cell densities for the corresponding normal ones.

Thereby, they confined themselves to a definite first part of the decay curves immediately after irradiation. Fig.5 displays these results (courtesy of Dr. Schamhart).

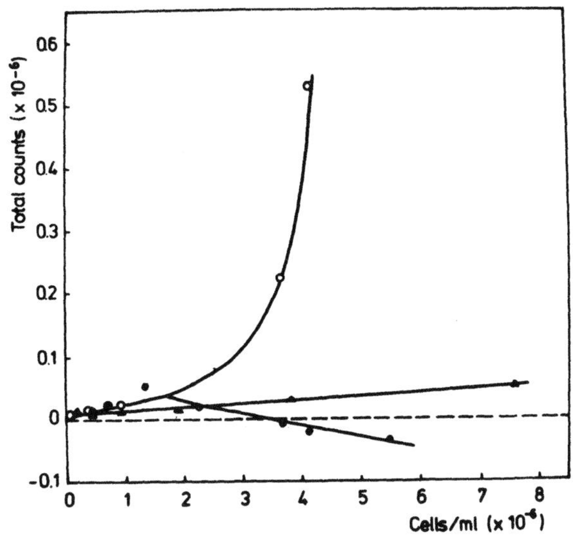

Fig.5: Total counts within the first seconds after exposure of cell suspensions to white light illumination. With increasing cell density, HTC cells (ooo) that are malignant and the corresponding normal hepatocytes (●●●) show principally different behaviour. The H35 cells (ΔΔΔ) are only weakly malignant.

At the same time Schamhart has shown that the relaxation dynamics of the normal cells agree better with the hyperbolic decay than that of the corresponding malignant ones, which display more rapid decay.

Before presenting own recent results on human cell cultures, the experiments shown in Fig.5 should be discussed.

This system consists of an ensemble of radiating cell layers in a cubic quartz cuvette within a colourless nutrition fluid. The total surface area of the cuvette is 6F, the diameter $d = \sqrt{F}$.

If p is the contributed photo count rate of one cell, and ρ is the cell density, we then obtain an increase di of the measured photon intensity by the contribution of a cell layer of thickness dx at a distance x from the counter according to

$$di(x) = \rho F p(1-W(x))dx \qquad \text{equ. 20)}$$

W(x) is the probability that the radiation is absorbed within the system on the way of length x between the layer and the counter that is located at point 0. First of all, there is no reason to expect for W(x) a value deviating from the Beer-Lambert law:

$$W(x) = 1 - \exp(-\mu x)\ \exp(-\rho\lambda x) \qquad \text{equ. 21)}$$

μ is a constant absorption coefficient of the device that is always the same for all the experiments. λ represents the absorption coefficient per unit of cell density for the cells within the medium. It is expected to be or order 10^{-6} cm^2.

After insertion of (21) into (20) and integration, we then obtain

$$i(0) = \frac{\rho F p}{(\lambda\rho+\mu)}\ (1 - \exp(-\lambda\rho+\mu)d) \qquad \text{equ. 22)}$$

where i(0) is the measured radiation intensity.

The result of Fig.5 describes

i(0) as a function of ρ

that exhibits a principally different behavior for normal and malignant tissues. From equ.22) we obtain

$$\frac{\partial i(0)}{\partial\rho} = \frac{i(0)}{\rho}\ (1 - \frac{\rho}{\rho+\mu/\lambda}) + \frac{\rho F p}{\lambda\rho+\mu} \cdot \lambda d \cdot \exp(-(\lambda\rho+\mu)d) \qquad \text{equ. 23)}$$

Since both terms of the RHS of equ.23) are positive definite, this model can never explain, firstly

$\frac{\partial i}{\partial\rho} < 0$ which is observed at higher cell densities of normal cells.

Secondly, it is not possible to explain

$\frac{\partial i}{\partial\rho} >> 0$ for malignant cells at higher cell densities, too, since for

$\rho \to \infty$ we have according to equ.23) $\frac{\partial i}{\partial\rho} \to 0$.

Fig.6 demonstrates the differences between the theoretical model due to the most reasonable assumptions and the real behaviour.

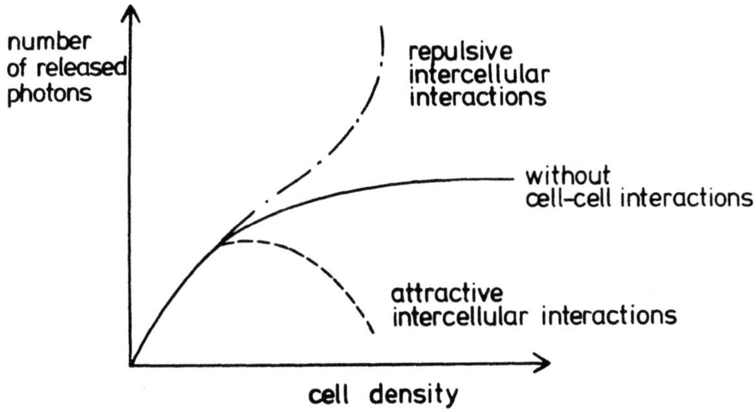

Fig.6: Theoretical calculation of the dependence of the photon intensity i (0) on the cell density for the cases that
(1) no interactions between the cells play a role (——),
(2) the interactions become aggregative (---), and
(3) the interactions become disaggregative (-.-.-.).

There are in principle only two possibilities to explain these significant deviations from expected results.

The first is a dependence of p on the cell density. This would mean that the production of photons alters very sensitively with mutual long-range interactions between cells. Malignant cells would produce more photons with decreasing mutual distance. The contrary would be valid for normal cells. This interpretation is supported by reports according to which tumour tissues may show a higher count rate than normal ones. We could not confirm this so far. However, this argumentation is supported by own observations of a dependence of photon intensity on differentiation. We generally observe a lower count rate of the unperturbated tissue with increasing differentiation.

However considering this argument one should realize that the experiments of Fig.5 are based on a count rate p of the order of about one photon per hour (or even less). If one prefers to envisage photochemical reactions as the source, for instance the alteration of enzymatic activity with the change of mutual distance of the cells, an explanation of even nonlinear (!) effects in terms of those in this case extremely rare events would be quite phantastic.

Hence, we prefer the second possible interpretation of this effect, namely the alteration of λ with varying cell density. Although the first possibility

$$\frac{\partial p}{\partial \rho} \neq 0$$

is not excluded by this and may really play a role, it appears more likely to explain the effect of Fig.5 in terms of

$$\frac{\partial \lambda}{\partial \rho} \neq 0$$

Since the cell densities used in the experiments are rather low compared to that of a solid tissue, a change of λ would indicate a very sensitive dependence of optical properties of living cells (as entities) on mutual long-range interactions. Since from Fig.5 we have consequently

$$\frac{\partial \lambda}{\partial \rho} > 0 \text{ for normal cells, and}$$

$$\frac{\partial \lambda}{\partial \rho} < 0 \text{ for malignant ones,}$$

normal cells exhibit an increasing absorption of weak mutual photons with increasing density, while malignant ones increase the reflection probability.

Roughly speaking, while normal cells improve the basis of mutual communication in the tendency of forming cell colonies by means of photon interaction with decreasing diestance, the contrary holds for malignant cell populations.

Although it is principally impossible to decide whether the effect of Fig.5 is due to

$$\frac{\partial p}{\partial \rho} \neq 0 \text{ or, alternatively, to } \frac{\partial \lambda}{\partial \rho} \neq 0$$

or possibly due to both of these alternatives, in any case there has to be concluded that

(1) there is a sensitive dependence of biophoton emission from living cells on mutual long-range interactions at a distance from at least about ten cell diameters on,

(2) in view of the very low intensities ($p \ll 1\ s^{-1}$) and very large distances between the single cells ($\rho v \ll 1$, where v is the volume of a cell), these non-linear effects corresponding actually to "stimulated emission and absorption of photons at very weak intensities" can only be explained in terms of coherence properties of the interacting photons. Taking into account the interaction distance of at least 10 cell diameters, the maximum coherence volume of biophotons is, according to these results, at least thousand times the cell volume.

At the same time, this test provides a powerful tool of differentiating normal and malignant tissues on the decisive level of intercellular interaction.

Since from the "most reasonable" model of equ.22) one would expect that the decay behaviour of the single cell (p(t)) corresponds exactly to the population, a further examination of these coherence effects concerns the characteristics of the decay functions. Therefore, we studied recently the relaxation dynamics of human cells after white-light illumination under the same conditions as Schamhart et al. have chosen. We compared human amnion cells with corresponding malignant ones, namely wish cells.

Fig.7 shows a typical example, where the decay functions of amnion cells and wish cells at a cell density of $3 \cdot 10^6$ cells/ml have been observed under the same conditions.

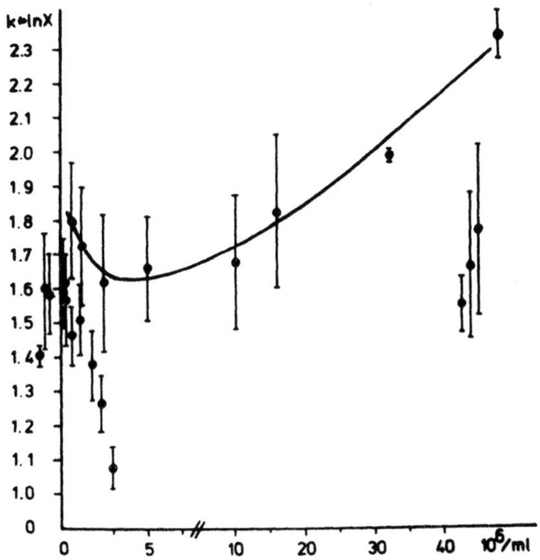

Fig.7: The decay parameter of the hyperbolic approximation that is adjusted to the relaxation dynamics of photon emission of different cell suspensions after exposure to weak white-light illumination in dependence on the cell density. The lower curve displays the behaviour of normal amnion cells. The opposite behaviour is shown by the corresponding malignant wish cells (upper curve). The three measurements at the right side of Fig.7 correspond to the nutrition medium alone.

These measurements were carried out with different cell densities ρ, where ρ was altered unsystematically, in order to avoid systematical errors. Then the best fitting of the hyperbolic decay according to equ.24) was calculated by use of a computer program.

$$i(0,t) = A\ (t+T)^{-\frac{1}{\kappa}} \qquad \text{equ. 24)}$$

T is a constant corresponding to the time delay between the first measuring point and the end of escitation. It is 3 seconds and was kept constant for all the measurements. The values A and $1/\kappa$ were determined for all the decay curves. $1/\kappa$ (ρ) is plotted in Fig. 7.

Again it can be seen that normal cell populations display a behaviour controversal to the malignant ones. While in the case of normal cells κ increases with increasing cell density (see, for clearness also equ.5)), κ decreases nonlinearly with increasing density for malignant cells, approaching more and more the chaotic exponential rescattering.

These results do not only confirm the importance of coherence in biology even, and in particular, in the case of weak effects: they are obviously fundamental in solving at least one of the most crucial problem and related questions. A more profound discussion of this last topic has been presented elsewhere (e.G. 25,37).

Finally, it may be worthwhile to note that non-exponential decay is also observed sometimes in condensed matter physics. As has been shown by Fain (52), however, in those cases non-random processes, including instabilities of the environmental conditions, are significant. This confirms the opinion that nature exhibits some tendency to coherence at low intesities. As Ngai et al. (53) emphasized, in these cases of coherent rescattering, a single linear exponential form is as unphysical as a superposition of them.

The connection to the usual exponential decay on the basis of perturbation expansions has been presented in ref. (54), while a summary of the given arguments appears in ref.(55).

REFERENCES

(1) A. Kelner: Proc.nat.Acad.Sci. (Wash.) **35** (1949), 73.
J. Setlow: 'The Molecular Basis of Ultraviolet Radiation and Photoreactivation'. In: Current Topics in Radiation Research, II (M. Ebert and A. Howard eds.), North-Holland Publishing Co., Amsterdam 1966.
(2) D.P. Häder: 'Photomovement'. In: Encyclopedia of Plant Physiology. New Series, **Vol. 7** (W. Haupt and M.E. Feinleib eds.), pp. 268-309, Springer, Berlin 1979.
(3) E. Bünning: The Physiological Clock. Third edition. Springer, Berlin 1973.
(4) T.T. Karu, G.S. Kalendo, V.S. Letokhov, and V.V. Lobko: Nuovo Cimento **6** (1982), 828.
(5) P.P. Calmettes and W.M. Berns: Proc.Natl.Acad.Sci. USA **80** (1983), 7197.
(6) D. Slawinska and J. Slawinski: Photochem.Photobiol. **37** (1983), 709.
(7) A.I. Zhuravlev (ed.): Ultraweak Luminescence in Biology. Moscow Society of Naturalists, Moscos 1972.

(8) H.H. Seliger: In: Chemilumenscence and Bioluminescence (M.J. Cormier, D.M. Hermes and J. Lee, eds.), Plenum Press, New York 1973.
(9) A. Boveris, E. Cadenas, R. Reiter, M. Filipkowski, Y. Nakase, and B. Chance: Proc.Natl.Acad.Sci.USA 77 (1980), 347.
(10) E. Cadenas, H. Wefers, and H. Sies: Eur.J.Biochem. 119 (1981), 531.
(11) M. Rattemeyer, F.A. Popp, and W. Nagl: Naturwissensch. 68 (1981), 572.
(12) W.B. Chwirot, R.S. Dygdala, and S. Chwirot: Cytobios, accepted for publication.
(13) F.A. Popp, K.H. Li, and W. Nagl: Z.Pflanzenphysiol. 114 (1984), 1.
(14) F.A. Popp, B. Ruth, W. Bahr, J. Böhm, P. Graß, G. Grolig, M. Rattemeyer, H.G. Schmidt, and P. Wulle: Collective Phenomena 3 (1981), 187.
(15) H. Precht, J. Christophersen, H. Hensel, and W. Larcher: Temperature and Life. Springer, Berlin 1973.
(16) I. Milczarek, J. Slawinski, and E. Grabikowski: Zeszyty Naukowe Akademii Rolniczej w Szeczecinie 39 (1973), 249.
(17) F.A. Popp: In: Electromagnetic Bio-Information (F.A. Popp, G. Becker, H.L. König, W. Peschka eds.), Urban & Schwarzenberg, München-Wien-Baltimore 1979.
(18) F.A. Popp: In: Proceedings International Symposium on Analytical Applications of Bioluminescence and Chemiluminescence (E. Schram and P. Stanley eds.), Brussels 1978, State Printing & Publishing Inc., Westlake Village, California 1979.
(19) J. Slawinski, E. Grabikowski, and L. Ciesla: J. Luminescence 24/25 (1981), 791.
(20) I. Prigogine: Pers.Communication.
(21) D.F. Mandoli and W. Briggs: Proc.Natl.Acad.Sci. USA 79 (1982), 2902.
(22) F.A. Popp, W. Nagl, K.H. Li, W. Scholz, O. Weingärtner, and R. Wolf: Cell Biophysics 6 (1984), 33.
(23) V.V. Lobko, T. Karu, and V.S. Letokhov: Biofizika 30 (1985), 366.
(24) W.B. Chwirot, R.S. Dygdala, and S. Chwirot: submitted for publication.
(25) W. Nagl and F.A. Popp: Cytobios 37 (1983), 45; 71.
(26) W.H. Louisell: Quantum Statistical Properties of Radiation. J. Wiley, New York 1973.
(27) K.H. Li: Laser + Elektrooptik 13 (1981), 32. F.A. Popp: ibid., p.34.
(28) K.H. Li, F.A. Popp, W. Nagl, and H. Klima: In: Coherent excitations in Biological Systems (H. Fröhlich and F. Kremer, eds.), Springer, Berlin-Heidelberg-N.Y. 1983.
(29) F.A. Popp and W. Nagl: Polymer Bulletin 15 (1986), 89.
(30) H. Fröhlich: Int.J.Quantum Chem. 2 (1968), 641.
(31) I. Prigogine, G. Nicolis, and A. Babloyantz: Physics Today 11 (1972), 23.
(32) T.W. Wu and S. Austin: J.theor.Biol. 71 (1978), 209.
(33) R.K. Mishra, K. Bhaumik, S.C. Mathur, and S. Mitra: Int.J. Quantum Chem. 16 (1979), 691.
(34) K.H. Li: Physics of Open Systems. Rep.Prog.Phys. (1986), in press.
(35) F.A. Popp: In: Information und Ordnung (G. Schaefer ed.), Aulis-Verlag, Köln 1984.
(36) W. Nagl and F.A. Popp: Cytobios 37 (1983), 45; 71.

(37) F.A. Popp: Biologie des Lichts. Paul Parey, Berlin 1984.
(38) R.J. Glauber: In: Quantum Optics (R.J. Glauber ed.), Academic Press, New York 1969.
(39) F.A. Popp and W. Nagl: Cell Biophysics, in press.
(40) F.T. Arecchi: In: Quantum Optics (R.J. Glauber ed.), Academic Press, New York 1969.
(41) J. Perina: Coherence of Light. Van Nostrand Reinhold Company, London 1971.
(42) I. Ersak: Yad.Fiz. 9 (1969), 458.
(43) L. Fonda, G.C. Ghirardi, and A. Rimini: Rep.Prog.Phys. 41 (1978), 587.
(44) E.B. Davies: Helv.Phys.Acta 48 (1975), 365.
(45) F.A. Popp: Proceedings of the ISG Conference on Selforganization and Dissipative Structures, London 1985, in press.
(46) B. Misra and E.C.G. Sudarshan: J. Math.Phys. 18 (1977), 756.
(47) C.B. Chiu, E.C.G. Sudarshan, and B.Misra: Phys.Rev. 16D (1977),520.
(48) A. Peres: Am.J.Phys. 48 (1980), 931.
(49) M. Bunge and A.J. Kalnay: Nuovo Cimento 77B (1983), 1.
(50) W.B. Chwirot and S. Dygdala: Cytobios, accepted for publication.
(51) D. Schamhart, A. Slawinski, and R. van Wijk: Cancer Res., in preparation.
(52) B. Fain: Phys.Rev. 24A (1981), 933.
(53) K.L. Ngai, A.K. Rajagopal, R.W. Rendell, and S. Teitler: Phys.Rev. 28B (1983), 6073.
(54) K.H. Li and F.A. Popp: Phys.Lett. 93A (1983), 262.
(55) F.A. Popp: Proc. on the Intern. Conference on Lasers '85, Dec. 2-6, 1985, The Society of Optical & Quantum Electronics, Las Vegas, USA, in press.

EVOLUTION: NATURAL SELECTION OR SELF-ORGANIZATION?

M.W. Ho
Developmental Dynamics Research Group
Open University
Milton Keynes MK7 6AA
UK

and

P.T. Saunders
Department of Mathematics
King's College
Strand, London WC2R 2LS
UK

ABSTRACT. When proper account is taken of the non-randomness inherent in matter and the self-organizing properties of process, natural selection becomes redundant and misleading as an explanation. This is exemplified in current studies on the origin of life, on development and pattern formation, and on the nature of heredity. As a metaphor which arose from a particular perception of life, natural selection serves only to reinforce the prejudice which gave rise to it.

1. INTRODUCTION

The scientific community of the 19th century inhabited a Newtonian clockwork universe in which physical cause and effect were linearly related, and bodies did not change their motion unless compelled to do so by external force. By contrast, the phenomenon of life was a total enigma. Darwin's special contribution was to offer a naturalistic alternative to both vitalism and divine creation: life evolved by the natural selection of random variations.

What is the theory of natural selection? It states that given the presence of heritable variations in individual organisms, their propensity for Malthusian (i.e. geometric) increase, and the limitation of resources in the environment, there will be competition for survival and reproduction in which the fittest variants will eventually win. This leads to an improvement in the adaptation of the species over a number of generations; and if the process is continued (and especially if there is a change in the environment), there will be a gradual transmutation of species. To this Darwinian account, the neo-Darwinists of the present day add the theory of Mendelian genetics, according to which heredity is mediated through discrete entities, the 'genes', and variations are caused by random mutations in those genes.

A further crucial assumption is involved, though few neo-Darwinists have stated it as explicitly as does Maynard-Smith (1): that

C. W. Kilmister (ed.), Disequilibrium and Self-Organisation, 231–242.

the theory of natural selection is both necessary and sufficient to account for the evolution of life on this planet to date.

True Darwinian natural selection is competitive: there is ´positive´ selection for the most advantageous variant. This distinguishes it from the ´negative´ weeding out processes due to lethal or semi-lethal mutations, diseases or extremes of temperature, which have been designated ´internal selection´ and ´environmental selection´ respectively by neo-Darwinists themselves. In practice, however, the competition paradigm prevails, and is the sense in which natural selection is most frequently used. This then constitutes the neo-Darwinian framework which purports to explain all of biological evolution.

But is natural selection a real mechanism? While it explains how an adaptive variation could be selectively preserved; it says nothing about how the variation arose in the first place. As Samuel Butler (2) put it, ´Mr. Darwin has given us an Origin of Species with the "Origin" cut out´. This and other limitations of the theory were apparent to Darwin himself as well as to some of his contemporaries, though they did not in the least dampen the great enthusiasm with which the theory was received. Much of the reason for its instant success is that the theory was cut from the very fabric of Victorian English society. Barzun (3) remarked that Darwin´s philosophy was ´imbibed, whether he knew it or not, from the economic, social and metaphysical speculations of his time. It did not arise spontaneously from his facts. Malthus and Spencer, combined with Lamarck and Positivism, provided him with the needed assumptions and attitudes, both seemingly so simple and so common in their kind as to pass for matter of fact. What brought him rapid victory and prolonged sway over his age was thus the ability of the age to recognize itself in him..´ In other words, competition, and the struggle for survival, were so much the order of the day that everyone believed they understood the theory and, moreover, that it must be true. Ultimately, natural selection - a metaphor borrowed from life - was most easily seen to be an apt explanation for life itself.

Today, special creation is hardly a serious issue (except in the United States). Furthermore, we no longer exist in the Cartesian universe of mechanical determinism. Physics has undergone a deep conceptual revolution with Einstein´s theory of relativity, and with quantum mechanics. Recently, the study of dynamics and nonequilibrium thermodynamics in chemistry (4) has revealed a host of physicochemical phenomena that exhibit a fundamental kinship with living organization. The time is ripe for a unification of physics and chemistry (and mathematics) with biology; not in the sense of reducing one to the other, but in the sense of ´carrying the understanding of physics and chemistry upstream to (the) comprehension of the "ordered complexity" of life process´ (5).

Traditionally, the study of evolution has consisted largely in showing how organisms are adapted to the environment as the result of natural selection. The justification for this resides in the word ´random´ in ´natural selection of random variations´. The definition given in the Oxford Dictionary is as follows: ´not sent or guided in a special direction; having no definite aim or purpose; made, done, occurring, etc. at haphazard´. This certainly implies that little if anything can be said of the variations; hence natural selection is considered to be the only direction-giving or creative agent in evolution. The discovery of Mendelian genetics has reinforced this view, as the genetic mutations, assumed to be responsible for generating variations, can indeed be shown to arise at random. (The obvious caveat is that even though the genetic mutations may be random, the phenotypic variations are not.) So pervasive is this way of thinking that even the most enlightened neo-Darwinists fail to conceive of explanations other than the two opposing categories of chance (randomness) versus natural selection (for example, see Lewontin´s (6) otherwise excellent critique of neo-Darwinian theory).

Whilst neo-Darwinists have taken randomness for granted, evidence has been accumulating for a long time that non-randomness is at the very core of life and evolution (7). The most recent demonstration of directed evolution is in the origin of biomolecular handedness (8). According to classical accounts, it was entirely a matter of chance that living organisms are composed of L-amino acids and D-sugars. In other words, the original accident was ´frozen´ or preserved by natural selection for an efficient homochiral biochemistry. (Note that even here, natural selection is already used in the sense of natural stereo-biochemical process.) However, the universe is not symmetric; the parity-violating, weak neutral current interaction gives rise to an energy difference between the mirror-symmetric isomers. This results in a slight stabilization of L-amino acids and the L-peptides relative to the corresponding enantiomers, which is sufficient to break the chiral symmetry of an autocatalytic racemic reaction in an open, non-equilibrium system.

Non-randomness (9) exists at the level of the organism as in pre-biotic chemistry, as we shall presently show. Thus, natural process is the real basis of self-organization in living systems. ´Self-organization is to be understood as the spontaneous emergence of coherence or structure without externally applied coercion or control´ (10). For this reason, the term ´natural selection´ is superfluous and, in many cases, obscurantist, misleading and ideological. We stress that non-randomness implies local determinism but not global finality; it is fully compatible with creativity and novelty. What non-randomness does offer is a non-arbitrary description of process; after all, even chaos, or unpredictability can be precisely described (11). This is the sort of description we propose for evolution in place of the arbitrary one based on the natural selection of random variations. The unfalsifiability of neo-Darwinism has already been demonstrated elsewhere (12), and we shall not be discussing that issue here.

2. CAN MATTER ORGANIZE ITSELF?

One of the most captivating questions about evolution is how life itself emerged. From the scientific point of view, life is a particular organization of matter which exhibits metabolism and reproduction. The question can thus be rephrased in the words of Pasteur: can matter organize itself (13)? The answer given within the neo-Darwinian orthodoxy is: not without natural selection. This stems from the assumption that the major features of the living state are extremely improbable, and require special explanatory principles. The scenario usually painted of the origin of life is the competitive replication of 'information molecules', i.e., nucleic acids. Thus, Eigen and Schuster (14) specifically invoke natural selection as 'the only understood means of creating information, be it the blueprint for a complex living organism... or be it a meaningful sequence of letters the selection of which can be simulated by evolutionary model games.' In a similar vein, Crick (15) and Orgel (16) both consider the origin of the genetic code to be mainly the result of accidents 'frozen' by natural selection, though neither of them rules out entirely the possibility that stereochemical interactions may have some role to play.

On account of the undue emphasis placed on the 'genes' as both the determinants of biological form (and, consequently, on 'genetic mutations' as the source of evolutionary novelty), neo-Darwinists seem unable to conceive of life except in terms of self-replicating nucleic acids. This has led them to postulate that the first RNAs, which they believe arose by chance, must either have had enzymatic functions or else coded for enzyme proteins which enabled them to reproduce. By itself, however, competitive replication of RNAs cannot result in anything but 'selfish replicators' (17), like the 'selfish DNA' which have recently been found in both eukaryotic and prokaryotic genomes. Metabolism and other cellular functions simply cannot emerge.

The answer to the question of whether matter can organize itself must surely be an unqualified yes. The physics and chemistry of matter is such that ordered structures can arise spontaneously in open systems (18). For example, highly ordered hexagonal convection cells can suddenly appear in a horizontal layer of fluid which is heated from below to beyond a critical temperature. Similarly, temporal and spatial oscillations associated with the Belousov-Zhabotinskii reaction will give rise to a whole range of coherent structures depending on the geometry of the medium (19).

Evidence for self-ordering in prebiotic evolution comes from extensive experiments in the laboratories of Fox and others (7). In the synthesis of proteinoids by thermal copolymerization of amino acids, highly non-random peptides are formed; in other words, the amino acid sidechains themselves provide the stereochemical 'information' necessary to determine the amino acid sequence of the polymers that are produced. Moreover, a variety of enzymatic, photochemical and even hormonal activities are found to be associated with these polymers. In contact with water, proteinoids readily form microspheres which exhibit many of the properties of modern cells, including fusion, fission and excitability. Cationic proteinoids also catalyse the polymerization of nucleotides.

Basic proteinoids complex readily with both DNA and RNA. With RNA, they form globular ribosome-like structures which exhibit selective interactions with other polynucleotides. A very suggestive observation is that under certain conditions, polymerization of each of the aminoacyl-adenylates (an intermediate in protein synthesis) of glycine, lysine, phenylalanine and proline can be catalysed most effectively in complexes of the proteinoid with the homopolynucleotide corresponding to the correct codon (20). This and other experiments demonstrate the stereochemical basis for the possible origin of the genetic code within protocells composed of proteinoids. Further, the process must have involved a two-way ´information flow´, or complementary recognition, between protein and nucleic acid. Thus the appropriate terms of description in prebiotic evolution are physicochemical stability, molecular complementarity and autocatalysis. To speak of ´competition´ or ´natural selection´ in this context is simply to introduce an obscuring and misleading metaphor.

3. THE DYNAMICS OF FORM

The physics and chemistry of self-organization are relevant to evolution in yet another respect, the generation of biological form. Within the neo-Darwinian framework, form is explained as a conglomeration of past random variations, accumulated by natural selection and preserved by heredity. There is consequently a tendency to decompose organisms into bits, or traits controlled by genes, for each of which some selective advantage must be postulated. It is here that the greatest confusion between cause and effect arises. Natural selection must, by definition, act <u>a posteriori</u> (assuming that it acts at all); it cannot call an adaptive character miraculously into being. The problem of the origin of the adaptive form is quite separate from that of its preservation, as Samuel Butler and many others have emphasized. A number of neo-Darwinists, both recently and in the past, have recognized the significance of what they call ´developmental constraints´ to natural selection, acknowledging that the developing system does not allow all possible forms to occur. This is a nod in the right direction, but it is hardly enough.

Forms arise from dynamic processes which can be described independently of natural selection. The dynamics are not merely constraints to form; for without them, there can be no form at all. Moreover, they are similar in kind (though not necessarily in detailed mechanism) to those generating forms in the physical world (21, 22) under dissipative conditions far from thermodynamic equilibrium (23). This kinship between the biological and physical world was the main thesis of D´Arcy Thompson´s classic work, <u>On Growth and Form</u> (24). There, he pointed out how a drop of fusel oil in paraffin develops the same shape as a jelly fish. Even earlier, Leduc (25) had placed fragments of calcium salts in some colloidal solutions and produced filamentous structures with regular constrictions which look like segmented organisms. And we are all familiar with the striking similarity between the spiral waves produced by the Belousov-Zhabotinskii reaction in a petri dish and the aggregation waves of the slime mould amoebae.

An understanding of the processes by which forms are created will tell us how they can evolve without invoking natural selection to cover up and justify our ignorance. The project to exhaust the physical and mechanical causes of biological form goes back at least to D´Arcy Thompson, who specifically rejected facile Darwinian explanations. There has been much progress since, which enables us to make inferences about evolution that are distinct from those based solely on selective advantage (21, 26). We shall consider two examples given by Saunders (21), one concerning mimicry and the other animal coat markings.

A voluminous literature has accumulated on the phenomenon of mimicry, the resemblance between organisms not only from closely related species, but also from different phyla. Mimicry is usually taken to be one of the most convincing demonstration of the power of natural selection. A selective advantage is postulated, and in some cases, this can be verified by experiment. But even if mimicry does confer a selective advantage, it tells us nothing as to how the forms themselves have come into being, only why they have persisted. The palatable Viceroy mimics the Monarch butterfly, which tastes bad, thus deriving protection from predators to some extent. The resemblance in the colour pattern of the wings, however, was generated by the similarity of underlying patterning processes in the first place, and not by natural selection. This is most clearly illustrated in cases of so-called pseudo-mimicry (kindly brought to our attention by Dick Vanewright). Anetia cubana and Lexias aeropus are in different butterfly families and live on opposite sides of the globe: the former in Cuba and the latter in Indonesia. Yet they resemble each other as closely as the most impressive mimics.

Because the patterning processes are dynamic, superficial resemblances can also occur between widely different organisms. Mivart (27), a major contemporary critic of Darwin, drew attention to the ´bird´s head processes´ which can be found in a marine bryozoa. It is difficult to imagine what selective advantage there might be for a bryozoa to mimic a group of birds. Similarly, it should occasion no suprise that an orchid can resemble in some respects a wasp; contrary to what Maynard Smith (28) implies, it is natural selection and not development that is irrelevant in the origin of this particular case of superficial convergence in form.

It is well-known that the protection afforded by mimicry breaks down if the frequency of the mimic exceeds that of the model; hence it would have been much more advantageous for the Viceroy to develop its own bad taste instead of copying the Monarch´s wing pattern. Clearly, the overriding ´cause´ for mimicry is the dynamics producing form convergence independently of any consideration of selective advantage.

Murray (29) has proposed a Turing type reaction-diffusion model for animal coat markings. This supposes that there is some morphogen with a single threshold concentration above which the colour will be dark and below which it will be light. It turns out that the precise pattern produced is dependent on the size of the embryo at the time the pattern is laid down. If the domain is too large or too small, there is no pattern at all. Small variations in the boundary conditions can significantly alter the pattern but almost always leave the general features unchanged: spots are replaced by similar spots and not by stripes or solid colour. These results show that processes similar in essence to those described by the equations are sufficient to account for some important features - such as the tendency for both very large and very small mammals to be uniformly coloured, and the existence of individual variation in coat markings in herds, without regard to natural selection.

We can now see that it is not simply a matter of taste whether one takes natural selection to be the primary explanation for evolution. On the contrary, it profoundly affects the way we study evolution. Neo-Darwinists will continue to take the greatest pains in measuring the selective advantage of real or imaginary characters or, worse yet, invent "Just So" stories to account for their existence. On the other hand, people who are aware of the need to study real processes will concentrate on investigating those instead, from the physics and chemistry of prebiotic evolution, to the dynamics of form generation. Moreover, the understanding gained in one area will have a significant impact in others. For example, the dynamics of form will serve as a firmer basis for a rational morphology (30, 28) and hence for phylogenetic systematics (31, 32), than hypothetical pathways of adaptive transformations.

The recognition of process in form generation has another important consequence. It makes explicit the role of genes in development as inntroducing consistent biases into the patterning dynamics (28, 22) rather than as acting as some ´central directing agency´ which can make all conceivable things happen. It also explains how environmental perturbations to those processes can have the same effect, or phenocopy those arising from genetic mutations (33), for the dynamics define a limited number of ways in which the system can be perturbed.

4. HEREDITY IS PROCESS

We now turn to the problem of inheritance and its relevance to development and evolution. Neo-Darwinists have traditionally maintained that Weismann´s barrier (which ensures that any modifications induced during an individual´s life time by the environment will not be inherited) makes the study of development separate from evolution. We have seen that even if Weismann´s barrier were absolute, the processes whereby form is generated during development would still have the primary deterministic influence on evolution. Within the past five years, however, recombinant DNA research has shown that Weismann´s barrier is far from absolute (34). In all higher organisms, a frequent route of communication between soma and germline is via the reverse transcription of processed mRNA from somatic cells into DNAs which are then reinserted into the germline genome. The processed ´pseudogenes´, as they are called, are present for both unique structural genes and for shorter sequences which are highly repeated and dispersed throughout the genome. For example, the Alu sequences, which are repeated some 500,000 times in the human genome, are processed pseudogenes of the RNA molecules which make up the cytoplasmic signal recognition particles involved in membrane translocation of protein.

Even more striking are the changes in germline DNA which can be induced by the environment within a single generation. The best studied example is the induction of heritable changes in some varieties of flax on treatment with different fertilizers (35). The stable lines, or ´genotrophs´ produced, differ from one another in morphological characters as well as in the amount of nuclear DNA, the redundancy of rRNA genes and isoenzyme patterns.

Thus the findings from recombinant DNA research point to an extremely dynamic feedback relationship between so-called soma and germline. Genomic DNA, once thought to be relatively static and unchanging, turns out to be very fluid. A substantial part of the genome of higher organisms consists of multiple copies of mobile genetic elements that can transpose, or jump, from one location to another. Transpositions, DNA rearrangements, gene amplifications and deletions occur frequently as part of normal development and in response to stress in both somatic and germcells. Consequently, on an evolutionary timescale genomic DNA undergoes turnover via cycles of sequence rearrangement, amplification (or deletion), and sequence divergence. There has been speculation that DNA turnover, under its own dynamic, can result in rapid speciation quite independently of natural selection (36). This is an interesting suggestion as it provides an easy solution to the problem of rapid evolutionary change which has plagued Darwin and many neo-Darwinists to the present day. But perhaps it is too easy a solution. For it creates the paradox of how, in the face of such fluidity of the genome, the process of development can remain so relatively stable, generation after generation. Nor can the difficulty be overcome by supposing that the apparent fluidity is due to normal control mechanisms which ensure that the right genes become expressed at the appropriate place and time. For then we run into an infinite regress when we ask what controls the control mechanisms themselves.

All this suggests that our concept of heredity is in need of reformulation (22, 33). Instead of positing heredity in some material (DNA sequences) which must be passed on almost unchanged from generation to generation, we must view heredity as a process of the developing system, which includes, among other things, the environment. After all, the most distinguishing characteristic of an organism is its ability to modify the environment and to respond to it. In this way, the problem of infinite regress disappears. In place of a linear chain of command from DNA to phenotype, there is a complex of interlocking feedback processes. These include cytoplasmic patterning reactions, intercommunication between nucleus and cytoplasm, intercommunication between cells, inductive relations between cell layers and tissues, and finally, the necessary relationships between the organism and the external environment whereby the internal processes can be orchestrated and co-ordinated. It is this whole complex of relations and processes, rather than the nuclear genes alone, which is reponsible for the stability and reproducible of inheritance. The causation of inheritance is thus non-localized and diffused, and is the property of the whole system.

It is easy to see why many genetic mutations will be neutral (37) in effect, because they are buffered by other reactions and especially by the external environmental parameters which remain unchanged. On the other hand, certain environmental changes will have discernible effects either by altering the expression of genes, or in some cases, the DNA itself. And, especially if the new environmental regime is maintained, the altered gene expression state or DNA rearrangement will appear to be stably inherited (22, 33). In this light, development and evolution are intimately connected, and not at all separate, as neo-Darwinists claim.

In what way can natural selection be said to act? The relationship between organism and environment forms an undivided whole. Any attempt to distinguish ´interaction´ from ´selection´ is purely arbitrary (38). We can see that here, as elsewhere, the introduction of the concept of natural selection serves only to obfuscate and confuse. Of course, organisms can perish from lethal mutations as from extreme environmental conditions. But (as suggested by Brian Goodwin), this is no different from saying that certain changes in the system or the environment can affect the stability, or persistence, of the system under consideration.

5. SOCIAL DARWINISM AND SOCIOBIOLOGY

Darwin´s theory of natural selection sprang from the same philosophical roots of mechanical materialism which inspired in turn Adam Smith´s free market political economy and the scientistic positivism of Auguste Comte. Darwinism was the culmination of this movement to which it gave a name. Barzun (3) remarks, ´Matter´ and ´force´, when applied to human beings, found some dangerous simple applications... And when the idea of force is embodied in the notions of Struggle and Survival of the Fittest, it should be expected that men will use these revelations of science as justifications for their own acts....´

This completes the positive feedback loop between the dominant sociocultural ideology of the day and a scientific theory to which it gave birth; and which, as we have seen, consists of little more than an empty metaphor. Nevertheless, it led via the Social Darwinists and theories of racial inequality to the rise of the Nazi regime.

But history is repeating itself now with Sociobiology. As a moderate within the camp remarks, ´..Some hard-selling proponents of sociobiology have been intent on using their relatively small portion of evolutionary biology as a means of taking over the human social sciences..´(39)

Sociobiology is based on the claim that most, if not all, of the social behaviour of animals is largely genetically determined, and has arisen as the result of natural selection. In the case of humans, the list of so-called characters includes ´aggression´, ´competitiveness´, ´male superiority´, ´incest taboo´ and so on. These are fine examples indeed of Whitehead´s fallacy of misplaced concreteness. Neither the genetic determination of the characters, nor indeed their very existence has been demonstrated for animals or for humans. But no matter, ingenious adaptive stories are invented to account for them.

Sociobiology is not just a parlour game for dons, however. Like Social Darwinism, it has considerable sociopolitical implications. As

Bateson (39) points out, 'The emphasis on selfishness and the struggle for existence in evolutionary biology has had an insiduous confirmatory effect on the public mind....' and again, 'The political dangers of representing all human social relationships in terms of competition is that the expectation is self-fulfilling.' Nevertheless, Bateson does not see the need to give up sociobiology. Instead, he proposes to sanitize it, so as to make it acceptable to as many people as possible, mainly by de-emphasizing competition and stressing cooperation (or co-adaptation) in nature. Of course, this is merely a rehearsal of the many arguments which have been pitched against Social Darwinism in its day, and history has demonstrated that they are woefully inadequate. Within Darwinism as within neo-Darwinism, competition is primary. Where cooperative behaviour does occur, it has to be explained as a phenomenon arising almost paradoxically out of competition -- as by kin selection or reciprocal altruism. Consequently, any attempt to construct a neo-Darwinist sociobiology in which cooperation is anything more than a relatively minor aspect of behaviour is doomed to failure. Besides, nothing is to be gained by replacing one metaphor with another.

The major weakness in the critiques of Sociobiology, as in those of Social Darwinism, is the almost universal and unquestioned acceptance of the foundations on which they are based: neo-Darwinism and Darwinism respectively. Our predecessors were combating vitalism on the one hand, and Fundamentalism on the other, and therefore could see no real alternative to mechanical determinism. We, however, live at a time when a new alliance between physical chemistry and biology has not only led to the enrichment of dynamics and thermodynamics with the mysteries of life (40), but should in turn result in the reenchantment and revitalizing of biology with process and creativity (41).

Acknowledgment

We would like to thank the following for comments on an earlier draft of the manuscript: Godfrey Vesey, Sidney Fox, Stephen Mason, Nick Furbank and Brian Goodwin.

REFERENCES

1. Maynard Smith, J. In Towards a Theoretical Biology, vol. 2, (ed. C.H. Waddington), p. 82, 1969.
2. Butler, S. Evolution Old and New, London, 1911. Darwin himself was conscious of the problem; the subtitle of the Origin is 'The Preservation of Favoured Races in the Struggle for Life' (emphasis supplied).
3. Barzun, J. Darwin, Marx, Wagner, 2nd ed., Doubleday & Co., New York, 1958.
4. Prigogine, I. From Being to Becoming, Freeman, San Francisco, 1980.
5. Piel, G. In Behavioural Evolution and Integrative Levels (G. Greenberg and E. Tobach, eds.), p. 9, 1984.
6. Lewontin, R. C. The Genetic Basis of Evolutionary Change, Columbia University Press, New York, 1974.
7. Fox, S. In ref. 41, p. 15.
8. Mason, S. Nature **311**, 19, 1984.
9. A better term here may be 'non-arbitrariness'. 'Randomness' within neo-Darwinism has the connotation of being arbitrary.
10. Godfrey Vesey points out that in using language such as 'self-ordering' and 'self-organization', we are in part returning to the Aristotelean view that teleology is internal to matter. However, we definitely reject a teleology that proposes organisms are shaped by adaptation to some external purpose or function, whether it be natural selection or some omnipotent creator that is postulated to account for it.
11. Berry, M. In Colloque: Les Systemes Dynamique (S. Diner, ed.). Springer, 1984.
12. Saunders, P.T. and Ho, M.W. Nature and System **4**, 179, 1982.
13. Cited by Fox, S. and Dose, K. Molecular Evolution and the Origin of Life, Marcel Dekker, New York, 1977.
14. Eigen, M. and Schuster, P. The Hypercycle, Springer-Verlag, Berlin, 1979. In fact, the authors are using natural selection as an extraneous metaphor. If, as they claim, the dynamic of molecular interactions they describe is sufficient to account for cellular metabolism and replication, then natural selection is irrelevant.
15. Crick, F.H.C. J. Mol. Biol. **38**, 367, 1968.
16. Orgel, L.E. J. Mol. Biol. **38**, 381, 1968.
17. Bresch, C., Nieser, U. and Harnasch, D. J. theor. Biol. **85**, 399, 1980.
18. Glansdorff, P. and Prigogine, I. Thermodynamic Theory of Structure, Stability and Fluctuations, Wiley, New York, 1971.
19. Winfree, A.T. Science **181**, 937, 1973.
20. Nakashima, T. and Fox, S.W. J. Mol. Evol. **15**, 161.
21. Saunders, P.T. In ref. 41, p.243.
22. Ho, M.W. Rivista di Biologia **77**, 147, 1984.
23. Nicolis, G. and Prigogine, I. Self-Organization in Non-Equilibrium Systems, Wiley, New York, 1977.

24. Thompson, D´A. W. On Growth and Form, Cambridge University Press, Cambridge, 1914.
25. Leduc, S. Theorie Physico-Chymique de la Vie, Paris, 1910.
26. Goodwin, B.C. In ref. 41, p.219.
27. Mivart, St. G. On the Genesis of Species, Macmillan, London, 1871.
28. Maynard Smith, J. New Scientist, 14 March, 1985, p.38.
29. Murray, J.D. J. theor. Biol. **88**, 161, 1980.
30. Webster, G. In ref. 41, p.193.
31. Lovtrup, S. In ref. 41, p.159.
32. Nelson, G. Syst. Zool. **27**, 324, 1978.
33. Ho, M.W. In ref. 41, p.267.
34. Pollard, J.W. In ref. 41, p.291.
35. Cullis, C.A. In Evolutionary Theory: Paths into the Future (J.W. Pollard, ed.), Wiley, London, 1984.
36. Dover, G.A. and Flavell, R.B. (eds.). Genome Evolution, Academic Press, London, 1982.
37. Kimura, M. Nature **217**, 624, 1968.
38. Ho, M.W. and Saunders, P.T. In Towards a Liberatory Biology, (S. Rose, ed.) p. 85, Allison and Busby, London, 1982.
39. Bateson, P. In Limits to Science (S. Rose, ed.), Blackwells, Oxford, 1985 (in press).
40. Prigogine, I. and Stengers, I. Order out of Chaos, Bantam Books, Toronto, 1984.
41. Ho, M.W. and Saunders, P.T. (eds.). Beyond neo-Darwinism: An Introduction to the New Evolutionary Paradigm, Academic Press, London, 1984.

THERMODYNAMICS AND COMPLEX SYSTEMS

P.T. Saunders
Department of Mathematics
King's College
Strand, London WC2R 2LS
UK

and

M.W. Ho
Developmental Dynamics Research Group
Open University
Milton Keynes MK7 6AA
UK

There is a striking similarity between thermodynamics and the study of biological evolution. Both are concerned with the time development of complex systems, and in both cases there is a definite trend to be explained. One of the many workers who have commented on this was Waddington (1), who wrote: "A few authors have tried to formulate a concept of a general parameter which will always change in one direction during evolution, as entropy always increases in physical systems, and as Fisher seems to have thought his fitness would always increase. If such a parameter could be defined, one could deduce from it the nature of the 'evolutionary force' which keeps evolutionary processes on the move ...."

In this paper we show that there are indeed parallels to be drawn between evolution and thermodynamics, although fitness and entropy do not in fact play analogous roles. We can use the relatively well understood subject of thermodynamics to give us insights into evolutionary theory, partly by discovering which ideas and results can be carried over, and partly by noting the points at which the differences become important. We suggest that the same strategy can be used in the study of other complex systems. For the sake of brevity, the biological detail has been kept to a minimum; more details can be found in refs (2), (3) and (4).

1. ORDER, ORGANIZATION AND COMPLEXITY

To begin, we have to decide what to take as the the fundamental quantity that increases in evolution. One candidate might be order, on the grounds that this is what decreases in thermodynamical systems, but this is unsatisfactory because it leads to the prediction that evolution should produce organisms which more and more resemble perfect single crystals.

C. W. Kilmister (ed.), Disequilibrium and Self Organisation, 243–253.

On the face of it, organization might seem a better choice. Unfortunately it turns out to be so difficult to define that we doubt that it is best thought of as a quantity at all. A number of authors (e.g. 5,6) have proposed definitions in terms of information content and redundancy, but these ignore too many aspects of our intuitive understanding of the word to be suitable. We suggest that it would be better to consider organization to be not a quantity but a structure on a set of components. This is much the same idea as a topology on a set of points, and indeed a topological space is a simple example of an organized system -- though from our point of view a trivial one, as there is no question of function.

We propose to regard complexity as the key parameter. It is a quantity which many biologists have already identified as increasing, and indeed a number of them have attempted to account for the trend. They do not, however, appear to have tried to define complexity, as they did not consider it necessary to have a firm definition. We need to be more precise, and so for the discussion in the first part of this paper we shall define the complexity of a system to be the number of different types of components it contains. This is based on von Neumann´s definition of the complexity of an automaton (7), except that he used the total number of components. This is not suitable here, because the mere repetition of structures is so readily accomplished by organisms that it is not a significant marker in evolution. As with entropy in thermodynamics, we shall then give a second definition of complexity based on probabilities, and we shall show that for our purposes the two definitions are equivalent. The reader should note the proviso "for our purposes": a recurring theme throughout this paper will be that in research of this kind it is often best to work with concepts that are defined adequately for our needs but no more rigorously than is necessary. Otherwise we run the risk of gaining apparent precision at the expense of actually moving further from the idea we are trying to capture.

2. COMPLEXITY AND EVOLUTION

It is easy enough to see where Waddington´s analogy goes wrong. He was looking for a parameter which would play the role in evolution that entropy does in physics. He believed that this parameter should represent that which drives evolution, which he took to be fitness. But a physical system which interacts with its environment, as an evolving organism clearly does, is not "driven" by entropy. Instead, it will seek a minimum of an energy function. This suggests that the analogue of fitness is energy, not entropy, and that we may think of organisms as seeking maxima of fitness just as thermodynamic systems seek minima of energy. (In fact, this is more than an analogy. As we point out in (8), saying that an organism is fit is very much the same as saying that a system is at or near a stable equilibrium, i.e. a minimum of potential energy.)

Complexity is a more appropriate analogue of entropy. It is certainly a quantity which is generally accepted to increase during evolution. Indeed it is much easier to see an overall trend towards increasing complexity than towards increasing fitness: most people would agree that a human being is more complex than a bacterium, but it is not at all clear which is fitter, or even that the two can be compared in this respect.

There are many parallels between complexity and entropy. For example, many physicists feel that they ought to be able to derive the second law of thermodynamics from other physical laws. Yet it is difficult to see how this can be possible when the second law possesses an arrow of time and the laws from which they believe it should follow do not. Attempts to derive the second law frequently turn out to depend on initial conditions: if we begin in a state which we take to be ordered then we may expect to pass to less ordered states.

Biologists have a similar problem with complexity. More complex organisms are not necessarily fitter. And if complexity does not increase with fitness then we have much the same problem as in physics: the principle of natural selection does not pick out a direction of time, which makes it hard to see how it can account for such a definite trend. Maynard Smith (9) has even suggested that if complexity has, on the whole, tended to increase in evolution, this may only be because the first organisms were so simple that there was only one direction in which the complexity could go -- in other words, it is a matter of initial conditions, just as many physicists claim is true for entropy.

But if to be more complex is not necessarily to be more fit, why does complexity generally increase? The answer lies in the nature of organisms as self-organizing systems. Essentially, it is always easier to add components to such a system than to remove them. This is a commonplace observation from everyday life, but it is not difficult to see why it should be so. The components of a self-organizing system will tend, over a period of time, to become more and more integrated and interdependent. This will make it progressively more and more difficult to remove any one of them without damaging the system. So while neither additions nor deletions may occur readily, the former will be much more likely to succeed than will the latter.

One particularly clear example of this asymmetry has been pointed out by Maynard Smith (9), who comments that while processes (such as gene duplication) are known whereby the genetic material of an individual can increase, it is hard to see how it could decrease "since most losses will involve losses of functions essential for survival."

Another reason why complexity tends to increase is that if a species is faced with an environmental challenge, it is far more likely that a suitable adjustment can be made by adding some new feature than by removing an existing one. It may even be easier to acquire a completely new trait than to modify an existing one; according to Dollo's Law, when a species adapts for a second time to the same environment it generally 'chooses' a different way of doing it, even though it may possess vestigal components of the original adaptation.

We expect, therefore, that complexity should increase in any evolutionary change (if it changes at all) just as entropy does not decrease in any thermodynamic process. In both cases, however, it is the nature of the process rather than the direction of the change which matters.

We can carry the analogy further. If a thermodynamic system can exchange energy with its surroundings, then its equilibrium is typically given not by a maximum of the entropy, S, but by a minimum of the Gibbs free energy

$$G = H - TS$$

Here H, the enthalpy, is a measure of the energy of the system, which we may take as the analogue of fitness.

Now with the identifications we have made so far, we see that maximizing G instead of H (note that we have to reverse signs from the thermodynamic model) corresponds to the idea that while a species will tend to evolve towards maximum fitness it will have to balance against this a tendency not to increase unduly its complexity. In the above expression, the effect of increasing T is to increase the relative importance of complexity in determining the future course of the system. This suggests that T should be a measure of the organization, since the more closely integrated the components are, the more 'inertia' a system will have, i.e., the less quickly it will respond to a fitness gradient. A highly complex, highly organized system will typically find it difficult to undergo alterations; this is sometimes described as constituting a loss of evolutionary potential, which implies a quantity with the appropriate dimensions (i.e. of energy).

We shall not pursue the analogy between temperature and degree of organization here, but we remark that from quite different considerations Goodwin (10) has defined for biological systems a 'talandic temperature', which he takes to be a measure of the organizational potential. It is surely significant that two entirely unrelated approaches should lead to the same conclusion.

It might be objected that while entropy never decreases, complexity occasionally does, but this is not a real problem. Consider, for example, two equal boxes, one empty and one filled with gas, and with a membrane separating them. If the membrane is removed, we expect that after a period of time the gas will be more or less equally distributed between the two boxes. But we do not expect that at every instant the net flux will be towards the box with the lower gas concentration. Evolution is a sufficiently slow process that we can observe almost every change individually, so we may well see some instances of decrease in complexity. But so long as the trend is consistently towards greater complexity, the analogy holds.

Almost the only cases in which there is a definite trend towards decreasing complexity are in organisms such as viruses and other parasites which live in environments more complex than themselves (9). These cannot really be considered as independent systems, and so the analogy is with processes like refrigeration, in which part of a system can undergo a decrease in entropy at the expense of a greater increase somewhere else.

3. COMPLEXITY AND PROBABILITY

We have seen how if there is a useful analogy to be drawn between thermodynamics and evolution, then the analogue of entropy is complexity. But another, more direct link has been proposed between the two quantities. Kolmogorov (11) and Chaitin (12) have defined the complexity of a sequence of numbers to be the amount of information required to specify it. Now the expression for the amount of information is $-\Sigma p_i \log_2 p_i$, which is, apart from the sign, the same as the definition of entropy in statistical mechanics. For this reason information is sometimes referred to as "negentropy".

This is encouraging, if only because it makes us more confident in the aptness of the analogy, but it remains to be seen if it leads to anything useful. In the first place, it is one thing to write down an expression for the complexity but quite another actually to evaluate it. The problem is to define the p_i. An early attempt was made by Dancoff & Quastler (13), who produced estimates of the information content of organisms based on the probabilities of particular atoms or molecules being found in their actual locations and orientations. This procedure has the serious drawback that it makes the information content of an adult human either 2×10^{28} or 5×10^{25}, depending on whether we take atoms or molecules as our basic unit, and yet gives no reason for preferring one to the other. It does not appear to have led to useful results.

It is not hard to see where the trouble lies. If we are going to define the complexity of something in terms of the amount of information required to specify it, we have to say who or what is doing the specifying. There will therefore be no unique definition of the complexity of a system; much will depend on which of its properties we happen to be interested in. Of course this is no different from our everyday use of the word: as Ashby (14) points out, a brain is highly complex to a neurophysiologist, but not to a butcher.

For most purposes, the complexity of a system is most naturally described in terms of the information required to construct it in the way that it is actually constructed. This is the definition we shall use here, though we still have to decide what we mean by the ´instructions´. A superficially attractive choice, and probably the only one which would allow us even in principle the possibility of actually evaluating the complete set of p_i, would appear to be the DNA, and Gatlin (15) and others have estimated the information content of the genome using either codon or amino acid frequencies. Such results can contribute to our understanding of molecular evolution, but they are not suitable for dealing with more general problems. The reason for this is simply that the link between the genes and the phenotype is nowhere near close enough. Most mutations appear to be neutral, i.e. a change in the genome will more often than not have no effect on the organism. Laboratory stocks of organisms are generally more uniform genetically but less uniform morphologically than the so-called wild type. Humans and chimpanzees are at least as close genetically as some sibling species of Drosophila which are almost identical. Conversely, effects such as cytoplasmic and cortical inheritance remind us that not all the instructions are in the DNA.

In our view, we can only obtain a useful definition of complexity if we are prepared to weaken our requirement in another direction. We have therefore proposed the following (3): Let C be the complexity of an organism at any stage of its evolution. Suppose that some change occurs, and that the probability of this change was p. Then the increase in complexity will be $\delta C = \log_2(1/p)$. The total complexity will accordingly be $-\Sigma \log_2 p_i$, with the summation over all the changes that have taken place. Note that this differs slightly from the familiar form because we are adding things to the system, not considering arrangements of a fixed set of components. We therefore work with the total amount of information rather than with an average.

This definition differs from earlier ones in that it is a differential relation. At most, it allows us to measure the change in complexity at any stage of the evolution of an organism. It is not, however, integrable; the complexity of an organism depends on its entire evolutionary history. We cannot make measurements on two arbitrarily chosen organisms and say which is the more complex.

At first these might appear to be important shortcomings, but in fact they are not, because they are irrelevant to the situations in which we might actually wish to use our definition. There is no real gain in being able to assign a number, which we may call the complexity, to an organism if that is all we are going to do, and whether an oak tree is or is not more complex than a sparrow is not really something we wanted to discuss anyway. For many other purposes, as the following example shows, our apparently limited definition is actually more suitable.

A major concern of many of those who have tried to measure the information content of organisms seems to have been to discuss the apparent paradox that the adult contains so much more information than the egg. Riedl (16) has suggested that we can solve the problem by distinguishing between what he calls the ´law content´ and the ´redundancy content´. The law content of two identical cells is the same as that of either alone. Only the law content of the adult (together with some extra information to organize the repeats) need be contained in the egg. Riedl estimates the average number of repeats of events in a human being to be 10^{19}, and this, he claims, is how to reconcile Dancoff & Quastler´s estimate of the information content of the egg, 10^6, with that of the adult, 10^{25}.

While the use of numbers may appear to make the argument more rigorous, it actually serves only to limit its validity. For repetitions are by no means the only way in which complexity (in the Dancoff & Quastler sense) can be increased. The atoms and molecules in an organism are arranged through physical and chemical interactions, not set individually into predetermined positions (17). The majority of the information required to specify the shape may be thought of as coming from the laws of physics and chemistry. Now we cannot assign a numerical value to the amount of information these laws contain and consequently we cannot carry out a calculation of the kind Riedl suggests. This does not, however, prevent us from discussing and resolving the problem of the egg and the adult. On the contrary, by not insisting on quantitative estimates which we do not need, we very much strengthen our result.

Much the same argument applies in the study of any self-organizing system. The whole idea of self-organization is that the result possesses a spatial and/or temporal structure which is not obviously implicit in the initial state. In fact, the structure arises out of an interaction between the initial or boundary conditions and the laws which govern the process. In trying to analyze such a system, therefore, we must be careful to decide whether our analysis is best carried out at the level of the initial conditions or that of the finished structure; the complicated link between the two means that we will obtain quite different results depending on which choice we make (18).

Another parallel between evolution and thermodynamics has now arisen, in that we have given two definitions of complexity, one phenomenological and the other in terms of probabilities, and we have to check that these can be reconciled. In fact, there are two points to be cleared up. In the first definition each different type of component makes the same contribution to the complexity, whereas in the second there is a weighting by probability of occurrence. We can, however, readily modify the first definition to include this without affecting any of the arguments.

There is also the difficulty that according to the second definition every change constitutes an increase in complexity, whereas on the first definition the addition or loss of one of a number of similar components would leave the complexity unchanged and the loss of all components of a type would decrease it. Again, we could improve matters by altering the first definition to allow a small contribution to complexity from repetitions (as in ref. 16) and since decreases are relatively infrequent (for the reasons given) the definitions are sufficiently equivalent for discussing general trends in evolution, which is all they are intended for.

4. THE PRINCIPLE OF MINIMUM INCREASE IN COMPLEXITY

For neo-Darwinists, evolution is the natural selection of random variations. The course of evolution is determined by selection. There is no need to investigate how the variations arise or which ones are likely to occur, because it is confidently assumed that "if the appropriate genetic variants to face an environmental challenge are not already present in the population, they are likely to arise soon by mutation" (19).

Organisms, however, are produced not by mysterious forces but by natural processes. These processes are clearly capable of producing entities of great complexity, but they are not omnipotent. They can only produce a certain, albeit large, range of things, and some of these are much more readily produced than are others.

Clearly any variant must be viable, and if it competes directly with another form it must also have some selective advantage. But natural selection cannot act on a variant which does not exist. Whether a particular feature appears in evolution therefore largely depends not on its relative fitness once it occurs so much as on the probability that it occurs at all. If we denote this probability by p, then the occurrence of the character increases the complexity by an amount $\delta C=\log_2(1/p)$. Larger values of p imply smaller values of C. In other words, evolution should proceed in accordance with a principle of minimum increase in complexity.

We now have yet another parallel with thermodynamics, viz. with the principal of minimum increase in entropy, which describes the behaviour of systems not far from equilibrium (20). What is more, we can find many examples of the operation of this principle. We outline them briefly here; for more detail see ref. (4).

a) Repeated structures: One way in which organisms can and do conform to the principle of minimum increase in entropy is the frequent occurrence of repeated structures. It is unlikely that the repetition or near repetition of a single form should so often be the optimal solution to a problem, but it is certainly readily accomplished. There are many examples both of actually repeated structures or of what are called "serial homologies": different features of an organism with the same basic structure. See ref. (16).

b) Williston´s Law: It is often remarked that when an organism possesses many similar parts there is a tendency for these to be reduced in number and for those that remain to become more specialized; this trend is called "Williston´s Law". There are also, however, increases: for example in the case of the number of florets in a flower. And we also have to explain how the many similar parts evolved in the first place.
Now repeated structures can occur in a number of different ways, but almost all the processes that have been suggested have the property that changing the number of occurrences is relatively easy: so much so, in fact, that the problem is generally more to explain why the numbers are so often the same in each individual. Out of a number of conceivable ways of coping with some environmental challenge, therefore, a change in the number of some component may not be the optimal solution but it is almost certainly to be the one most likely to occur in the first place. This is surely why such changes are so common.

c) Parallel Evolution: Parallel evolution is the development of similar characters in different lineages of common ancestry. One of the best known examples is the existence in Australia of marsupials which closely resemble many of the placental mammals found elsewhere.
The neo-Darwinist explanation of the phenomenon involves the assumption that the two lines experienced much the same selection pressures in much the same sequence. There is seldom evidence for this, and in some cases, like the trend in fossil reptiles to develop mammalian characters, the environments appear to have been quite different.
We can, however, provide an explanation for parallelism by using the principle of minimum increase in complexity. Suppose there are a number

of possible ways in which a species could evolve and suppose that each of these has a probability p_i of occurring. Now we can hardly expect to know all the possible changes, still less the probabilities. If, however, we know that a certain change occurred in one line, then by a Bayesian argument we infer that this is the change with the greatest p_i, and so this is the one we expect to observe in other lineages as well.

d) Mimicry: The Monarch butterfly is foul tasting, and so tends to be avoided by birds. The Viceroy is palatable, but it looks so much like the Monarch that the same birds tend to avoid it. But while it is easy to see the advantage that this mimicry gives to the Viceroy, that is not the end of the story. For the Monarch has a beautiful and complicated wing pattern, and one is bound to ask why the Viceroy has gone to all the trouble of copying it, rather than just evolving its own bad taste. Indeed, on strict neo-Darwinist grounds it is the latter adaptation that should have been preferred because it confers a greater advantage: the gain in fitness due to mimicry decreases as the number of mimics increases (because the birds will tend to associate the pattern with the mimic instead of with the model). There is, of course, no such limitation on the advantage gained by being unpalatable.
It turns out, however, that the majority of butterfly wing patterns, complicated though they appear to us, are but variations on a common theme (21). Consequently for one butterfly to mimic another is much easier than it looks. So the Viceroy´s choice of adaptation was in accordance with the principle of minimum increase in complexity, rather than on strict grounds of optimality.

5. TRENDS IN COMPLEX SYSTEMS

The starting point for this work was the observation that while complexity seems to increase in evolution it is hard to account for this within neo-Darwinism. We have found a surprisingly close analogy between biological evolution and thermodynamics, and this includes the problem of trends. Neo-Darwinism is a simple optimization theory, and as such contains no arrow of time.

One can ask a similar question of any model of a complex system. Are there irreversible processes going on within the system, and, if there are, is the model capable -- even in principle -- of accounting for them? Thus, for example, that there are definite trends in economics indicates that a simple equilibrium model is inadequate. It is a sobering thought that an attempt to manage an economy on the basis of such a model is as likely to succeed as is a project to construct a perpetual motion machine -- and for much the same reason.

6. CONCLUSIONS

One of the most ambitious tasks in theoretical science is the building up of what von Bertalanffy (22) called ´general systems theory´. It would clearly be of great value to be able to call on a body of results applicable to many different kinds of systems, yet at the same time we are bound to wonder whether this is not too much to hope for.

Surely the answer lies somewhere in between. There are statements which we can reasonably hope to make about systems in general, but they are not quite the same as those we can make if we restrict our attention to much smaller classes of entities. The results will generally be less categorical: to tell us what we should expect to observe rather than what we must. They are therefore to be applied with caution, as there will probably be exceptions to almost all the rules we derive. Even then the results can be useful, for the observation that a system is behaving in an atypical fashion should lead us to ask why, and thus direct our attention to what may be its most significant features. The results are also far more likely to be qualitative than quantitative, but this is not necessarily a disadvantage. Most of the questions we ask about complex systems are really qualitative anyway, even though we may use quantitative methods to try to answer them (23).

The work described here shows the value of one particular approach to the study of complex self-organizing systems, viz., the drawing of an analogy between the system we are interested in and one which we already know quite a bit about. We have seen how closely the analogy could be drawn, and how far it could be carried: from the phenomenological definition of entropy right through to the principal of minimum increase in terms of probablities.

At the same time, we have seen how we must not push analogies too far, and run the risk of carrying over properties which the model system has but the one we are trying to learn about does not. Part of the secret of studying such broad subjects as complexity and self-organization is to be willing to employ approaches which are tentative and imprecise, but at the same time to recognize them as such and to be correspondingly careful in their use and in the claims we make for our results.

REFERENCES

1. Waddington, C.H. In Towards a Theoretical Biology, vol. 2, (ed. C.H. Waddington), Edinburgh University Press, Edinburgh, 1969, p114.
2. Saunders, P.T. and Ho, M.W. J. theor. Biol. **63**, 375, 1976.
3. Saunders, P.T. and Ho, M.W. J. theor. Biol. **90**, 515, 1981.
4. Saunders, P.T. and Ho, M.W. In Evolutionary Theory: Paths into the Future (ed. J.W. Pollard), Wiley, London, 1984, p121.
5. Von Forster, H. In Self Organizing Systems (eds M.C. Yovitz and S. Cameron), Pergamon, London, 1960, p31.
6. Atlan, H. L'Organisation Biologique et la Theorie de l'Information, Hermann, Paris, 1972.
7. Von Neumann, J. Theory of Self Reproducing Automata (ed. A.W.Burks), University of Illinois Press, Urbana, 1966.
8. Ho, M.W. and Saunders, P.T. (In this volume)
9. Maynard Smith, J. Studium Generale **23**, 266, 1970.
10. Goodwin, B.C. Temporal Organization in Cells. Academic Press, London, 1963.
11. Kolmogorov, A. Problemy Peredaci Informacii **1**, 3, 1965.
12. Chaitin, G. J. Ass. comp. Mach. **13**, 547, 1966.
13. Dancoff, S.N. and Quastler, H. In Information Theory in Biology (ed. H. Quastler), University of Illinois Press, Urbana, 1953, p263.
14. Ashby, W.R. Cybernetic Medicine **9**, 1, 1973.
15. Gatlin, L. Information Theory and the Living System, Columbia University Press, New York, 1972.
16. Riedl, R. Die Ordnung des Lebendigen, Paul Parey, Hamburg, 1975.
17. The classic work on this topic is of course D'Arcy Thompson's On Growth and Form (Cambridge University Press, 1917). See also (24) and (25).
18. It is precisely for this reason that the neo-Darwinist program fails, since it makes the claim that the evolution of organisms can be adequately explained by studying the genes, which are at most part of the initial conditions.
19. Dobzhansky, Th., Ayala, F.J., Stebbins, G.L. and Valentine, J.W. Evolution, Freeman, San Francisco, 1977.
20. Prigogine, I. Acad. R. Belg. Bull. Classe Sci. **31**, 600, 1945.
21. Nijhout, H.F. J. exp. Zool. **206**, 119, 1978.
22. Von Bertalanffy, L. General System Theory, Braziller, New York, 1968.
23. Saunders, P.T. Rivista di Biologia **77**, 325, 1984.
24. Thom, R. Stabilité Structurelle et Morphogénèse. Benjamin, Reading, 1972.
25. Saunders, P.T. In Beyond neo-Darwinism: An Introduction to the New Evolutionary Paradigm (eds M.W. Ho and P.T. Saunders), Academic Press, London, 1984.

THE TRANSITION FROM TOWN TO CITY: METROPOLITAN BEHAVIOUR IN THE 19TH CENTURY

Rainer E. Zimmermann
c/o Institut für Philosophie, Wissenschaftstheorie, Wissenschafts- und Technikgeschichte
Technische Universität
D - 1000 Berlin 12

ABSTRACT. A scenario is discussed which describes the transition of Paris from a town to a large metropolitan area in the period of migration between 1821 and 1830. The influence of perception on the styles of behaviour as exhibited by the inhabitants is studied. The model shall give some more insight into the formation of large cities which can be thought of as being basic structures determining the history of Europe in the 19th and 20th centuries. The different paradigms of science are inspected then, and their origin and nature are closely investigated. In looking for a possible unified application of both, it is discussed how topological concepts can be carried over into the domain of social processes in order to develop a common methodological basis for both fields of science.

1. INTRODUCTION: THE RELEVANCE OF LARGE CITIES

The modelling of urban structures has gained a lot by the advent of qualitative theories of self-organization such as those given in the works of Thom (1), Prigogine and his school (2), and Eigen and Schuster (3). Catastrophe theory has been applied to urban structures by Amson (4) and Wilson (5). This equilibrium approach is discussed shortly in the next section. Prigogine's coworkers have introduced models for the development of urban structures which are based on their evolutionary paradigm (6). The explicit history of the system is significant within the frame of this non-equilibrium approach. Combining this latter ansatz with aspects of population dynamics, mainly based on the theory of hypercycles of Eigen and Schuster, I have shown at another place (7) that the spontaneous creation of new life-styles is mandatory for securing the dynamics underlying evolution. In particular, it was found that the interactions of certain groups in a city are strongly influenced by the formation

C. W. Kilmister (ed.), Disequilibrium and Self-Organisation, 255–284.

of group complexes, consisting of a given group and its associated catalyst group in a pairwise fashion. Under certain initial conditions it is possible then that an originally unstable state of the urban system is rendered stable by the spontaneous creation of a new group exhibiting a new life-style.

These results are generalized in this paper here in view of two aspects: the rôle of communicating information within the system on the one hand, and the modified perception in a city by its inhabitants on the other, this latter aspect being especially related to the significance of vertical structures in a city, and the subjectively perceived relative velocities.

The idea is that during the period of migration between 1821 and 1830, Paris (as urban system) performed a transition into a new state which we call "metropolitan system". This transition characterizes an unstable period in the general history of the city. But in order to explain why the achieved new state remained stable in the following decades, it is necessary to invoke an additional influence on the system as a whole. Since the inhabitants of a city define the substance of the urban system, it is obvious to look for that additional influence in the life-styles of the various groups populating Paris at the time. We find that especially the "labyrinthine" effect of a large city as result of the clustering of high buildings and street crossings plays a central rôle. This is related to the fact that, in contrary to a town, a large city is some kind of condensed space in the sense that it is characterized by a critical increase in the relative frequencies of passers-by, of the density of vertically significant built structures (stressing the three-dimensionality of metropolitan space), and of the relative velocities of typically urban means of transport as perceived by the inhabitants. Benjamin (8), and more recently, Klotz (9) have discussed these aspects in detail.

The emergence of large cities in the early 19th century plays also a very important rôle with respect to the history of culture the results of which still today dominate our behaviour. The radical changes of life-styles due to the agglomeration of urban centres, typically divided into "areas of working" and "areas of living", call forth an almost complete re-structuring of the cultural scenario associated with the Europe of late feudalism. Apart from the statement that in France this development was obviously related somehow to the great revolution of 1789 (this was not the case in England), the true roots of this change can be traced back to the economical functions emerging from the situation of the industrial revolution, this being a purely self-referential process, indeed. The concentration of such economical functions is certainly one important reason for the agglomeration of centres, the question is whether this reason suffices. The later stabilization of metropolitan areas is not at all trivial, however. In fact, it could well happen that agglomerated centres vanish again (or reduce at least) after a further significant change in its underlying economical structure. In some extreme cases, such as those of North-American towns which evolved into quasi-metropolitan structures when certain raw materials were found in the vicinity (gold or oil, e.g.), this could be actually observed to happen. Of course, European cities (and the majority of American cities) served mainly the purpose of manufacturing and distributing goods rather than further the raw materials for them. But, on the other hand, thinking of the self-referential aspect of the process, one could well imagine that attaining a certain critical size is a condition for advancing industrialism. There is in fact an important parallel (which did not lead to an industry - therefore?): In ancient Greece (in the 7th century B.C. that is), the formation of (a then new type of) cities is closely related to the emergence of trade and derived economical functions, as Vernant has shown (10). Greek cities however, lost their influence with the advent of the Roman reign in Europe, when the flow lines of trade were arranged differently. So, the original economical functions these cities had, and which created them in first place, could not prevent their relapse into a state of marginal importance. Hence, it can be suspected that in the case of modern European cities, there might have been further influences which established stability and eventually made them a sort of cultural institution (in the sense of Gehlen (11)).

But - there is a basic difference in the fact that trade in ancient Greece was not backed by industrialism. However, we can easily re-construct similar situations of shifting trade lines for European cities which in spite of their reinforcement by industry could not avoid a similar relapse of the associated urban system.

So, the basic idea is that the critical size of a city is an additional factor for stability (in some sense). With respect to the changes of perception mentioned above (being also related to the size of the city), we find that a special group, the "flaneurs", did not only participate in the creation of the scenario, but at the same time initiated its stability, in focussing the new kind of perception arising, and weaving a new kind of life-style out of it. This confirms former (more preliminary) results which indicated that the Paris scenario of Pre-Balzac-type is unstable, if migration is added, but that the creation of a new life-style (associated to the "flaneur") leads to stability (12).

2. THE NON-EQUILIBRIUM PERSPECTIVE

First, let us review shortly one of the equilibrium models: The "abstract city" of Amson (13) comprises two basic interactions, a cohesive coercion (civic attraction), given by a potential $\phi = k \ln (1/r)$, and a relocation inducement (civic pressure) of the form $- dp/dr$. The population density ρ is coupled to the civic pressure p by an appropriate equation of state of the well-known type $p = K\rho^{\gamma}$, where K is called "rental coefficient" and γ "rental exponent", which are characteristic parameters which determine the behaviour of the inhabitants with regard to attraction or relocation. If $M(r)$ is the total "mass" of inhabitants, defined in a natural way, of the form

$$M(r) = \int_0^r 2\pi r'\rho(r')dr' ,$$

then we can write down the equilibrium equation for the system which is

$$k\rho(r)M(r) + r\ p'(r) = 0 \quad (r \geqslant 0),$$

where the prime indicates a derivative with respect to r. A very interesting solution of this is for the choice $\gamma = 2$ the socalled "Bessel city" which provides for the formation of satellite towns of a larger city. This concept can be generalized in defining an "urbanitic system" C which consists of a distribution of civic "mass" in a civic space (chosen to be the Euklidean plane) together with the parameters from above. Then we can define also a positive measure on C which we call "opulence" Ω of the system, referring to some wealth of the inhabitants which can be appropriately described - though not clearly measured - by observable activities of these inhabitants, modes of consumption etc. Then we can replace the former equation of state by a more general one which is said to constitute an "urbanitic law" of the form $p = f(\Omega, \rho)$ - this being simply a constraint on the possible states of the system. The interesting case here is the "saccadic city" satisfying the relation

$$(p + \alpha\rho^2)(1/\rho - \beta) = K\Omega,$$

this actually turning out to be an equation of the van der Waals-type (14). Here the coefficient α is a kind of density response factor, reflecting the state of condensation of the system, and β gives a measure for the individual space available in a city. This equation expresses the fact that the system can exhibit discontinuous changes in density in response to continuous changes in opulence and rental. It is very obvious that such a situation can be straightforwardly related to the cusp catastrophe in the sense of Thom (15). In fact, what we do have is a city which consists of two urbanitic sheets according to the typical picture of a cusp.

There are some particular cases which exhibit clearly how the kind of instability encountered depends on the actual path taken by the process when traversing the critical region of the cusp. If e.g. the opulence of a city decreases, the city may suffer a saccadic increase in density (at constant rentals), but does not return to its original density, if an inverse motion occurs. The same is true for a decrease in rental leading to a decrease of density. Of course, in history, such "saccadic" jumps are always delayed by the natural inertia of urban systems. They are "relative jumps" (as comp-ared to the usual time scale of history).

This equilibrium situation can be visualized now as a special case of a more general non-equilibrium situation. Nicolis and Prigogine (16) have explicitly shown that the cusp is contained in a typically autocatalytic model with the rate equation

$$\dot{X} = - k_2 X^3 + k_1 AX^2 - k_3X + k_4B,$$

where A and B are the initial and final products of the associated chemical reaction scheme and X is a transient product. It can be shown that the stability diagram is of the cusp-type (17), and that the steady-state diagram resembles roughly the van der Waals graph. In fact, the steady-state solutions are

$$X_o^3 - aX_o^2 + kX_o - b = 0$$

with suitably chosen new coefficients (18). This can be easily compared with the equation of state for a saccadic city as displayed above, which is recovered, if

$$a = 1/\beta, \; k = p/\alpha, \; b = (p + K\Omega)/\alpha\beta \; .$$

The behaviour variable is then equal to ρ. If the general variable X is re-defined by

$$Z := X - \frac{1}{3} (k_1/k_2) A,$$

then this gives us the standard form of a cusp,

$$- dZ/d\tau = Z^3 + uZ + v,$$

with u, v and τ being (re-written) expressions in A,B and k's.

In the perspective of non-equilibrium processes this means that van der Waals-type situations differ from more general situations by the fact that we can unambiguously determine the thermodynamic branch of available solutions and place the multiple steady states at a finite distance from this branch. This is e.g. not generally true for limit-cycle or symmetry-breaking transitions (19). Only for the case of the cusp is the actual singularity point identical with the bifurcation point. The crucial difference between the two is that in general, bifurcation points indicate co-existing branches of solutions, and for one of them the process has to "decide". In the cusp case however, it is possible for the process to avoid the region of instability in first place, because the multiple steady states belong to the same branch of solutions. This can also be seen by the fact that the cusp path is basically a hysteresis-like loop which is related somehow to a kind of reversibility (of part) of the process involved. For our studies of urban systems this has the important consequence of enabling us to think of a useful approximation of such systems by means by means of catastrophe theory, although we would expect that a city is an open system in the strict sense. In so far, the rôle of the cusp in both equilibrium and non-equilibrium models motivates a new study in the classification of singular points and the criteria for a system eventually attaining one type or the other, this being an open question.

3. THE SIGNIFICANCE OF PERCEPTION

How to incorporate perception now in the model? This is a difficult undertaking indeed. But we know at least something about perception in a city: The difference between an outer observer who is looking at a city and a co-moving observer (who is actually inside a city and therefore part of the system) is the following: The outer observer basically describes the dynamics of the system by local time derivatives $\partial/\partial t$. The co-moving observer describes the dynamics adequately by convective time derivatives of the general form

$$d/dt = \partial/\partial t + (v \cdot \nabla) .$$

Here, the correcting term reflects the eigen-velocity v relative to the other constituents of the system. This velocity can sometimes be approximated by the average velocity within the system (20). Now, let us take the following view: The observer judges his own velocity with respect to the average velocity he perceives when moving in the system. Though this concept can be generalized to all kinds of motions, it suffices here to deal with the actual motions of bodies in a city (as a given local space) such as the motions of horse-drawn vehicles, passers-by, horses etc. Note now that the velocities perceived will depend on the density of the traffic. The more bodies the observer perceives in his vicinity, the more the actual velocities will be biased to higher values. The same is true for the observable dimensions of space in a city: The more packed a region is with buildings, and the more street crossings and junctions are available, that is the more complicated the perceived network of traffic situations becomes, the more the velocities are biased (to higher values). These effects have been studied in another context by Gibson for instance (21).

In a city with a population density above a certain critical size, these effects are catastrophically enhanced (22). Vertical structures are far more relevant therefore in cities than in smaller towns. The actual height of the buildings leads out of the common two-dimensionality of urban models and stresses the three-dimensionality of metropolitan space. This is what we call "labyrinthine effect" (according to Benjamin (23)). If ℓ is a typical mean free path for an observer in a city (giving a measure for his relative mobility), and h is the average height of buildings in his vicinity (as given by a piece of metropolitan space bound by what we might call an individual horizon), then we define the "space contrast" C(S) of a section S (the vicinity as just mentioned) in some non-linear way, say of the form

$$C(S) \sim h^2/\ell^2 \ .$$

We define also the "labyrinthine coefficient" κ which is basically given by the number of crossings c and junctions j with respect to S. This is a purely combinatorial quantity (which has in fact non-trivial implications on the topology of metropolitan space - we come to that later), so

$$\kappa = c + \frac{1}{2}\, j,$$

where the junctions have a weight 1/2 because of self-explicative reasons. The height h can be neglected for κ, because it has no immediate effect on the possible number of bodies passing. If $\langle v \rangle$ is now the actual average velocity in S, then the perceived velocity $\langle u \rangle$ can be thought of as a biased $\langle v \rangle$, namely

$$\langle u \rangle = \kappa C(S)\ \langle v \rangle,$$

where the factor gives the total bias of $\langle v \rangle$. Note that the bias is dimensionless (because we have to end up with a velocity again, in terms of km/h say). We can incorporate perception of an observer in the system now by replacing the usual local time derivatives by convective ones, putting u for v, with $\kappa C(S) =: F(x,y,t)$ being a function of the coordinates. The coordinate z can be suppressed here, if an explicit value for h is included in the expression.

The following table displays some typical values, bias and average velocities can attain in an urban system of the early 19th century.

	$c + \frac{1}{2}j$ (1)	h^2/ℓ^2 (2)	(1) x (2)	<v> kmh^{-1}	<u>
Country	1,5	10^{-4}	$1{,}5\times10^{-4}$	25	0,004
Town	4,5	0,64	2,88	9	25,92
City	13,5	225	3037,5	7	2×10^{4}

Table 1: Typical values of urban systems around 1825 (Data taken partly from M. Bloch (24)).

Note that this *does not* mean that the observer perceives real velocities of this order, or that he could measure or estimate in one way or another actual relative velocities in the system, and would end up then with several thousands of kilometers per hour. The values indicated refer only to the consequences for his perceptive physiology which arranges his bodily functions (such as the rate of his reactions, the capacity of his storing visual or other information and so forth), and therefore creates an individual scenario for his behaviour which is closely related to the urban scenario he is moving in. As this is subject to other fields of (microscopic) investigation, we assume here that the given values reflect (macroscopically) the *impressions* the observer has when moving in the city or given scenario. The microscopic picture is discussed at other places (25).

4. A PARIS SCENARIO

Remember now the scenario for Paris as preliminary described in a former work (26). Again we have the number densities of "pre-Balzac" inhabitants n_1, and of the rural population occupying space in Paris, n_2. We have the associated "catalysts", the intellectuals m_1, and the aristocrats (with rural possessions) m_2. The re-lated complexes of interaction are z_1 (for the pair n_1, m_1) and z_2 (for n_2, m_2), respectively. The density of information η (as thought of as being measurable output of communication somehow) is included now in the model explicitly. All variables (referring to the urban system) depend on coordinates x,y,t with coordinate z being incorporated in the convective time derivatives in the evolution equations.

As we visualize the formation of a large city as a process of agglomeration of centres, we apply basically equations of the Keller-Segel-type (27), of the form

$$\dot{n} = -\nabla(D_1 \nabla \eta) + \nabla(D_n \nabla n),$$

$$\dot{\eta} = P(\eta) + D_\eta \nabla^2 \eta,$$

where $P(\eta)$ is the net production of information in the system, and where we have suppressed indices and net production as well as spontaneous creation terms for n (of the Verhulst-type, compare equation 3.3 of the reference in note 26). Here D_1 refers to the coupling of information to the system, of "chemotactic type": $D_1 \sim n/\eta$. As is well-known, equations of this type lead to the agglomeration of groups in the population. We give now the full set of equations for some time t prior to the time of creation t_o, which refers to the period in which a new group shows up in the scenario:

$$\dot{n}_1 = k_1 n_1 (N_1 - n_1 - \beta_2 n_2) - d_1 n_1 - \frac{n_1}{n_1 + n_2} m_2 k_2 + $$

$$+ D_1 \nabla^2 n_1 - D_{12} \nabla^2 \eta \quad (4.1),$$

$$\dot{n}_2 = k_2 n_2 (N_2 - n_2 - \beta_1 n_1) - d_2 n_2 - \frac{n_2}{n_1 + n_2} m_1 k_1 + $$

$$+ D_2 \nabla^2 n_2 - D_{21} \nabla^2 \eta \quad (4.2).$$

This is the first pair of equations. Note that the coefficient of net production for group n_2 is governed by k_2 which is thought of as being considerably larger than k_1 (because of the inflow of rural citizens during the epoch in question). On the other hand, the "capacity" N_2 is roughly of the same order than N_1, because it is assumed that newcombers to the city will in general accept poorer living conditions (at least, for some initial period). The niche overlaps β can be taken as coupled parameters, in the fashion $\beta_1 + \beta_2 = 1$, say, for the custom's barriers of Paris do not (at first) allow for the agglomeration of slum satellites. In contrary, newcombers will tend to settle in ghettos inside the city (an effect comparable to the "Mietskasernen" of Berlin, some hundred years later). The death rate (including re-migrating) d_2 will be possibly larger than d_1 (due to the general conditions of health and hygiene in such ghettos), but it will certainly not upset the high inflow characterized by k_2. Hence, the "net production" of n_2 is considerably larger than that of n_1. Of course, both n_1 and n_2 are thought of as being considerably larger than the respective densities of the catalysts. The cyclic terms express the relative frequency of interactions as established by direct contact in one way or the other. This is different from the flow of information as displayed by the last terms. The latter has always some polarizing effect on the average. In this sense, the diffusion terms reflect the random element in the system. The sign convention in the equations can be interpreted such that diffusion always tends to smooth the system, while flow of information enhances the coupling among the constituents of the system, and thereby encourages the formation of structure. It is thought that additionally, information diffuses within the system in a random manner, as it is carried by those individuals who migrate (more or less randomly) within the system. The flow of information is given then by

$$\dot{\eta} = (n_1 + n_2)f(\eta) - k(\eta)\eta + D_\eta \nabla^2 \eta \quad (4.3),$$

where the "carriers" of information are coupled to the production term while the annihilation term is taken to depend on the information available. Note that the information originating within groups m_1 and m_2 is included in the interactions due to the formation of complexes. This is so, because it is assumed that the main task of the catalysts is to create and communicate information. Hence, the formation of complexes represents still another sort of "flowing" information. The remaining equations have a simpler structure, therefore, of the form

$$\dot{m}_1 = f_1 z_1 - \frac{m_1}{m_1 + m_2} f_2 z_2 \quad (4.4),$$

$$\dot{m}_2 = f_2 z_2 - \frac{m_2}{m_1 + m_2} f_1 z_1 \quad (4.5),$$

$$\dot{z}_1 = m_1 n_1 s_1 - \ell_1 z_1 + D_{z1} \nabla^2 z_1 \quad (4.6),$$

$$\dot{z}_2 = m_2 n_2 s_2 - \ell_2 z_2 + D_{z2} \nabla^2 z_2 \quad (4.7).$$

Here the coefficients $f_1, f_2, s_1, s_2, \ell_1, \ell_2$ are appropriate rate parameters, and D_{z1}, D_{z2} diffusion coefficients, respectively. The equations (4.1) through (4.7) of the scenario have to be completed for times $t>t_o$. We use t_o here, as indicated above, as injection time, when the group n_3 (the flaneurs) is spontaneously created out of the scenario together with its catalyst m_3 (the "bohème"). The enlarged system of equations replaces then the equations displayed above.

We do not give the complete set of equations once more, but write down the additional equations only, and indicate new terms entering the former equations. The equation for n_3 becomes

$$\dot{n}_3 = k_3 n_3 (N_3 - n_3 - \beta_1 n_1 - \beta_2 n_2) - d_3 n_3 - \frac{n_3}{n_1 + n_2 + n_3}(m_1 k_1 + m_2 k_2) + D_3 \nabla^2 n_3 - D_{312} \nabla^2 \eta \quad (4.8).$$

The effort, n_3 has with establishing itself within the given scenario, is reflected in the two niche overlaps in the equation which can only be counteracted by the various interaction terms, because k_3 is very small, too. In fact, the diffusion becomes here of central importance. The catalyst's equation is

$$\dot{m}_3 = f_3 z_3 - \frac{m_3}{m_1 + m_2 + m_3}(f_1 z_1 + f_2 z_2) \quad (4.9).$$

The complex is described by

$$\dot{z}_3 = m_3 n_3 s_3 - \ell_3 z_3 + D_{z3} \nabla^2 z_3 \quad (4.10).$$

The rest of the equations has to be completed in the sense of these three equations: All sums have to take into account n_3 or m_3, now (this refers to the cyclic terms and the sum in the flow of information). In the evolution equations for n_1 and n_2, we have to incorporate the probabilities for the creation of n_3 in view of the particular subgroups,

$$-w_{13} n_3, \quad \text{and} \quad -w_{23} n_3,$$

respectively. These terms have the function of a spontaneously created new niche overlap entering the system. All ten equations constitute the description of the system for $t>t_o$.

5. DISCUSSION

The first we can do is to look for the equilibria of the system. In this section we refer to times $t<t_o$ as scenario S_o, and to times $t>t_o$ as scenario S. Then for S^o we find the following cases: If trivially, n_1 and n_2 vanish, then the flow of information and the complexes vanish as well. However, it is not possible then to say anything about the catalysts m_1 and m_2. If, on the other hand, we put m_1 and m_2 equal to zero in first place, then the complexes vanish, but for n_1 and n_2 (as well as η) we find

$$n_{1o} \sim (1 - \beta_1\beta_2)^{-1}(N_1 - \beta_2 N_2 - d_1/k_1),$$

$$n_{2o} \sim N_2 - \beta_1 n_1,$$

$$\eta_o \sim N_2 + \tfrac{1}{2}(N_1 - \beta_2 N_2 - d_1/k_1).$$

We have applied here the mild plausibility conditions on the system as introduced at the beginning of the last section. Note especially, that d_2/k_2 is very small as compared to d_1/k_1.

On the other hand, for the general case of non-vanishing number densities for the groups, we realize that equations (4.4) and (4.5) imply a contradiction, if taken for themselves: Eliminating the z gives $m_1 m_2 = (m_1 + m_2)^2$, which can only be true, if one of the m is negative (we would not expect that). But, if we take these equations together and combine them with equations (4.6) and (4.7), we find (up to constants) that $m_2/m_1 \sim (n_1/n_2)^2$. This is quite a strange property of the equations for m: The first case implies that there are no equilibria for the m, and that therefore the whole scenario must be unstable. The second case implies that the quotients of the respective number densities govern the equilibria rather than the number densities themselves. This hints towards the significance of the relative composition of the scenario, where absolute quantities are not especially important. If we apply these results to the equations for the n, and eliminate the m, then we find that we can recover a condition on the quotients of the n again, of the form

$$(n_2/n_1)^2 \sim (n_1 + cn_2 + d)/(n_2 + an_1 + b)$$

which can be re-written to give the cubic equation

$$y^3 + ay^2 - cy - 1 \sim 0,$$

where

$$a := (1 + \beta_2)/\beta_2, \quad c := (1 + \beta_1)/\beta_1,$$

$$y := n_2/n_1.$$

It is interesting to note that a quotient of number densities which has a stabilizing effect (in the sense of providing an equilibrium for the system) depends mainly on the relative niche overlaps of the groups. We can show that in general this cubic equation has three solutions (which are real), because the characteristic expression $D = q^2 + p^3$ for

$$q := \frac{1}{2}(2a^3/27 + ac/3 - 1) \text{ and}$$

$$p := \frac{1}{3}((-3c-a^2)/3)$$

is always negative for possible values of β_1 and β_2.

These results imply that we have either no equilibrium for S_o, or relative equilibria for the quotients of number densities. We can visualize this situation with a kind of "flowing" non-localizable equilibrium which will be discussed later.

True equilibria in the strict sense , we can only obtain for very special conditions. For instance, if the net production rates are equal (of n_1 and n_2), and if in particular, $k_1 = k_2$, then we obtain an equilibrium for all n (which simply means that in this case the n are equal, or start at least under equal conditions). If, on the other hand, the inflow k_2 vanishes, then an equilibrium is attained, if the birth rate of n_1 decreases (!), or, equivalently, the death rate decreases. These results are obviously very unlikely.

Let us turn now to S: The trivial cases are equal here to those of S_o. If the catalysts vanish, then we have a similar situation, only that the expressions become a bit more complicated (because n_3 shows up now). We will not display these expressions again. It is interesting, however, that for n_2 we get *exactly* $N_2 - \beta_1 n_1$ again, which we had for S_o. That is, the coupling of the equilibria between n_1 and n_2 is still the same and only governed by the actual value for n_1. This means that n_2 is mainly influenced by n_3 *via* n_1! This is exactly what we have expected. For the cyclic equations of the complexes, we have no difficulty now with a contradiction when eliminating the z and looking at the m. The scenario S has therefore, it appears, another stability behaviour.

Looking at the actual stability analysis now, we find

the following: The characteristic matrix of the equations governing S_o is of 7x7-type with 49 elements of which 25 are non-zero. The matrix for S_o is displayed on the following page. Note that for comparatively mild conditions on the values of the equilibria (which determine the functions F, G, and H showing up in the matrix), the left half of the matrix (below the main diagonal) can be made zero. Then the resulting secular equation is simply the trace of the matrix. Of course, not only is the evaluation of the matrix tedious work, but also is the resulting secular equation very difficult to solve (because it is of the seventh order). But, as we look for the formation of structures, it suffices to check whether it has any positive roots at all. According to the law of Descartes, the number of positive roots of the secular equation can be estimated in looking for the change of sign occurring in the coefficients of the equation: Is the number of changes of sign N_s, and the number of positive roots N_p, then $N_p \leq N_s$, and differs from N_s by an even number, only. For the secular equation we find then:

$$\sigma^7 + \alpha\sigma^6 - \beta\sigma^5 - \gamma\sigma^4 - \delta\sigma^3 - \varepsilon\sigma^2 + \phi\sigma - \eta = 0,$$

with appropriately chosen coefficients. Obviously, N_s is 3, so the equation has either 3 or 1 positive roots. In any case, the onset of instability is secured, therefore.

It is a principle of complexity now that the higher the degree of the secular equation associated to a system, the higher is the probability that it has at least one positive root. Obviously, the secular equation for S is of a still higher degree than that for S_o, because it is given by a 10x10 matrix. So we will expect that the number of positive roots (their expectational value rather) increases. Indeed, this is clearly shown when comparing the equation of 9th order with that of 10th order: The former has $N_s = 4$ so that the number of positive roots is either 4 or 2. The latter however has $N_s = 6$ so that there are either 6 or 4 or 2 such roots.

So it appears as if we have even more instability after t_o than we had before which is somewhat contradicting our interpretation (namely that the arrival of n_3, m_3 leads to a stabilization of the former scenario).

Why do we speak of stabilization at all? The crucial point here is the following: We can visualize the pair (n_3, m_3) as a perturbation on the system, showing up at t_o. If this perturbation survives (that is, if it can co-exist in the scenario, at least), then it enriches the structure of the system. But it also increases the order of the new secular equation. So, although it makes S_o more stable in a sense, it also creates more instability at the same time. What it is doing actually is to "create *meta*stability" (or rather: to prolong the metastable status of the system).

n_1	n_2	η	m_1	m_2	z_1	z_2
$\sigma - G_1 - q^2D_1$	H_1	q^2D_{12}	0	I_1	0	0
H_2	$\sigma - G_2 - q^2D_2$	q^2D_{12}	I_2	0	0	0
$-f(\eta_o)$	$-f(\eta_o)$	$\sigma - G_3 - q^2D_\eta$	0	0	0	0
0	0	0	$\sigma - G_4$	$-H_4$	$-f_1$	I_4
0	0	0	$-H_5$	$\sigma - G_5$	I_5	$-f_2$
$-m_{1o}s_1$	0	0	$-n_{1o}s_1$	0	$\sigma + \ell_1 - D_{z1}q^2$	0
0	$-m_{2o}s_2$	0	0	$-n_{2o}s_2$	0	$\sigma + \ell_2 - D_{z2}q^2$

Table 2: Characteristic matrix of S_o (the indices of the various abbreviated functions give the line of the matrix, important for the subsequent analysis are only the quantities σ and q - the latter being the wave number of the spatial perturbation amplitude). To recover the equation of seventh order displayed above, re-scale the matrix so that the left half becomes zero, multiply the trace and re-name the functions by Greek symbols. The elements of the main diagonal are those with a σ.

This state is achieved by increasing the complexity of the system. Indeed, the permanent production of stability and instability, and the permanent competition between them, characterizes simply the necessity for the system to re-process itself. This is nothing but the central property of *autopoietic systems* (28). And in so far, the autopoiesis establishes a useful representation for the dialectic structure of evolutionary processes: A given scenario is a kind of "thesis" in which all of the system's history is "suspended" (in the sense of Hegel). The immanent instability of the system (reflecting intrinsic conflicts of the groups constituting it) is the associated "anti-thesis". The product (or challenge) of this instability is the emergence of a new structure due to fluctuations. The competition between thesis and anti-thesis reflects also the immanent dilemma of the system which I have discussed at another place (29): to assemble its structural components around an average behaviour (in order to secure optimal stability) on the one hand, and to admit the creation of new structures, being initiated far from the average behaviour, on the other hand. The resolution of this dilemma is autopoiesis: The system tries to incorporate its stable and unstable tendencies in re-processing itself towards a new state of meta-stability, in the sense that the attained state is synthesis and new thesis at the same time. For social systems (which we are discussing here) this re-processing is done by the system's cultural institutions. In fact, re-processing means here producing *aesthetical categories* (we come to this oncemore).

How the new structure is treated within the system, depends on the outcome of the local competition after the injection time t_o. This decides about damping out or co-existing (dominance is very rare). We realize that from this follows that there is nothing like "progress" in the classical sense. Because new structures emerge from local fluctuations (as dissipative structures, Prigogine would say), the following process is only another "trial" of the system involved. Its outcome can be an error or not. The guiding principle is that of resilience (30), a local principle only, and steadily leading to higher complexity in the system.

Once again: In the discussion of the model we have stressed that we are dealing with meta-stability rather than with stability in the strict sense. Re-processing the system means re-processing its type, but not its detailed structure. Hence, type and stability are ascertained in a global sense, but locally stability breaks down in favour of creating new structures. In exactly this sense, the urban space we discuss here (the metropolitan system) is an institution in the anthropological sense as has been mentioned in the introduction. Critical number and size is a *space-time concept* for structured (social) systems. Such concepts are vital therefore for the system's re-processing.

And we can say more: If we visualize the evolution of systems as a series of transient states (or local scenarios) within the unfolding of complexity, then we can approximate a single state by an almost closed system in between transitions. In exactly this sense, catastrophe theory can be applied to such scenarios, as we have seen in section 2. This is the first close relationship to catastrophe theory, another one will be discussed in section 7.

We have one point left: what is the actual effect of the convective derivatives on the system? In the stability analysis these additional terms which we have introduced earlier add to the left-hand-side of the equations. Is $F = F(x,y,t)$ the total bias, then we get terms like

$$F(x,y,t)(q_1 v_x + q_2 v_y),$$

where the q are the associated wave-numbers for the spatial part of the chosen perturbation (in the usual normal mode ansatz). Remember that this is given in the form

$$n = n_o + \varepsilon\, \tilde{n} \exp\,(q_1 x + q_2 y) \exp \sigma t\ .$$

Then, after insertion and expansion of the polynomial terms according to Taylor, we have on the left-hand-side

$$\sigma\, \tilde{n} + F(x,y,t)(q_1 v_x + q_2 v_y),$$

where the second term is the correction from above. The sign of this additional expression shows us that it is adding to instability. Hence, the higher the bias of the velocity perceived (that is, the more complex the city network), the higher the probability for instability.

6. BIFURCATION OF PARADIGMS

In this section we discuss a number of points which are implied by the foregoing results, and which shed some more light on further levels of research.

In Table 3 the development of Paris with respect to its inhabitants is displayed for a period of about 250 years. This development is compared with that of London for the same period (31). We realize that Paris exhibits a very particular behaviour in the period of migration, when the number of inhabitants is increased by two millions. This is an increase of about 300% as compared to the level of 1820, which takes place in a period of only nine years: This can be truly called "saccadic" increase. The only comparable city of that time, London - until 1820 always ahead of Paris -, cannot follow so fast. Berlin, by the way, starts to become a city about this period, but a similar transition will not take place before the late 80's of the 19th century. As we have discussed in this paper, for Paris, this period is characterized by a complete change of urban structure, and of the behaviour of its inhabitants.

Year	Paris (inh. in est. absolute numbers)	London (inh. in est. abs. no.)
1594	180 000	-
1602	-	250 000
1700	500 000	674 350
1789	650 000	-
1801	-	864 845
1820	1000 000	-
1821-1830	3000 000	1200 000

Table 3: Inhabitants of Paris (in absolute numbers) as compared to those of London. Data according to Sombart (31).

One problem remains however: First of all, it is necessary to study the explicit differences between Paris and London in more detail, because they (and their historical roots) are not trivial at all. It is also important to investigate the nature of relevant fluctuations occurring in the system. It is interesting to note e.g. that significant architectonic structures in Paris (like "Les Halles", the passages, or railway stations) emerge (in their "modern" form) between 1811 and 1821 (being re-built about 1850 (32)), while in London this does not appear in such a striking manner. A similar development starts in Berlin not before 1880 (33). So, Paris appears to have expressed the "architectonic language of space" (34) for the first time.

The conceptual difficulty we have here is related to a similar problem we encountered when dealing with the notion of (perceived) velocities: We have put more *meaning* into the equations than they actually do reflect by their symbolic form. This is to say that "behind" these equations, there is more meaning with reference to the social implications of the physical phenomenology than one would recognize when simply looking at the equations. It is exactly at this point of the investigation, where the paradigms (of humane sciences and those of the sciences of nature) bifurcate in the following sense: If we wish to express a model in mathematical terms, then we have to neglect its wider context, because we end up with a purely formal description. We deal with the "output" of phenomena, only. If, on the other hand, we wish to deal with the meaning of the observed phenomena, then we cannot apply formal mathematics strictly. The difficulty is therefore, to find a sort of "intermediate" theory which is capable to reflect both paradigmatic domains, and to take into account the fact that we are dealing with human beings, who, by their active participation in the structuring of the system, should be adequately part of the system's interpretation.

In fact, the velocity concept introduced here, gives us a chance to actually promote this idea. As will be discussed in a later section, a first trial to achieve a unified approach could be to apply topology to the systems involved. The point here is that the convective term in the time derivatives can be interpreted as covariant derivative in the sense of connections on topological spaces (35). This can be made clear in the following way: A linear connection ∇ of the vector bundle (E, M, π) is a real bilinear map of the kind $V(M) \times \Gamma(E) \rightarrow \Gamma(E)$, where M is a topological space (or a manifold) and E is the fiber space over M, π being the natural projection. Also, $V(M)$ is the Lie algebra of vector fields on M (which is actually nothing but the space of first order homogeneous differential operators!), and $\Gamma(E)$ is the space of cross-sections of E. If now X is a vector field, and Ψ is a cross-section, then $\nabla_X\Psi$ is the image of (X, Ψ) under ∇, and this is the covariant derivative of Ψ by X which is again a cross-section of E. This setting has a close relationship to cohomology theory (36), and has therefore deep roots in topology. The important aspect here is that we know on the other hand that the convective terms in the equations are also related to the (perception of the) observer who participates in the system as one of its (singular) parts. Obviously, therefore, the state of an intrinsic observer is closely related to the topology of the space the observer perceives, and which in turn is defined by the system.

Moreover, the complexity of a system is also related to topology (as we shall see later). Hence, the *concept of meaning* in a social system may be eventually disclosed by discussing this complexity in topological terms, because the latter are formalizable, though in a qualitative sense. This can also be shown when thinking of the network structure of cities: It is possible to define certain geometric properties of the network of streets (as curves on the plane) such as curvature and winding. According to the type of network, this can give a measure for the *redundancy* of the city's labyrinthine structure. As this redundancy is nothing but the excess of references which can be verified within the complexity of the system (37), we have here a close relationship between "space" and "behaviour in space", this being a concept of important anthropological meaning (38). Of course, space now is not coordinate space in the classical sense: *it is space filled with meaning*. And, this meaning is projected on the topological properties of space. The investigation of such relationships is not a very difficult task (or: not so difficult as one would expect). In fact, the situation of the scientist is similar to that of the detective, as I have explained by means of the "detective's metapher" (39): It is the detective in fact, who evaluates information which comes from both paradigmatic domains. Only when he tries to embed (objective) data into a context (of meaning), can he be successful. In the same way, the scientist might eventually be able to combine these different paradigmatic principles in order to achieve a better understanding of complex processes in nature (of which he is a part)(40).

7. A UNIFIED APPROACH

As we have seen now, the main difficulty for a formal treatment of processes in the social domain arises from the different paradigmatic foundations of the respective fields of interest. In this section the nature of this difference shall be investigated in some more detail. The idea is that topology might be a useful tool for eventually bridgeing the methodological gap between the two domains of science. In the standard approaches to self-organizing systems far from thermodynamical equilibrium (41, 42), topological ideas have in general not been discussed further than was necessary for a treatment of the stability of differential equations. A significant exception is the work of Thom (43). In this section we discuss first various topological concepts and their relations to autopoietic systems, and inspect then some notions of Thom with respect to their applicability.

In cosmology, for instance, discussion usually starts with a given 4-manifold characterized by a metric with components g_{ab} such that the fundamental form is

$$ds^2 := g_{ab}\, dx^a\, dx^b.$$

On the pair (M,g), where M means the manifold and g the metric, Einstein's equations are postulated to exist.

Basically, Einstein's equations consist of expressions in the g_{ab} up to their covariant derivatives of second order. For more details see any book on relativity theory(44). It depends on the explicit boundary conditions of the Einstein equations whether the problem solved is of local or of global nature. If we introduce black holes e.g., we choose a local model. The cosmological problem of finding a metric in the large is of global nature. It was a persistent difficulty in the past to determine how local is local on space-time. A way out shows the concept of asymptotic flatness, originally introduced by Penrose (45). For a detailed review on this problem see the work of Geroch in the volume of Esposito and Witten (46). We choose here the notion of weak asymptotic simplicity as given by Hawking (47) which suffices for our purposes: We map M on a manifold M' with boundary ∂M. This can be achieved by a conformal transformation, characterized by some conformal factor Ω such that on M, $g' = \Omega^2 g$ (in components), and on ∂M, $\Omega = 0$. If additionally, every null geodesic in M has past and future end points on ∂M, and Einstein's equations hold on M (such that M is empty or has an electromagnetic field near ∂M), then this implies that ∂M consists of two disjoint null hypersurfaces $\Xi^{\pm}$ with topology $R^1 \times S^2$. Then (M,g) is said to be weak asymptotically simple, if there is a pair (M',g') which is asymptotically simple such that a neigbourhood of $\Xi^{\pm}$ in M' is isometric to a similar neigbourhood in M. And M' is asymptotically simple, if the above mentioned conditions hold concerning the conformal factor. The basic idea of this is that we expect flatness of space-time, if curvature falls off with the distance from the source of gravitation. But as it is very difficult to observe the behaviour of a *limes* of curvature (or similar quantities), it is more practical to study the infinity of space-time as if it were very nearby. This can be achieved by a conformal transformation of the kind discussed above. Instead of studying space-time now at infinity, we can study the conformal space-time at its (finite) boundary $\Xi^{\pm}$.

What we have found here is a useful treatment of *systems*: Instead of trying hard to explain the difference between a system and one of its subsystems, and instead of trying to express what it means that the sum of the subsystems does not equal the totality of the system (but is "less" than that), we simply differ between systems and subsystems in defining a conception of local and global. We can hope therefore, to eventually found the very notion of system on manifold properties.

A similar idea led the Berlin school of Gestalt psychologists of the 1920's to involve topological concepts in the interactions of groups of persons. But because the concepts used were not very elaborate at the time, they basically achieved not more than a formal re-phrasing of field theory in terms of psychology.

We would expect that other physical systems which are somewhat more complicated than those discussed in relativity theory, can nevertheless be dealt with by the same topological concepts just developped. Of course, the formation of matter in the universe will become more and more s local concept (in fact, all of the entropy paradox rests on the conceptual difference of local and global systems). But on the other hand, the more we study local and complicated processes (such as those in astrophysics), we deal with sets of differential equations. But we know also that differential equations can be related directly to the topology of a system (as discussed earlier). The notions of differential topology give us a straightforward correspondence therefore, between the more "static" picture of topology and its "flowing" consequences as visualized by the dynamical processes which are described by differential equations. So, in principle, we can justify the general idea with a view to fairly classical applications of dynamical systems.

In between (social and physical systems) we find chemical and biological systems which share their paradigmatic context with physical systems. Such a context we call "objective paradigm" which includes the idea that in principle all models arising from this concept can be judged in terms of appropriate measurements which are "objectively" obtainable from experiments, in the sense that anyone can obtain them provided he performs his activities due to the same set of boundary conditions securing the reproducebility of these experiments. This is - as we know - quite different in social systems, where new aspects emerge which are related to a new concept called *meaning*.

As we have seen in the last section, the concept of meaning is a direct consequence of the reciprocity of the individual within the social totality. We will shortly describe how this reciprocity develops.

Social reciprocity differs from mere cooperation in the fact that it is mediated by an individual through the population by means of communication as immanent pre-supposition of its own being and social being (of the system). This is a very difficult notion which involves other concepts such as complexity, contingency, redundancy, and sense (meaning). All these have been introduced recently into the framework of the theory of (autopoietic) social systems (48). At their basis lie the ideas of self-reference and self-observation of systems (49) which have been discussed earlier within cosmology by Wheeler (50). In Wheeler's picture the universe as a whole is a self-referential system which creates communicating participators (rather than passive observers). The latter give meaning to the universe by their very existence. The consequence is an "anthropic principle" which secures that the physical properties of the universe are mild enough in order to enable the creation of living beings (51).

In social systems, self-reference and self-observation are the constitutive elements of consciousness, and meaning for such systems is created by themselves.

This can be seen in the following way: Social systems are very complex such that for an individual (or a group) it is necessary to select interrelationships within the system, because it is not possible to verify all relations which are (combinatorically) available. Selection then means contingency, and contingency means risk (52). This is the reason why similar units (of social nature) can constitute very different systems. The selection leaves open a number of "excess references" towards further possibilities of living and acting (53). This is the aspect of *redundancy* in a system, which we have discussed already in this work. Excess of references is one form of the phenomenon "meaning". So, the concept of meaning is nothing but the ultimate result of reciprocity, the historical evolution of which I have discussed at another place (54). Communicating excess is the application of symbolical techniques which are deeply rooted in the creation of language (55).

Meaning and *context* are closely related to each other. Context may be interpreted as a reference frame for meaning in a system. Different social groups exhibit local meaning (by their practical actions) within the context of the groups' coupling to other groups of the system. Local groups can be visualized then as parts of a page in a book. A page represents a state of the system, the collection of pages (for all times t) is the book. You possibly can't understand a page (or even a chapter) of the book, if you don't know the book. Hence, it is quite pointless to analyze a certain short period of history (or biography) without trying to re-synthesize this period within the complete social context of the time. This idea is referred to as "text paradigm" of social systems (56).

During the evolution of the universe the emergence of meaning within some historical context is a very recent phenomenon. If we measure the scale of the earth's coordinate time in units of 10^7 years, then we notice that meaning emerges virtually in the last moments of evolution, after a long time of "objectifiable" processes immediately succeeding the Big Bang scenario (we had 2000 such units: galaxies forming after about 100 such units, Earth after about 1500 units, men exist not even three such units). It is interesting to note that this "unfolding" of history corresponds to a true unfolding of the geometry of the phenomenal manifold, if we translate this into topological language. Indeed: as the manifold of the social system will have a large number of dimensions, n say, we can expect for the historical part of evolution a space similar to the product $M^4 \times M'^n$, where M^4 refers to space-time and M'^n is the n-dimensional manifold associated to the social system. The emergence of meaning (or, of a historical context, or, of a social system at all) can be visualized therefore by an embedding

$$\phi: M^4 \to M^4 \times M'^n$$

so that space-time becomes part of a larger manifold (which has not existed beforehand?). What kind of embedding we have here is left open for the time being, because this involves arguments on the rank of the derived linear function ϕ_* on which we will not elaborate here (57). It is to be noted however, that at present, science does not take this structure into account at all. The sciences of nature can be thought of as a study of the range of the projection

$$\pi_1: M^4 \times M'^n \to M^4,$$

while the humane sciences can be thought of as the study of the range of the projection

$$\pi_2: M^4 \times M'^n \to M'^n .$$

This is obviously equivalent to stating the difference of paradigms used in the respective fields of research.

8. TOPOLOGY OF SOCIAL SYSTEMS

We are now in a position to come back to our original problem: linking system properties to topology.

Why do we deal with a manifold at all, and not just with an R^n? The answer is straightforward: As we deal with human behaviour, we would not expect a linear scaling of the "coordinate axes". We will expect rather that all variables involved depend on all others and undergo non-linear variations. This is a very important aspect, which implies that the manifold we are dealing with will have curvature of some kind. In relativity theory, local (small) curvatures mean sources of gravitation, global curvature implies the totality of gravitational effects (superposed on all of space-time). Geometrical interactions in relativity are presented by the interaction of sources (and test particles) on the general gravitational background.

For a social system, the individual person is the source of local interactions, and social space is curved by these interactions. It is useful in fact, to think of a person in social space as being a singularity in a field associated to that space. Hence, an individual person is to the social system what a particle is in physics to the field: a pure singularity at which the macroscopic theory breaks down in favour of the microscopic theory (for a social system this is the point where sociology stops and psychology sets in, where the singular and the universal - in the sense of Sartre - are reciprocally coupled to each other).

Topological concepts have been used in a number of times for economical models, usually assuming a part of R^n to represent "commodity space", i.e. the space of demand for the sonsumers. But in general, these models were of static nature, and handled preference relations and utility functions in order to achieve a description of equilibrium states of (simple) economies according to the concept of Pareto stability (58). In particular, the picture of an economy was mainly that of persons demanding linearly and independently for a certain measurable amount of commodities. The point here is that for a social system, we will generally deal with actions which are not explicitly measurable and moreover coupled to each other in a non-linear way. So, if we can define a metric on our manifold at all, it is not probable that it will be Euklidean. Allen and his coworkers (59) have used metrics for the decisions made by persons with respect to urban settlement. Different from earlier works, he uses weighting factors for the "coordinates" in commodity space and an additional measure of the information available to an individual which introduces a high degree of non-linearity into the model. However, commodity space itself is still of the R^n-type.

We have seen already that meaning arises from a certain excess of references which in turn is a result of the complexity of social systems. Thom has shown (60) how information and complexity can be linked to topology (and to entropy). The idea is the following: Given two manifolds X and Y, there is a locally closed hypersurface with singularities H (called "bifurcation set") in the function space L(X,Y) such that each connected component of L(X,Y)-H defines a structurally stable form of maps. Are f and g two forms which can be continuously deformed into each other, then they can be joined by a differentiable path which crosses H transversally (61). This introduces a metric d(f,g) in a straightforward way. Complexity can be defined then simply as the integer $q(f) = d(f, f_o)$, where f_o is the topologically simplest (ground) form. The choice of an appropriate ground form can be comparatively easy determined for the case that $n > 2m + 1$, whenever $m = \dim X$ and $n = \dim Y$ (62). This is the case for the large number of dimensions we might have to handle when dealing with a social system and its subsystems. A simplest form is then the embedding of X in Y (properly speaking: in the set of all homotopy classes of maps from X to Y, for any two such embeddings are isotopic, and their topological type is unique).

Within this context, entropy can be thought of as the sum of measures of open sets X_i of points having topologically the form of F, if F is given and structurally stable. Is $\mu_i(F)$ a measure of dimension i, then entropy of F is the sum

$$\mu(F) = \sum_i \mu_i(F)/i! \quad .$$

If, on the other hand, B is the function space of a dynamical system, which can be represented by its spectrum as a sequence of subsets $(X_1, X_2, \ldots)$, where the X_i are Riemannian manifolds, then en-tropy is simply the total measure of elements of form F in B (63). There are many practical consequences of this argument (but also a conceptual difficulty, because Thom relates entropy to topological complexity such that the larger complexity of a form, the larger its entropy - provided we think of topological complexity as being related to that complexity we introduced for systems, this is a contradiction, because more complexity means more order then, and *less* entropy therefore). It is also possible to link these ideas with the concept of energy (64) and probability (65). The latter relations are not particularly new if thinking of the thermodynamical theory, but the important thing here is that they are derived from topological concepts! This is exactly what we are aiming at. Together with the aforementioned interpretation of vector fields as being differential operators (66), and the correspondence between combinatorical concepts (critical points in view of their indices and the winding number of curves e.g.) and the functional aspects of systems as being reflected by the (formal) complexity of systems of differential equations (67), this serves as a basis for a further development of a unified theory of social systems.

9. FURTHER PROSPECTS

Is there any hope for eventually achieving the objective? I think yes: Social systems are "very" complex indeed, but hope founds on the fact that nevertheless each individual has to decide about many things in daily life (which actually creates his behaviour). This means that there takes place an on-line selective ordering at any moment of daily life which actually creates permanently a symbolical correspondence between meaning (of action) and excess of references available. This correspondence has to be modelled, and the promising direction seems given by a unification of topological concepts, the theory of natural self-organization, and hermeneutic strategies (68). The latter are particularly important, because we have found that the production of aesthetical structures serves as a link between the re-processing system and the meaning emerging out of this re-processing. It appears to be promising therefore, to relate the concepts discussed here to the explicit development of the cultural institutions of social systems. This has been done at another place (69) in order to clarify the socially relevant aspects of philosophy and literature within the theory of Sartre. So far this is a special case. But it can be suspected that the formation of structure is also of central meaning for a large number of fields in arts, if questions of innovation, stability, flow of information, social reciprocity, and so forth, shall be discussed (70).

10. NOTES AND REFERENCES

If notes are referred to several times, the sign "Ref. n" is used with n being the number of the note at the first occurrence of the respective reference. Numbers of notes are given here without brackets. Page numbers are given without "p." or "pp.". Volume numbers are *singly* underlined.

1 R. Thom: *Structural Stability and Morphogenesis*, Benjamin, Reading (Mass.), 1975.

2 G. Nicolis, I. Prigogine: *Self-Organization in Non-Equilibrium Systems*, Wiley-Interscience, New York, London, Sydney, Toronto, 1977.

3 M. Eigen, P. Schuster: 'The Hypercycle - A Principle of Natural Self-Organization. Part B: The Abstract Hypercycle.' *Naturwissenschaften* 65 (1), 1978, 7-41.

4 J. C. Amson: 'Equilibrium Models of Cities' (2 parts), *Environment and Planning* 4 (1972) 429-444, 5 (1972) 295-338.

5 A. G. Wilson: *Urban and Regional Models in Geography and Planning*, Wiley-Interscience, London, New York, 1974.

6 P. M. Allen: 'The Evolutionary Paradigm of Dissipative Structures' (Reprint), Annual Meeting of the AAAS, San Francisco, 1980.

7 R. E. Zimmermann: 'Spontaneous Creation of Life-Style: A First Attempt Towards Mathematical Psychohistory', in: Kilmister, C. W., R. E. Zimmermann, M. Bayerl, K. S. Islam (eds.), *Proceedings of the 2nd Int. Whitsun Meeting (ISG)*, Klosterneuburg, 1983, 237-251.

8 W. Benjamin: *Das Passagenwerk*. Gesammelte Schriften V1 & 2, Suhrkamp, Frankfurt/M., 1982.

9 V. Klotz: *Die erzählte Stadt*, Hanser, München, 1969.

10 J.-P. Vernant: *Les origines de la pensée grecque*, Presses Universitaires de France, Paris, 1962.

11 A. Gehlen: *Urmensch und Spätkultur*, Athenaion, Frankfurt/M., 1975 (espec. section 9 of part 1).

12 Cf. R. E. Zimmermann: 'Das Textparadigma der Geschichte - Psychohistorie als Hermeneutik und Analyse' Paderborner Studien 1/2, 1984, 100-103 (see also Ref. 7).
13 J. C. Amson: 'Equilibrium and Catastrophic Modes of Urban Growth', in: London Papers in Regional Science 4 (1972) 108-128.
14 Thom, Ref. 1, 56, 62.
15 ibd. 62-3.
16 Nicolis and Prigogine, Ref. 2, 170-8.
17 ibd. 172-3.
18 ibd. 171.
19 ibd. 174.
20 R. B. Bird, W. E. Stewart, E. N. Lightfoot: *Transport Phenomena*, Wiley, New York, London, Sydney, 1960, 73.
21 J. J. Gibson: *The Ecological Approach to Visual Perception*, Boston, 1979. Cited according to the German Translation, Urban und Schwarzenberg, München, Wien, Baltimore, 1982, 217.
22 Cf. Amson, Ref. 13.
23 Benjamin, Ref. 8, 55, 541, 559.
24 M. Bloch: *La société féodale*, Michel, Paris, 1939. Cited according to the German translation, Ullstein und Propyläen, Frankfurt/M., Wien, Berlin, 1982, 85.
25 Cf. R. F. Schmidt (ed.): *Sinnesphysiologie*, Springer, Berlin, Heidelberg, New York, 1973. See also Ref. 21, passim.

26 Zimmermann, Ref. 7, 245-9.
27 E. F. Keller, L. A. Segel: 'Initiation of Slime Mold Aggregation Viewed as an Instability', *J. Theor. Biol.* 26 (1970) 399-415.
28 N. Luhmann: *Soziale Systeme*. Suhrkamp, Frankfurt/M., 1984, 46-7, 78.
29 Zimmermann, Ref. 7, 242.
30 C. S. Holling: 'Resilience and Stability of Ecosystems', in: Jantsch, E. and C. H. Waddington (eds.), *Evolution and Consciousness. Human Systems in Transition*. Addison-Wesley, Reading (Mass.), 1976, 73-92.
31 W. Sombart: *Liebe, Luxus, Kapitalismus*. Wagenbach, Berlin, 1984 (first published 1922), 43sqq.
32 A useful documentation can be found in the essays by Rita Schober, contained in the German Winkler edition of Zola's *Rougon Macquart* series of novels. For more explicit research refer to the Pléiade edition by Gallimard, Paris. The Winkler edition is published in München.
33 J. Boberg, T. Fichter, E. Gillen (eds.): *Exerzierfeld der Moderne: Industriekultur in Berlin im 19. Jahrhundert (I)*. Beck, München, 1984.
34 Cited from Mayer, in: Ref. 33, 117.
35 R. Hermann: *Vector Bundles in Mathematical Physics*, Benjamin, New York, 1970 (2 vols.). Here: I, 88-9. See also the third chapter of S. Kobayashi, K. Nomizu: *Foundations*

of Differential Geometry (I), Interscience, New York, London, 1963.
36 J. L. Dupont: *Curvature and Characteristic Classes*, Springer, Berlin, Heidelberg, New York, 1978, 38.
37 Luhmann, Ref. 28, 73.
38 Gehlen, Ref. 11, 54sqq.
39 R. E. Zimmermann: 'Die Totalität und ihre Struktur: Existentielle Psychoanalyse und Geschichtstheorie', *Lendemains* 40 (1985) 75-85.
40 R. E. Zimmermann: 'Das Paradies der Damen: Zur Maschinisierung der Urbanität bei Zola.' Lecture given to the "Historian's Circle", Inst. Hist. Med., Free University Berlin on the 1st June 1985. Cf. also: sections 3.2.2 and 3.2.3 of my forthcoming thesis: *Die außerordentlichen Reisen des Jules Verne - Zum Technikverständnis im Frankreich des 19. Jahrhunderts*. Technical University Berlin, 1986, 433-474.
41 Nicolis and Prigogine, Ref. 2.
42 Eigen and Schuster, Ref. 3.
43 Thom, Ref. 1.
44 S. W. Hawking, W. Israel (eds.): *General Relativity. An Einstein Centenary Survey*. Cambridge University Press, 1979. See for a short review: J. Ehlers: 'Einstein's Theory of Gravitation', in: *Einstein Symposion Berlin*, Lecture Notes in Physics 100, Springer, Berlin, Heidelberg, New York, 10-35.
45 R. Penrose:'Structure of Space-Time', in: Lectures on Mathematics and Physics, Battelle Rencontres, 1967, 121-235 (see especially: 171sqq.).
46 R. Geroch: 'Asymptotic Structure of Space-Time', in: Esposito, F. P., L. Witten (eds.), Asymptotic Structure of Space-Time, Cincinnati, 1976, Plenum, New York, London, 1977, 1-105.
47 S. W. Hawking: 'The Event Horizon', in: Black Holes, Les Houches Summer School 1972, Gordon and Breach, New York, 1973, 5-55 (see especially: 14-5).
48 Luhmann, Ref. 28, 47-50, 93, 101.
49 ibd. 63, 101.
50 C. M. Patton, J. A. Wheeler: 'Is Physics Legislated by Cosmogony?', in: Isham, C. J., R. Penrose, D. W. Sciama (eds.), *Quantum Gravity. An Oxford Symposium*. Clarendon, Oxford, 1975, 538-605 (especially: 564-8).
51 R. Penrose: 'Singularities and Time-Asymmetry', in: Ref. 44, 581-638 (especially: 633-4).
52 Luhmann, Ref. 28, 47.
53 ibd. 93.
54 R. E. Zimmermann: *Anlage und Umwelt. Der sozialphilosophische Bezug*. Manuscripts of the Semester Sessions, Math.-Nat.-wiss. Verein, Berlin (MNV), 1985.
55 R. Gurney: *Language, Brain and Interactive Processes*, Arnold, London, 1973, 22-3.

56 H. Brunkhorst: 'Wie man sich zu dem macht, der man ist.' In: T. König (ed.), *Sartres Flaubert lesen. Essays zu "Der Idiot der Familie"*. Rowohlt, Reinbek, 1980, 27-43.

57 Cf. F. Brickell, R. S. Clark: *Differentiable Manifolds*. Van Nostrand Reinhold, London, 1970, 60.

58 S. Smale: 'Global Analysis and Economics I', in: M. M. Peixoto (ed.), *Dynamical Systems*. Bahia, 1971. Academic Press, New York, London, 1973, 531-544. - E. Dierker: *Topological Methods in Walrasian Economics*. Lecture Notes in Economics and Mathematical Systems 92, Springer, Berlin, Heidelberg, New York, 1974.- G. Debreu: *Theory of Value*, Wiley, New York, 1959. In German see: Springer, Berlin, Heidelberg, New York, 1976.

59 P. M. Allen, M. Sanglier, F. Boon, J.-L. Deneubourg, A. de Palma: 'Models of Urban Settlement and Structure as Dynamic Self-Organizing Systems.' U.S. Department of Transportation. Contract TSC-1640, 1980.

60 Thom, Ref. 1, 127.

61 R. E. Zimmermann: 'Katastrophentheorie und die Geometrie der Entscheidungen II', Blätter DGVM (Karlsruhe) 14 (3) 1980, 385-412.

62 Thom, Ref. 1, 128.

63 ibd. 133.

64 ibd. 134.

65 ibd. 135.

66 M. Henle: *A Combinatorial Introduction to Topology*, Freeman, San Francisco, 1979, 60.

67 Allen, Ref. 6.

68 After completion of this paper I noticed that S. Strogatz describes two immediate applications based on the correspondence of self-organizing processes in the sense of Prigogine to topology. See: 'Yeast Oscillations, Belousov-Zhabotinsky Waves, and the Non-Retraction Theorem' *Mathematical Intelligencer* 7 (2) 1985, 9-17.

69 R. E. Zimmermann: 'Poetik des sozialen Seins: Sartres Auffassung vom Unbewußten.' to be published in *Lendemains* 41 (1986).

70 R. E. Zimmermann: 'Totalität und Hermeneutik: Zur textparadigmatischen Grundlegung sozialer Systeme.' Manuscript ISG-070685-3/85, forthcoming.

CREATION, ORGANISATION AND PRESERVATION OF DISSIPATIVE STRUCTURES

Saiful Islam
Visiting Scientist
Physical Resource Theory Group
Chalmers University of Technology
University of Göteborg
S-412 96 Göteborg
Sweden

ABSTRACT. Three fundamental resources - energy, material and information - are necessary for the creation, organisation and preservation of ordered structures. A human society for example is an ordered structure which functions by using information and by transforming and exchanging resources, goods and services. Ordered structures are systems far from thermodynamic equilibrium. The tools and concepts of thermodynamics are used to model the creation of useful energy in the early Universe, the role of information in consumption behaviour and the optimal development of an economy under resource constraint.

1. INTRODUCTION

The first law of thermodynamics says that energy is conserved. It does not imply that all structures, once realized, will last forever. The maintenance of a structure requires the dissipation of energy and material - the degradation of high quality energy and material into low quality energy and material.

The second law of thermodynamics states that the quality of energy is not conserved. Quality of energy is determined by the distance of an energy source from equilibrium. The following equation shows that exergy |1-4| is a measure of the distance from the equilibrium:

$$E = S\ (T-T_o) - V\ (P-P_o) + \Sigma\ N_i\ (\mu_i-\mu_{io}) \qquad (1)$$

T, P and μ_i are the intensive parameters of the system - temperature, pressure and chemical potential. V, S and N_i are the extensive parameters - volume, entropy and number of different molecules. The o-subscribed symbols correspond to their values in equilibrium. Thus the exergy vanishes at equilibrium, or disorder is maximal at equilibrium.

The change in entropy in a system can be written as

C. W. Kilmister (ed.), Disequilibrium and Self-Organisation, 285–305.

$$T_o dS = - (dE + dW) \qquad (2)$$

If during the process through which equilibrium is attained, exergy consumed is E, work extracted is W and the change in entropy is ΔS (S_{eq}-S), the integrated form of (2) is

$$E = W + T_o \quad \Delta S \qquad (3)$$

If a process is reversible ($\Delta S = 0$), all exergy can in principle be converted into mechanical work. Mechanical work itself can be transformed into any other form of energy. It is a consequence of the second law of thermodynamics that entropy is produced in all processes that proceed at a non-zero speed. So that after a processs has taken place, the amount $T_o \Delta S$ has been dissipated - this amount is no longer available as useful energy. In other words exergy is the maximum (corresponding to reversible processes, i.e. when $\Delta S = 0$) amount of work that can be extracted from a system. We see that entropy production is a synonym for exergy consumption. Net exergy can be destroyed or consumed but not created, except on a cosmic scale |5-9|.

If the system evolves towards equilibrium without doing any work, eq. (3) gives

$$E = T_o \quad \Delta S \qquad (4)$$

ΔS is negentropy. Exergy is thus equilibrium temperature times negentropy.

Negentropy is information capacity |10|. A structured system has more information in it than a disordered system. Exergy and negentropy are related in that exergy can be used to create ordered structures and new information |11-12| . Let us consider solar radiation reaching the earth. It contains very little information. Since sunlight is hotter than the earth's surface, its exergy content is high (cf. eq. (1)). This exergy makes the creation, preservation, organization and orderliness of various terrestial dissipative structures possible.

Without the solar radiation, the earth would have cooled down with a continuous flow of heat from the interior of the planet into the outer space. Neither the solar radiation nor the earth's surface, if considered in isolation, has much negentropy. It is the contrast between the two interacting systems that makes the combined system rich in exergy or negentropy. Life would not emerge on the earth without the influx of solar radiation . Information in the genetic material of the primeval life grew into a rich and complex biosphere in the course of evolution. Out of the exergy or information capacity of the solar radiation, new information is generated and existing information and structures are maintained.

2. CREATION OF NUCLEAR EXERGY

According to the standard theory, the early Universe, due to high density was in thermal equilibrium within one hundredth of a second after the Big Bang. thus the Universe had already reached the state of heat death, its exergy was zero. But when we look at the sky we see examples of extremely ordered systems consuming high amount of exergy - the stars.

How could order be restored in a totally unordered Universe? Expansion helped the Universe to come out of the stalemate. Due to the expansion, the conditions for local equilibrium changed with the changing temperature and pressure. The universe expanded so fast that the mechanisms for sustaining equilibrium could not keep pace with the expansion. Thus order and disorder can grow simultaneously in the expanding Universe. This does not violate the second law of thermodynamics. In an expanding Universe the entropy S increases but its maximum attainable entropy S_{max} may increase faster due to the expansion which increases disequilirium. Thus the negentropy (S_{max} -S) and consequently exergy increase (cf. eq. 4).

The source of starlight is the nuclear reactor at the core of the star where mainly hydrogen is burnt into helium. To understand the predominance of hydrogen in the Universe we have to go back to the instant 0.01 seconds after the Big Bang.

Temperature of the Universe at 0.01 seconds after the Big Bang was 10^{11}K. At this temperature the electron and its antiparticle positron, and neutrino and antineutrino can keep protons and neutrons in equilibrium. This means that we cannot distinguish between protons and neutrons. Since exergy is contrast, we have no nuclear exergy. Reactions of the following types occur at this temperature.

$$\text{Antineutrino + proton} \rightleftharpoons \text{positron + neutron}$$
$$\text{Electron + proton} \rightleftharpoons \text{neutrino + neutron}$$

In thermodynamic equilibrium the relative abundance of two types of particles is determined by their mass and is given by the Boltzmann's statistical theorem:

$$(\text{neutrons/protons}) = \exp\left((m_p - m_n)c^2/kT\right)$$

The heavier neutron is thus less bountiful to start with. At a slightly lower temperature, it is easier for the heavier neutron to turn into proton, so that the reactions going from right to left are more favourable and thus the neutron proton ratio decreases further. About 14s after the Big Bang the temperature has dropped so far that the radiation has no more energy to produce electron positron pairs. The conversion of protons into neutrons requires the presence of electron positron pairs. So neutrons and protons at this point can be distinguished and the nuclear exergy comes into existence.

It started with 24% neutrons and 76% protons. At this point the nuclear exergy is tied to the neutrons which can decay into proton, electron and antineutrino, thereby releasing 1.294 MeV per neutron. As times goes on the Universe becomes cool enough for various stable nuclei to form. If the Universe had been expanding slowly, it would have happened. But the Universe expanded extremely rapidly. In this condition nuclei can only be formed in a series of two particle reactions. And here was the lucky coincidence which prevented the consumption of all nuclear exergy. The deuterium nucleus - half way toward helium - is extremely loosely bound. They were torn apart by the energetic radiation as soon as they were formed, so that the heavier nuclei do not have a chance to form. Nuclear exergy has been trapped in the form of protons. Only about one fourth of the baryonic matter has been synthesized into helium.

The era under consideration is characterized by high temperature but the density of matter is somewhat less than that of air. We consider the nucleon gas as a non-relativistic Maxwell-Boltzmann gas. The exergy per nucleon is given by |13|:

$$E = \sum_i P_i \frac{kT}{A_i} \ln \frac{P_i}{P_i^{(0)}} \qquad (5)$$

where i lables states and P_i is the actual probability distribution for one nucleon, and $P_i^{(0)}$ is the probability distribution for one nucleon at thermodynamic equilibrium.

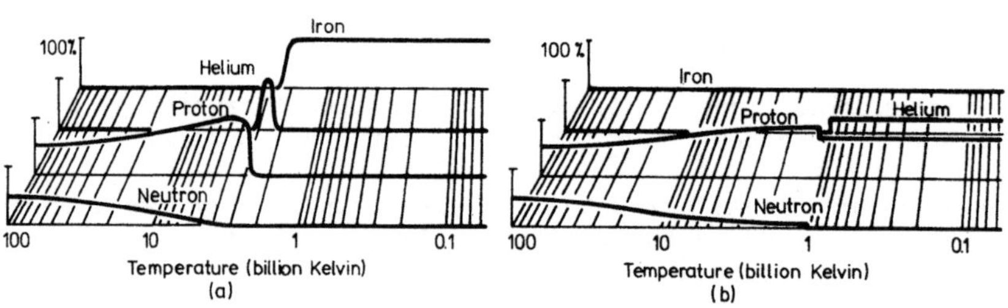

Fig. 1. In equilibrium the lower energy states are more probable. It is for this reason that as the Universe cools down (lower energy) the neutrons disintegrate into proton. As the cooling continues the protons and the neutrons fuse together to form heavier nuclei e.g. helium and ultimately iron. Iron is the lowest energy state of nucleons.

(a) shows which state (among the four chosen) would be most probable had the Universe been expanding slowly, making the above mentioned scheme possible. According to this scheme already 30 seconds after the Big Bang there should have been only iron nuclei.

(b) shows how the nuclear exergy was trapped in protons (predominantly) and in helium. The more the matter deviates from equilibrium, the more is the exergy.

Fig. 2. is thus a measure for the difference between (a) and (b).

In a simplified model |fig. 1| we consider that a nucleon may either be a free proton or a free neutron or it may be bound in ^4He or $^{56}Fe_{56}$. ^4He was formed abundantly during the cosmic nucleon-synthesis and Fe is a natural candidate by virtue of its being the most tightly bound atomic nucleus.

i	A_i	g_i	b_i	P_i $(T > T_o)$	P_i $(T < T_o)$
1	1 (H)	2	0.0	$1-\gamma\ (T)$	$1-2\ \gamma_o$
2	1 (n)	2	-1.30	$\gamma\ (T)$	0
3	4 (He)	1	6.73	0	$2\gamma_o$
4	56 (Fe)	1	8.41	0	0°

Table 1. Mass number (A_i), spin multiplicity (g_i) binding energy per nucleon (b_i) and the actual probability distribution of the four states. All helium is assumed to be synthesized at $T = T_o = 10^9$K with $\gamma_o = \gamma\ (T_o) = 13{\cdot}5$. Neutron fraction , $\gamma(T)$ is calculated numericallly.

The equilibrium distribution is given by

$$P_i^{(o)} = \kappa r \left(\frac{mc^2}{kT}\right)^{3/2} g_i\ A_i^{5/2} \exp\left(-A_i\ (\mu\ (T)-b_i)/kT\right) \quad (6)$$

where r, the photon to nucleon ratio, is taken to be 10^9. The nucleon mass m and the constant κ are given by

$$m = 0.24m_n + 0.76m_p = 938.6\ \text{MeV}/c^2 \quad (7)$$

$$\kappa = \frac{2\pi}{8\zeta(3)} = 0.26 \quad (8)$$

The chemical potential $\mu(T)$ is fixed by the normalization

$$\Sigma\ P_i^{(o)} = 1 \quad (9)$$

Eqs. (5) and (6) and the data in Table 1 give the exergy per nucleon before (10) and after (11) the helium synthesis as:

$$E\ (T) = \mu\ (T) + |b_2|\gamma(T) - kT\ \ln \frac{2\kappa r\left(\frac{mc^2}{kT}\right)^{3/2}}{\gamma(T)^{\gamma(T)}\ (1-\gamma(T))^{1-\gamma(T)}} \quad (10)$$

$$E(T) = \mu\ (T) - 2\gamma_o b_3 - kT\ \ln \frac{2\left|\kappa r\left(\frac{mc^2}{kT}\right)^{3/2}\right|^{(1-\frac{3}{2}\gamma_o)}}{(1-2\gamma_o)^{1-2\gamma_o}\ \gamma_o^{\frac{1}{2}\gamma_o}} \quad (11)$$

The creation of nuclear exergy is plotted in fig. 2.

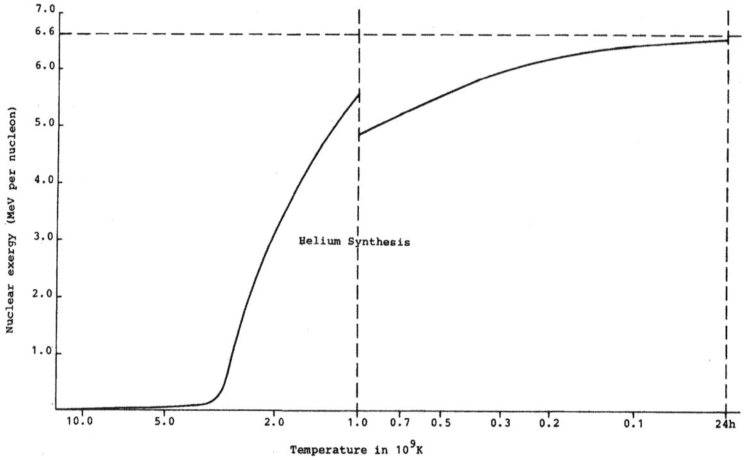

Fig. 2. The creation of nuclear exergy started around 14 s after the Big Bang. Most of the exergy was created during the first few minutes. The process was essentially completed within 24 hours after the Big Bang or on the first day of creation.

After about 700,000 years the Universe was cool enough for the atoms to be formed, i.e. electrons could settle down on the atomic nuclei. If the hydrogen atoms were uniformly distributed the nuclear exergy would not be available to do work. In the cosmic order gravity plays the most important role. Gravitational force is experienced by all particles of the Universe. In time gravity clumped matter into protostars and protogalaxies. The stars burn protons into heavier elements and release energy and this makes life possible.

Gravity is also in itself a source of exergy. After hydrogen and helium has been burnt in a star, further gravitational contraction with its consequent heating takes place and heavier nuclei are formed in the stars until the end product is iron. If the star is heavy, the gravitational attraction would be so high that the protons will capture electrons to become neutrons. Enormous temperature and pressure develop, allowing all sorts of nuclear reactions that cannot otherwise occur and heavier nuclei are formed. Pulsars are the rest of such neutron stars which became supernovae and blew out matter in the interstellar space. Many proton rich nuclei can be produced only through (p, γ) reactions, since the coulomb barrier in such a reaction is very high they can take place only in a highly energetic situation, as in a supernova explosion. The fact that our solar system contains heavy elements indicates that it is made of material which was blown out of stars, and there is evidence that it has gone through this star-interstellar dust-star cycle more than once.

The exergy per nucleon in a star thus is a function of its mass. At low mass the stars ends up as white dwarfs, in this case the nuclear exergy is augmented by a few keV per nucleon of gravitational exergy. For neutron stars the nuclear-gravitational exergy per nucleon is of the order of 40 MeV. For systems large enough to have a black hole as its final state the exergy per nucleon may be around 100 MeV/nucleon.

Modern elementary particle physics suggests that the proton is unstable. In this case matter itself represents a form of exergy with a very long half life (10^{31} years). But if the Universe is open, even this exergy will also be consumed in nuclear decay and the Universe will consists of very thinly spread electrons and positrons (so that they cannot annihilate each other), photons, neutrinos and gravitons.

Exergy is a measure of contrast. This contrast can be internal to the system (e.g. proton decay) or against an external environment (e.g. the formation of the iron nuclei). Iron is the most tightly bound nucleus. Energy is thus released when nuclei heavier than iron are torn apart (fission) and lighter than iron nuclei are fused together (fusion). Stellar nucleosynthesis uses the exergy trapped in the protons at the beginning of the Universe and the gravitational exergy of supernovae are trapped in the uranium and in other proton-rich nuclei (e.g. the tin isotopes with $A<116$). We thus also use exergy sources which are older than the solar system.

3. CONSUMPTION AS A SYSTEM FOR THE EXCHANGE AND CONTROL OF INFORMATION

From equation (1) it is obvious that a resource is characterized by its deviation from equilibrium. It is a law of nature that systems tend towards thermodynamic equilibrium. Spontaneous processes drawing a resource into equilibrium with its environment destroy that resource. For this reason all resources are to some extent scarce. Exergy and negentropy both measure the deviation of a resource from equilibrium. Either of them can be used as an overall resource measure. It is relevant to restrict negentropy to the physical level and use the term 'information' for knowledge. Exergy can be used to measure both useful energy and material which is not in equilibrium with its environment |11|.

More than ten thousand years ago humanity invented agriculture and started to abandon gathering/hunting as the main activities to bring resources into the society. With agriculture human societies moved away from a close balance with their natural environment into a situation where a considerable input of labour, organization and resources was necessary to keep their land away from such a balance. With time this input has increased in organized societies. One result of this development is inequality in the control and use of resources. About three thousand years ago an empire was organized as shown in an adapted version of Herrmann's |14| representation of an oriental society.

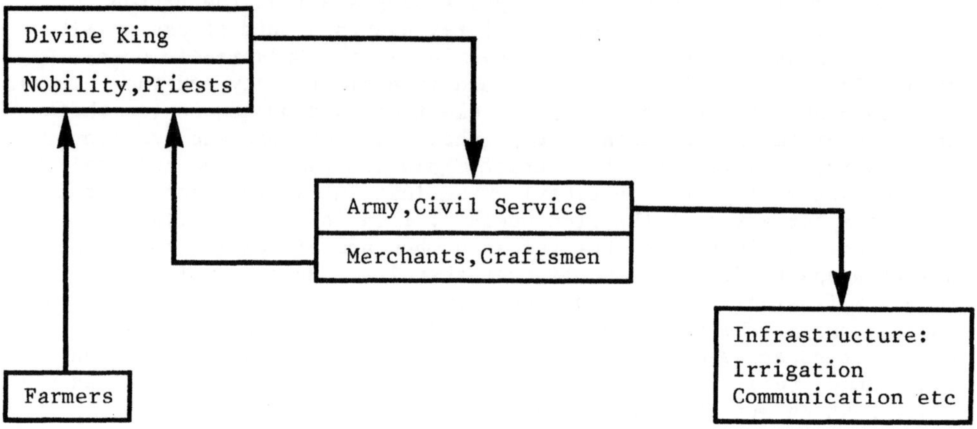

Fig. 3. Schematic representation of an ancient empire. The net flow of exergy is shown with arrows.

Fig. 3 shows that power increases in the vertical direction and it is directly correlated with the use of information by different groups. Information is the resource par excellence. From the earliest period the individuals and groups follows strategies which would bring them nearer to the centre of information generation and transmission. Possession of information and the ability to manipulate it is power. Priests were almost always more powerful than mere kings. Where the offices of the highest executive and the chief priest were united (the divine king) the power reached its maximum.

Exergy can be used to generate new information and information can be used to lower the exergy required to perform certain tasks. Technical development means that increased output can be obtained from a fixed amount of labour and resources, or alternatively that the same level of output requires less labour and resources. When technical development is embodied in new machines, it is called embodied technical progress. Disembodied technical progress is the increase in efficiency because of organizational and institutional changes.

In short, technical development increases efficiency thereby giving people more voluntary leisure. The upper class in a society wishes to remain near the centre of information generation. The power of the upper class lies in the fact that they can manipulate information. They also ensure that the world is interpreted as they wish. To retain this capacity in the face of a mass of new information they need to control their use of time. If they are tied to high frequency (repetitive) jobs like household chores or factory work they loose their manipulative power. The upper class for this reason always has occupations which Douglas and Isherwood |15| call low frequency activities. This makes it possible for them to be available at short notice for encounter with other powerful persons and groups. Accordingly a large part of their consumption is information based consumption. This class could as well be called the information class. A person in this class preferentially possesses and consumes commodities and uses services which will increase his ability to interact with the information system as a whole.

The concept of "availability for encounter with persons and groups of influence" can be illustrated by the following example. When Bismarck was playing a very important role in Europe, he spent a considerable amount of time feasting and hunting with his peers. His power base was the German aristocracy. He had to be personally available to them. A German Chancellor nowadays can hardly indulge in such activities. He has to be available to the parliament, his party, trade unionists, other politicians, the media, industrial magnates, in Brussels and so forth. The German Chancellor uses an airforce jet to cover distances of the order of a few hundred kilometers. To retain his office he must be available to as many people, groups and places of influence as possible. His consumption in terms of exergy is a few hundred times more than the more powerful Prince Otto von Bismarck.

Technological development thus increases the personal availability and enhances the consumption of the information class. This concept can be modelled with the following qualitative relationship

$$\text{Availability} = A \quad 1 - \exp\left(\frac{\alpha I+\beta T}{E}\right) \quad E \tag{12}$$

Where A, α and β are constants, E is exergy and I and T denote the information component and the technological component of consumption. When the use of information and the level of technology is very low, availability is independent of the input of exergy, since in this limit eq. (12) reduces to

$$\text{Availability} = A\ (\alpha I+\beta T)$$

When, on the other hand, the level of technology and the use of information is very high per unit of used exergy, eq. (12) becomes

$$\text{Availability} = AE$$

i.e. availability in this case is directly proportional to the input of exergy.

This is simply a qualitative way of showing that as technology develops availability increases. With the help of new technology more information can be processed which in turn increases the availability even more. This is illustrated by a modified version of Spreng's |16| triad. Fig. 4. shows the well-known fact that the people performing non-repetitive tasks use more exergy and more information. Speed is important for them, so they substitute exergy and information for time.

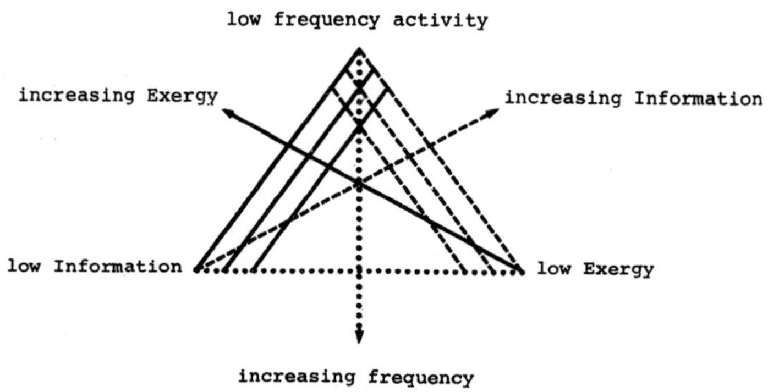

Fig. 4. Parellel lines connecting the vertices are isoexergy (continuous) lines and isoinformation (broken) lines. Spreng gives the value t=0, to the vertex "low frequency activity". This means that tasks are performed very quickly. On an isoexergy line, a task can be performed faster by using more information and similarly on an isoinformation line, more use of exergy increases the speed of performance. The triad shows that exergy, information and time are substitutes.

Technological development has resulted in more consumption for every one in the industrialized countries. But the consumption of the information class has increased disproportionately. Technological development makes it possible for them to be available at different places at ever increasing speed. Their consumption is mainly composed of university education, visits to concerts, dining with persons of influence, attending conferences, lecturing abroad, telephone calls, telex, computers, books and journals. People of the information class must let the less important persons know that they are busy and unavailable.

This apparent unavailability is in fact their very high availability in their technologically expanded social interaction.

Technical development definitely increases the voluntary leisure of all classes in a society but the middle and lower classes do not preferentially demand or consume those technical gadgets which would increase their capacity to acquire more information or interpret information independently of the manipulators.

Douglas and Isherwood classify consumption in three categories

1) consumption of staples
2) technology based consumption
3) information based consumption

A commodity is called staple if it is needed for high frequency tasks e.g. household applicances, motor car for a travelling salesman. Technology based consumption bundle is composed of goods that have been newly introduced in the market. People with higher income but not belonging to the information class tend to buy every new technical gadget that is introduced in the market - e.g. colour television, video. The information class do not buy these goods immediately, if such goods neither increase their availability for encounter with powerful persons nor their capacity to acquire and manipulate information.

With this classification there is no need to distinguish between luxuries and necessities. A television set, when first on the market, was a luxury item in the conventional sense, but now almost every household in the industrialized countries own a television set. This is also the case with refrigerator and vacuum cleaner. The acquisition of such goods is sometimes illustrated by the "spread of infection" model.

In this simple model, all households which are "susceptibles" end up buying the commodity. The interesting thing is that the "rate of infection" varies considerably for different goods in the same price category. The rate of spread of many new commodities and services in the same price category among different social classes cannot be explained by the budget-constraint alone.

The technical gadgets needed for high frequency activity - refrigerator, vacuum cleaner - have been acquired by all households. Although the telephone was introduced more than a hundred years ago, still it has not infected everyone (fig. 5). Some people argue that technical limitations prevented the service from spreading faster. This can hardly be true. If a service or a commodity is demanded then the producers see to it that it is made available, apart from reducing prices by economies of scale, credit facilities are used to bring the commodity to the consumer.

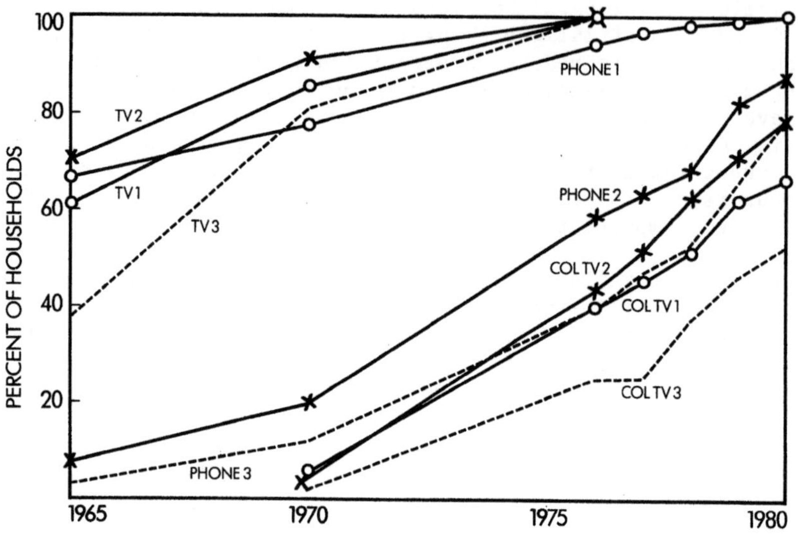

Fig. 5. Spread of telephone (PHONE), black and white television (TV) and colour television (COL TV) among the West German households. Upper, middle and lower income groups are indicated by 1,2 and 3. Telephone was introduced in 1877, television in 1948 and colour television in 1968. Data is taken from ref. 17.

People of the lower class did not need the telephone. Firstly because none of their friends and relatives had a telephone, secondly because they do not have control over their time. They perform high frequency tasks. They cannot choose to be available in a certain place at a certain time. But the people performing low frequency tasks need to cancel their appointments, make new appointments, be available at short notice in faraway places.

They have to talk to their lawyers, their publishers, their agents, their acquaintances of influence in different countries. They need the telephone, so they have the telephone. Not so in the case of television. Almost right from the beginning the middle income group owned more television sets than the information class. Television might increase the amount of received information but it is not superior to other forms of information e.g. books, newspapers, radio etc.

Services and commodities supplied are oriented towards the needs of the information class. Before the advent of the efficient high speed electronic communication era, mail was delivered three times a day. At that time the Cambridge dons and the business community used the postal service as the main channel of communication. One could get a reply to a letter sent to London from Cambridge on the same day. Now they have other channels of communication so that the letters are delivered only once a day. In Sweden letters are not delivered at the week-ends, registered letters have to be collected from the post office and the telegrams are slower than express letters. If the upper class **does not demand or consume a service, the service deteriorates.**

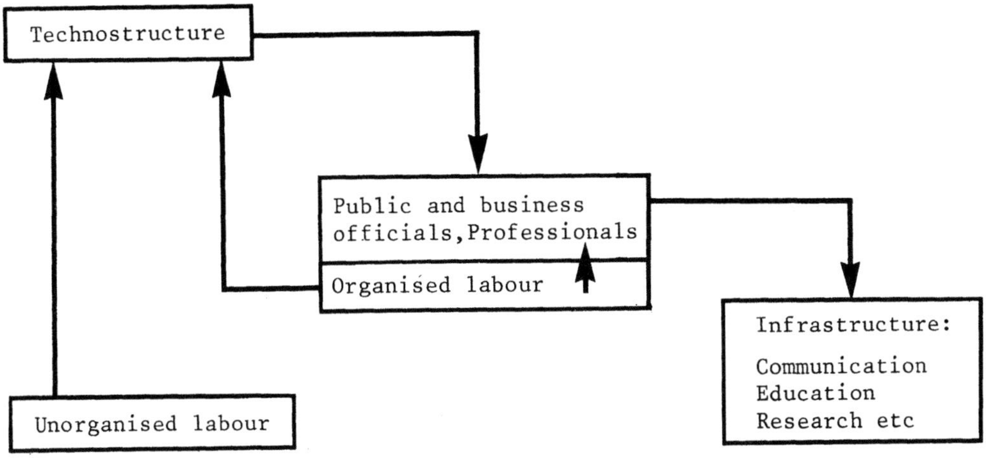

Fig. 6. Schematic representation of a modern society. Power increases in the vertical direction, the net flow of exergy is shown with arrows. Professionals include technology based small firms. Unorganized labour include small scale nontechnical economic activity. Compare with fig.3.

The power structure in any society with a considerable infrastructure, is similar to that of an ancient empire. In the modern high technology society, the divine king with his priests and courtiers has been replace by Technostructure (a term coined by J.K. Galbraith to denote the top people in science, technology, military, industry, party and administration). Organized labour and technology based small firms have replaced the merchants and craftsmen. Unorganized labour and the non-technical small economic units are the least privileged.

The differences within a society have their counterpart on a global scale (fig. 7). Poor countries have less information. The problem of freeing people from poverty is not how to feed them. Poor countries need technology to produce their own food. It is true that different societies have different capabilities to gain from technological development but certain economic activities generate more (technical and non-technical) inventiveness than others in a given society. The way to development is to identify and pursue economic activities which make people more inventive.

Most research is done to fulfill the goals of the technostructure - to increase their information manipulation capacity. Research is not oriented to the needs of the poor countries or to the needs of the consumers in the rich countries. An industrial policy that could control the industrial evolution through the selection of useful innovation would be a major achievement.

An increase in the degree to which individuals, groups and nations are linked to the information system as a whole will be a great advance as it will reduce the power of the manipulators. The capacity of people to interpret information which concerns them must increase, so that they can choose new innovation or discard old modes of production irrespective of whether the capital or the organized labour lose in the short run. The problems that concern us all cannot be solved by a small group.

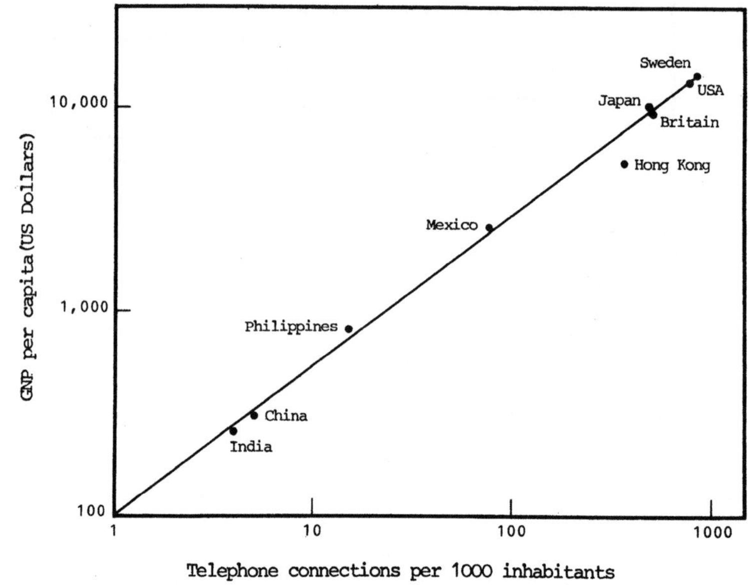

Fig. 7. The correlation between the number of telephone connections per thousand of the population and the GNP per capita in selected countries (1984). Data from International Telecommunication Union. Density of telephone connections can be taken as an indicator of wealth (cf. fig. 4).

The information revolution can go in either direction. It might increase the interpretative power of the people - which to some extent has already happened in the North-western Europe - or it could result in complete control by the technostructure. Information can both be created and destroyed on a very large scale - the potential for great good or great evil is therefore enormous.

4. OPTIMAL ECONOMIC DEVELOPMENT UNDER RESOURCE CONSTRAINT (12).

The stock of capital (K) in our economy is measured in terms of exergy. The technology 'stock' is measured in some dimensionless unit T. The resource intake into the economy is an exergy flow R.

There will always be some upper bound ($\bar{R}$) to the inflow of exergy which can be diverted to the economy. This upper bound exists not only because of some physical constraints. It may exist due to regulations for the preservation of some species or landscapes and due to regulations limiting the level of thermal, chemical, noise and radiation pollution. These regulations might be a result of the attitude of a society towards ecological, medical and aesthetical issues.

The control variables of the system are investment rates, I into capital and J into technology and the resource intake R. We consider it to be impossible to consume technology and we do not allow for the consumption of capital. The control variables thus satisfy

$$I \geq 0,\ J \geq 0,\ R \leq \bar{R} \tag{13}$$

The equations of motion in (K,T) plane are

$$\dot{K} = -\beta_K K + I,\ \dot{T} = -\beta_T T + \sigma J \tag{14}$$

β_K and β_T are depreciation rates for the capital and technology and σ is the technology increase per invested unit of exergy. These three quantities are assumed to be constant.

The production rate is expressed in terms of the exergy flow and is given by |18|.

$$P = E\ (K,T)\ R \tag{15}$$

E is exergy efficiency and is given by

$$E = E(x) < 1$$

where $x = \beta_x\ e^{\lambda(T)}\ \frac{K}{T} \geq 0$ (16)

The consumption rate is given by

$$C = P-I-J \tag{17}$$

The utility rate U(c) satisfies

$$p_e \equiv U'(C) > 0,\ U''(C) < 0 \tag{18}$$

The optimal development path is characterized by the property that it maximizes the total utility from present to all future. That is the functional

$$W = \int_{to}^{\infty} dt\ U(C)(t)) - U(C(\infty)) \tag{19}$$

is maximized. Where we have normalized the integrand in such a way that the total utility converges. Following Ramsay |19| we do not discount the utility of the future generations.

The problem of maximizing (19), given the equation of motion (14) and the constraints (13), is described by the Hamiltonian function

$$H(K,T,p_K,P_T;I,J,R,q_I,q_J,p_R) = U(RE-I-J) + p_K(I-\beta_K K) + +p_T(\alpha J-\beta_T T) + q_I I + q_J J + p_R(\bar{R}-R), \tag{20}$$

where q_I, q_J and p_R satisfy

$$q_I \geqq 0,\ q_J \geqq 0,\ p_R \geqq 0, \tag{21}$$

and

$$q_I I = q_J J = 0,\quad p_R(\bar{R}-R) = 0 \tag{22}$$

In (20) p_K and p_R are the prices per unit exergy of capital and of the natural resource; p_T is the unit price of technology, and q_I and q_J are price differences (see (25) below) to prevent the consumption of capital or technology. There is no market in the model, and all prices are shadow prices.

The equations of motion obtained from (20) and (14) are

$$\dot{p}_K = -\frac{\partial H}{\partial H} = -p_c\frac{R}{K}xE'(x) + \beta_K p_K$$

$$\dot{p}_T = -\frac{\partial H}{\partial T} = -p_c R\lambda(T)xE'(x) + \beta_T p_T \tag{23}$$

where we have used (18) and (16).

The conditions for optimality.

$$\frac{\partial H}{\partial I} = \frac{\partial H}{\partial J} = \frac{\partial H}{\partial E} = 0 \tag{24}$$

imply

$$P_K = p_c - q_I,\qquad \sigma p_T = p_c - q_J, \tag{25}$$

and

$$p_R = p_c(E+R\frac{\partial E}{\partial R}) = p_c(E(x)-xE'(x)), \tag{26}$$

Using (25) we can eliminate q_I and q_J in (21) and (22),

$$0 \leqq p_K \leqq p_c,\quad 0 \leqq p_T \leqq \frac{1}{\sigma}p_c,\quad p_R \geqq 0, \tag{27}$$

and

$$(p_c-p_K)I = (p_c-\sigma p_T)J = p_R(\bar{R}-R) = 0 \tag{28}$$

For simplicity we put

$$\beta_x = \beta_K = \beta_T = \gamma \tag{29}$$

The equations of motion (14) are then

$$K = -\gamma K+I, \qquad \dot{T} = -\gamma T+\sigma J \tag{30}$$

The own rates of interest for capital and technology obtainable from the equation of motion (23) are

$$-\frac{p_K}{p_K} = \frac{p_c}{p_K} \cdot \frac{R}{K} xE'(x)-\gamma$$

$$-\frac{p_I}{p_T} = \frac{p_c}{p_T} \cdot RxE'(x)\ \lambda'(T)-\gamma \tag{31}$$

The general interest rate, connected to the price p_c of consumption goods is

$$r = \frac{p_c}{p_c} = -\frac{d\ \ln U'(C)}{dC} \cdot \dot{C} \tag{32}$$

For utility rate, we choose the following risk-averse function

$$U(C) = \ln(C-C_o) \tag{33}$$

This gives

$$r = -\frac{\dot{p}_c}{p_c} = \frac{\dot{C}}{C-C_o} \tag{34}$$

During a period when it is optimal to invest in both capital and technology, i.e. when I and J are both non-zero, then according to (28),

$$p_c = p_K = \sigma p_T \tag{35}$$

In this case clearly the three interest rates coincide,

$$-\frac{p_K}{p_K} = -\frac{\dot{p}_T}{p_T} = r, \tag{36}$$

and hence from (34) we get the following relation between K and T:

$$\sigma K = \frac{1}{\lambda'(T)} \tag{37}$$

This is a relation of balance between capital and technology. When this balance holds, capital increases with increasing technology.

$$\sigma \frac{dK}{dT} = -\lambda''(T)/(\lambda'(T))^2 > 0 \tag{38}$$

If (37) is not fulfilled, but

$$\sigma K > 1/\lambda'(T), \tag{39}$$

then there is excessive capital; p_K falls below p_c, and because of (28), I vanishes, while J is still non-zero. Similarly when there is excessive technology,

$$\sigma K < 1/\lambda'(T) \tag{40}$$

p_T falls below p_c/σ, and J = 0, but I is non-zero.

We now study two different régimes.

I) For $R < \bar{R}$, eq. (28) gives

$$p_R = 0 \tag{41}$$

with eqs. (18) and (26) this implies

$$x = x^* \tag{42}$$

where x* is uniquely determined by (26)

$$x^* E'(x^*) - E(x\bar{A}) = 0, \; x^* > 0 \tag{43}$$

This equation together with (16) fixes the optimal resource intake:

$$R = \frac{\lambda}{x} Ke^{\lambda(T)} \tag{44}$$

Let $Z = \frac{R}{\bar{R}}$

For $R = \bar{R}$, II = Z > 1, E = E(*Z) (46)

$$P = \bar{R} E \tag{47}$$

$$p_R = p_e (E - xE') > 0 \tag{48}$$

The following policy rules can be obtained:

1. If either the capital stock (39) or the technology stock (40) is in excess, do not invest in this stock until the line of balance (37) has been reached. If the stocks are in balance, stay on the line of balance.
2. Before the resource ceiling $\bar{R}$ has been reached, keep a fixed optimal efficiency (43). This determines the optimal resource intake (44). The price of the resource in this region is zero (41). The scarcity of the resource is manifested in the limited availability (27) of relevant technology and capital necessary to divert the flow of exergy into the production process.

3. The resource has a price after the resource ceiling has been reached (48). Fix the resource intake at its maximum (47).

These principles determine everything except the overall investment level. To determine the initial consumption level and hence also the initial investment one has to study the full optimization problem. Once the initial values have been found however, the decision on how much to invest and how much to consume is governed by the interest rate (34).

4. The resource has a price after the resource ceiling has been reached (48). Fix the resource intake at its maximum (47).

These principles determine everything except the overall investment level. To determine the initial consumption level and hence also the initial investment one has to study the full optimization problem. Once the initial values have been found however, the decision on how much to invest and how much to consume is governed by the interest rate (34).

We note that like the consumption rate, the total investment rate has to be a continuous function of time, whereas the control parameters I and J (investment into capital and into technology, respectively) may have jumps.

We also note that the efficiency in the final state E_{∞} lies well below 1. The laws of thermodynamics requires the input of exergy for the maintenance of structures and organizations. To achieve an efficiency close to unity a very large stock of capital and technology is needed. This would increase the minimum saving requirement, to compensate for the depreciation, to a very high proportion of the total output. It is to be mentioned, however, that the economy can grow significantly even after reaching the upper bound of the resource flow by increasing its stock of capital and technology (45-47).

5. CONCLUDING REMARKS

In this paper I analyzed systems as different as the Universe, a society and an economy using the tools of thermodynamics. The investigation of systems as complex as a society or an economy using the tools of classical mechanics has not been very successful. The reason is that the information content of a dissipative structure changes with time, so that such structures are inherently indeterministic. The theory of self-organization which uses the tools of thermodynamics, statistical mechanics, combinatorics and information theory might be well suited for understanding large-scale dissipative structures.

This work was supported by the Swedish Council for Planning and Coordination of Research, the Science Research Council and the Energy Research and Development Commission. I am thankful to Professor Karl-Erik Eriksson for introducing me to the theory of physical resources, to Baron Isherwood for helpful suggestions, Amanda and Paul Lomas for linguistic corrections. Gun Fornell for typing various versions of the paper and Gisela Kolek and Doris Tschentscher for drawing the figures.

References

1. Z. Rant: Forschung Ing-Wesens 22, 36 (1956).

2. R.B. Evans: A Proof that Essergy is the only Consistent Measure of Potential Work. (Thesis, Dartmouth College, Hanover, NH, 1969).

3. M. Tribus and E.C. McIrvine: Scient. Amer. 225, 179 (1971).

4. B. Eriksson, K.-E. Eriksson and G. Wall: Basic Thermodynamics of Energy Conversion and Energy Use. Institute of Theoretical Physics, Göteborg, 1978.

5. R.C. Tolman: Relativity, Thermodynamics and Cosmology (University Press, Oxford 1934).

6. G. Gamow: Nature 162, 680 (1948).

7. S. Weinberg: Gravitation and Cosmology (Wiley, New York, 1972).

8. P.C.W. Davies: Physics of Time Assymetry (University of California Press, Berkeley, 1974).

9. P.T. Landsberg: Thermodynamics and Statistical Mechanics (University Press, Oxford, 1978).

10. L. Brillouin: Science and Information Theory (Academic Press, New York, 1962).

11. K.-E. Eriksson, S. Islam, E. Tengström: Resurser i natur och samhälle (Background report number 14 of the Committee on the Natural Resource and the Environment, Swedish Ministry of Agriculture, Stockholm, 1983). English version (Resources in nature and society) is available as report number 1983-6 of the Physical Resource Theory Group, Göteborg, 1983).

12. K.-E. Eriksson, S. Islam, S. Karlsson and B. Månsson : Resources and Energy, 6, 235 (1984).

13. In an earlier paper (K.-E. Eriksson, S. Islam and B.-S. Skagerstam: Nature 296, 540 (1982)) there are a few mistakes, consequently we overestimate the amount of nuclear exergy created (fig. 2 of the paper in Nature). These mistakes have been corrected in this paper.

14. J. Herrmann: Spuren des Prometheus (Urania-Verlag, Leipzig 1982).

15. M. Douglas and B. Isherwood: The World of Goods (Penguin, Harmondsworth 1980).

16. D.T. Spreng: On Time, Information and Energy Conservation. Institute of Energy Analysis, Oak Ridge, Report no. ORAU/IEA-78-22 (R) (1978).

17. Institute der Deutschen Wirtschaft: Zahlen zur wirtschaftlichen Entwicklung der Bundesrepublik Deutschland (Deutscher Instituts-verlag, Köln 1979-82).

18. R.U. Ayres and S.M. Miller: Journal of Environmental Economics and Management 7, 353 (1980).

19. F.P. Ramsay: Economic Journal 38, 543 (1928).

INDEX

GPSR Compliance

The European Union's (EU) General Product Safety Regulation (GPSR) is a set of rules that requires consumer products to be safe and our obligations to ensure this.

If you have any concerns about our products, you can contact us on ProductSafety@springernature.com

In case Publisher is established outside the EU, the EU authorized representative is:

Springer Nature Customer Service Center GmbH
Europaplatz 3
69115 Heidelberg, Germany

Batch number: 10607320

Printed by Printforce, the Netherlands